CW00602989

CITIES OF PLEASURE

This collection of cutting edge and accessible chapters explores various connections between urban living, sexuality and sexual desire around the world. The key themes featured address a number of topical issues including: (i) the controversies and debates raging around the evolution, defining patterns and appropriate regulation of commercial sex zones and markets in the urban landscape (ii) how have gay public spaces, districts and 'gay villages' emerged and developed in various towns and cities around the world? (iii) How do changing attitudes to and the usage of urban sexual spaces, as depicted in iconic television series such as *Sex and the City* and *Queer as Folk*, reflect the reality of working women's or gay men's changing life experiences?

This book was previously published as a special issue of the journal *Urban Studies*.

Alan Collins is Reader in Applied Economics in the Department of Economics, University of Portsmouth.

URBAN STUDIES MONOGRAPHS

In the contemporary world cities have a renewed significance. Trends such as globalisation, neo-liberalism, new technologies, the rise of consumption and consumerism, the re-definition of modes of governance, demographic and social shifts, and the tensions to which such changes give rise, have re-defined the critical role cities play, as well as the problems arising from city life. By focusing on specific issues this series aims to explore the nature of the contemporary urban condition.

Cities of Pleasure
Sex and the Urban Socialscape
Edited by Alan Collins

Clusters in Urban and Regional Development
Edited by Andy Cumbers & Danny McKinnon

Globalisation and the Politics of Forgetting
Edited by Yong-Sook Lee & Brenda S. A. Yeoh

CITIES OF PLEASURE

Sex and the Urban Socialscape

Edited by
Alan Collins

LONDON AND NEW YORK

First published 2006
by Routledge
2 Park Square, Milton Park, Abingdon, Oxon OX14 4RN

Simultaneously published in the USA and Canada
by Routledge
270 Madison Avenue, New York, NY 10016

Transferred to Digital Printing 2006

Routledge is an imprint of the Taylor & Francis Group, an informa business

Typeset in Times by Infotype Ltd, Eynsham, Oxfordshire
Printed and bound in Great Britain by Antony Rowe Ltd, Chippenham, Wiltshire

British Library Cataloguing in Publication Data
A catalogue record for this book is available
from the British Library

Library of Congress Cataloging in Publication Data
A catalog record for this book has been requested

ISBN 0-415-36012-9
ISBN 978-0-415-36012-8

CONTENTS

Sexuality and Sexual Services in the Urban Economy and Socialscape: An Overview

Alan Collins

Introduction

Given that towns and cities are essentially spatial concentrations of large numbers of people, then it is this very population density that serves as a defining characteristic. Such high densities mean that a more diverse range of people may be found within these urban locations than beyond their margins. If people with particular personal preferences are looking for other people with particular characteristics then the very presence of such diversity is likely to help the search and matching process to take place more efficiently. Clearly, large towns and cities are also characterised by higher-order services than smaller settlements. Among these higher-order services will tend to feature a larger and more diverse range of leisure opportunities, the pursuit of which is often combined with mating opportunities and rituals. Consider the large packs of men and women who roam, in circuit drinking mode, the huge number of pubs and bars of Newcastle, England (and of course many other towns and cities around the world). They are not only engaged in socialising and drinking, but are also instinctively aware that opportunities for sexual contact can be acquired through this route (albeit with an uncertain probability of success on any given occasion). Further, the actual pursuit (or thrill of the chase) can also offer considerable enjoyment to many individuals. Cities are thus places that offer spaces containing an extensive range and number of pleasure-based outlets and these often have considerable sexual dimensions. These may be characterised by no actual physical contact, but merely comprise the sexual *frisson* from the physical proximity of potential partners or various sexual goods, services and spaces. They may

Alan Collins is in the Department of Economics, Portsmouth Business School, University of Portsmouth, Richmond Building, Portland Street, Portsmouth, PO1 3DE, UK. Fax: 02392 844 037. E-mail: alan.collins@port.ac.uk. *The author wishes to thank Ronan Paddison and Isabel Burnside for their tenacity, faith and hard work in helping to make this Special Issue happen. Thanks are also due to Samuel Cameron for useful theoretical discussions and Bill Johnson for his help with producing the diagram. The usual caveat applies.*

also offer, of course, the real prospect or means to achieve actual sexual contact. Sexual desires can thus be viewed as important behavioural drivers that may push people to look frequently to urban locations for expending their pleasure time. Potentially, they may also tip the decisions of some determined seekers of sexual contact, such that they actually choose to reside in a large town or city, thus enabling them to make even more effective use of the greater sexual opportunities such locations present.

Why then can sexual desire be such an important influence on individual behaviour, such that it can strongly influence the intensity of use of urban leisure opportunities? From a medical perspective, participating in safe sexual activity may simply be contended to be generally conducive to mental and physical well-being and individuals generally prefer more sexual activity to less, albeit not necessarily with the same sexual partner. While there is increasing technological scope for the separation of recreational sexual activity, companionship and child production (reproduction), Baker (1999) contends that it is not the human psyche that is necessarily driving the exploitation of this separation, but rather natural selection. Humans are not the only species to seek sex far more often than is necessary to reproduce. From this standpoint, Baker suggests that advances in reproduction medicine and technology are merely contributing to an increasing existing trend towards single parenting becoming the norm. Further, he contends that such individuals may well want to share household overhead costs and enjoy the companionship and sexual opportunities of other single parents, in the same physical household, with same-gender or multigender permutations. This means that partner seekers will increasingly be able to contemplate partnering decisions in terms of seeking a range of more readily available, specialised or individual options. The alternative is to expend potentially higher search costs looking for the elusive single individual who would have to demonstrate satisfactory to excellent matches, possibly across all of the relationship dimensions noted above. Stylised dramatic representations of these alternative perspectives might be discerned in the relationship preferences suggested in the characters of Samantha and Charlotte in the US Home Box Office television drama series *Sex and the City*, based on the Candace Bushnell (1996) novel of the same name.

In sociobiological terms, sexual desire may be viewed as a device helping attempts to dominate the gene pool (Wilson, 1975; Barash, 1979). Humans, like all other life-forms, are instinctively driven to reproduce. Were it not so, the human species would have died out long ago. In the sociobiological terms of maximising the fitness of the gene pool, for the human male, the sub-conscious goal is to impregnate as many females as possible. To this end, the average ejaculate has many million sperms and little else. A sperm is merely a collection of genetic material and possesses no other cellular functions. The male ability to impregnate is quite durable and does not fully cease in later years. The human female must nurture the pregnancy and the resultant infant and thus typically devotes significant biological resources to the single egg produced each cycle. Typically, these will not become fertile again until the current offspring is older—i.e. stops breast feeding. Both the male and female are programmed to maximise the success of their contribution to the gene pool.

What then of the purpose of homosexual desire in this seemingly heterosexual vision of the continuing life-force of society? While Freudian psycho-analytical theory and neuro-hormonal theories (Ellis and Ames, 1987) offer alternative accounts for the genesis of homosexuality (and various degrees of bisexuality), these can be readily combined with sociobiological reasoning, as they are certainly not mutually exclusive bodies of thought. Only sociobiology, however, articulates an integrated sense of social and genetic purpose for homosexuality. Two key explanations emerge from Wilson (1975, 1978). First, he points to work suggesting that homosexual genes may possess superior fitness in heterozygous conditions, which can be maintained in evolution if they tend to survive into maturity better, produce more offspring, or both. The second and related theory pertains to 'kin selection' whereby it is argued that homosexuality developed normally, in a biological sense, through the course of various primitive societies. Its function was to assist close relatives, such as siblings. Accordingly, Wilson suggests that homosexuality has evolved as a distinctive beneficial behaviour that constitutes an important element in human social organisation. By this reasoning, homosexuals may be the genetic carriers of many of mankind's altruistic impulses.

Hence, sociobiologists would contend that all humans are genetically programmed to be reproductively successful. This translates into a desire to search actively for and mate with sexual partners. Yet society forms constraints on how this search process and mating are conducted, even though society ultimately wants the mating to take place (otherwise there is no society in the future). In a rural setting, cosy romanticised *gemeinschaft* notions of some close-knit village community providing more enduring and satisfying personal and sexual relationships does not necessarily sit comfortably with evolutionary biological objectives. Excessive rounds of mating within the village community will clearly not ensure gene pool fitness maximisation. Recognition of this simple fact has led rural communities throughout the ages to respond actively to these sociobiological imperatives. In so doing, rural societies have also recognised that the search costs of finding suitable sexual partners may be very high, as greater distances have to be traversed to find suitable mating prospects. Thus more sophisticated customs, routines and rituals to foster inter-village mating opportunities have been devised by rural societies around the world. Many of these have now fallen into disuse with changing demographic structure, rural depopulation and social change (see—for example, Sundt and Anderson, 1993; and Friedl, 1997). In essence, many deeply rural areas, particularly in developed economies, have evolved to become extremely thin physical markets for partner search. They effectively came to form 'dating deserts' prior to the partial mitigating effects of increasing private car ownership and the advent of Internet chat rooms and web dating sites. For similar reasoning, it is not surprising that arranged marriage transactions have also traditionally been more common in rural communities around the world.

As rural employment opportunities diminished, alongside rapid industrialisation and more liberal social change, urban living has emerged to form the dominant backdrop to the socialscape of people's lives in most developed economies. In this more *gesellschaft* context, both men and women have had to become more responsible for finding their own mating opportunities, although social customs in many societies might dictate less effort needing to be expended by younger age cohort women.

While the distance-generated search costs of suitable partner search, more typical of rural societies, may be less problematic in urban centres, the higher population density potentially raises the matching-sourced costs of suitable partner search. This relates to the likelihood that, in a city, far greater numbers of unsuitable partner prospects may have to be filtered out prior to finding an appropriate subset of suitable sexual partners. Such costs suggest the need for cheap screening mechanisms to eliminate quickly large numbers of people. Paradoxically, cheap screening devices may actually involve the use of expensive prices in clubs and other venues, in an attempt to contrive a 'better' search pool of potential partners. That said, many individuals may be more content to accept such matching-sourced search costs since for many people they are, to a significant degree, offset by a measure of in-search utility—i.e. the enjoyment and benefits derived from simply meeting and 'evaluating' (sexually or otherwise) new potential partners. Commercial ventures and enterprises, alongside public authorities, have emerged as significant players influencing the ease, or otherwise, with which individuals can match themselves up with suitable sexual partners in urban areas. These players exercise this influence through commodification of sexual services and opportunities, or via their ability to regulate sexual spaces and markets. Although following Christaller-type logic, a broader range of sex-based commercial premises is likely only to feature in higher-order towns and cities.

This Special Issue presents a collection of papers, from various disciplinary perspectives, which investigate the many ways these sexualities and urbanities connect and not just offering an exploration of how urban areas assist the process of partner search. Realisation of such connections is by no means new. Indeed, there has been significant early work in the 1920s, particularly in human geography, identifying the nature and role of sexual markets and sexual spaces in urban development and processes (see, for example, Park and Burgess, 1925/1984; see also the work reviewed in Heap, 2003). As the papers in this volume make evident, research on the broader reverse causal themes of how urbanism and urban spaces condition individual sexuality and sexual activities and *vice versa* have also generated a significant body of work in the social sciences. More recently, the potential and scope for embedding even more liberal sexual expression in urban life have also been recognised by policy-makers and researchers as a potential source of competitive advantage for some cities. This arises as many municipalities chase economic success by trying to affirm their cosmopolitan and progressive aspirations and credentials within their planning policies. In a parallel contemporary vein, many advanced economies have also witnessed a huge growth in the number of single-person households. This has arisen, in large part, as a consequence of the social acceptance of cohabitation (which offers relatively low partnership exit costs), delay in marriage and high divorce rates. While this has obvious urban physical manifestations in terms of boosting the demand for smaller dwelling units, it also has repercussions in terms of the urban socialscape. Higher numbers of partner-seekers generate greater demands for infrastructure in the urban economy to accommodate more time-diverting leisure and to provide more sexual partner search opportunities.

Clearly, the gravitational pull of larger numbers of single persons (of various sexual orientations) is a hallmark feature of cities which in itself may attract further waves of potential partner-seeking individuals. Since the pursuit of sexual contact can be a time-consuming activity and one that is often bundled together with other leisure activities to save time, this may influence individual residential choice decisions or, at the very least, the spatial extent of their travel-to-leisure ranges. Yet it is not only single people who are seeking sexual partners. Many individuals who are in existing relationships (straight or gay) may wish to make up for what they perceive are deficiencies in the amount and/or range of sexual activities in which they participate. Alternatively, they may simply wish to diversify their sexual consumption, to satisfy a demand for sexual variety (in partners or activities). This can more easily be camouflaged or hidden from existing partners when contained in the potentially more anonymous urban settings of large towns and cities. They may do this either through forming other relationships of varying degrees of commitment, furtiveness and durability, or via explicit commercial means, such as use of the prostitution services that are typically found in most large urban centres around the world. Yet physical proximity to such commercial sex markets has often presented for some city residents a significant source of urban land use conflict and negative externalities (see, for example, Hubbard, 1998). It has also presented local government with many challenges and opportunities in formulating urban policy. These are all contemporary themes explored by various contributors herein.

This collection of papers is thus a timely and useful device to bring together work from various research strands that currently feature in a diverse range of outlets. The papers are grouped around three overlapping themes: commercial sexual spaces, markets and community responses; the evolution and development of urban gay spaces; and, urban sexuality. In the next section, some theoretical exploration is offered of the boundaries and commonalities linking work in these areas, within a simple conceptual economic framework. The central themes and concerns in the three groupings of papers are then briefly considered. Concluding remarks are offered in the final section which also attempts to draw attention to various research gaps and areas that are not, for various reasons, adequately represented in this Special Issue but which nonetheless merit future research attention.

A Time and Place for Sex

While personal relationships can be viewed as providing individuals with, among other things, a source of recreational sexual services and marriage-type companionship, there is no reason to suppose such individuals desire the same levels or proportions of these two relationship elements. Many individuals may garner adequate or substantial companionship from their friends and family, which can offer a reasonable substitute for the companionship afforded by a potential personal relationship partner. Alternatively, some individuals may be in a particular phase of their life where they simply do not require high levels of companionship at all and are content or satisfied with their own company or the minimal companionship afforded by a restricted social circle. Turning to sexual services, clearly libidos vary widely and may change markedly over a lifespan. Furthermore, some individuals may want to explore a wider sexual repertoire than may be provided by one partner. Purely for graphical expositional simplicity, let the entire range of recreational sexual services be condensed and represented by a single characteristic—**S**. In a similar vein, let all aspects of marriage-type companionship be condensed and represented via a single characteristic—**M**.

In trying to satisfy these two needs, individuals (whether single or already partnered) are faced with resource constraints that attenuate their search for **S** and **M**. These resource constraints are in large part determined by their different endowments of wealth, physical attractiveness and various personality traits. They may also have differing levels of efficiency in using their personality traits and physical attractiveness to acquire **S** and **M**. For those already in relationships, additional resources may also be required to help camouflage or hide extra-relationship sexual consumption. Crucially, in our study context, another significant element featuring in individuals' resource constraints relates to the simple observation that acquiring **S** and **M** of various amounts and types takes time. For example, due to higher population densities in towns and cities, the time required for meeting or acquiring suitable partner matches for greater amounts of, or more diverse, sexual contacts, is likely to be significantly lower in highly urbanised settings and substantially higher in deeply rural settings. This also means the time-dependent search costs will also be likely to vary with towns and cities of different ranks or orders. In Figure 1, the resource constraints (or efficiency frontiers[1]) for individuals A and B, which apply in either 'rural' or 'urban' settings are labelled $\mathbf{E}_{\mathbf{rural}}$ and $\mathbf{E}_{\mathbf{urban}}$. These individuals are identical in every way apart from their personal preferences for differing levels of **S** and **M**. Thus the distinction between the two efficiency frontiers in Figure 1 arises purely from the urban–rural divergences in time-dependent partner search costs. The frontiers depict the maximum feasible amounts of **S** or **M** that may be obtained by individuals A and B in a given 'rural' or 'urban' location. In this example, the frontiers overlap precisely below the point marked *c*, although this need not be the case, as some may argue that rural settings may actually present intrinsically greater resources for acquiring more **M**. Figure 1 also uses the elementary economic concept of indifference curves to represent all the consumption bundles (combinations) of **S** and **M** that yield the same level of satisfaction (or utility) to an individual. In a rational choice

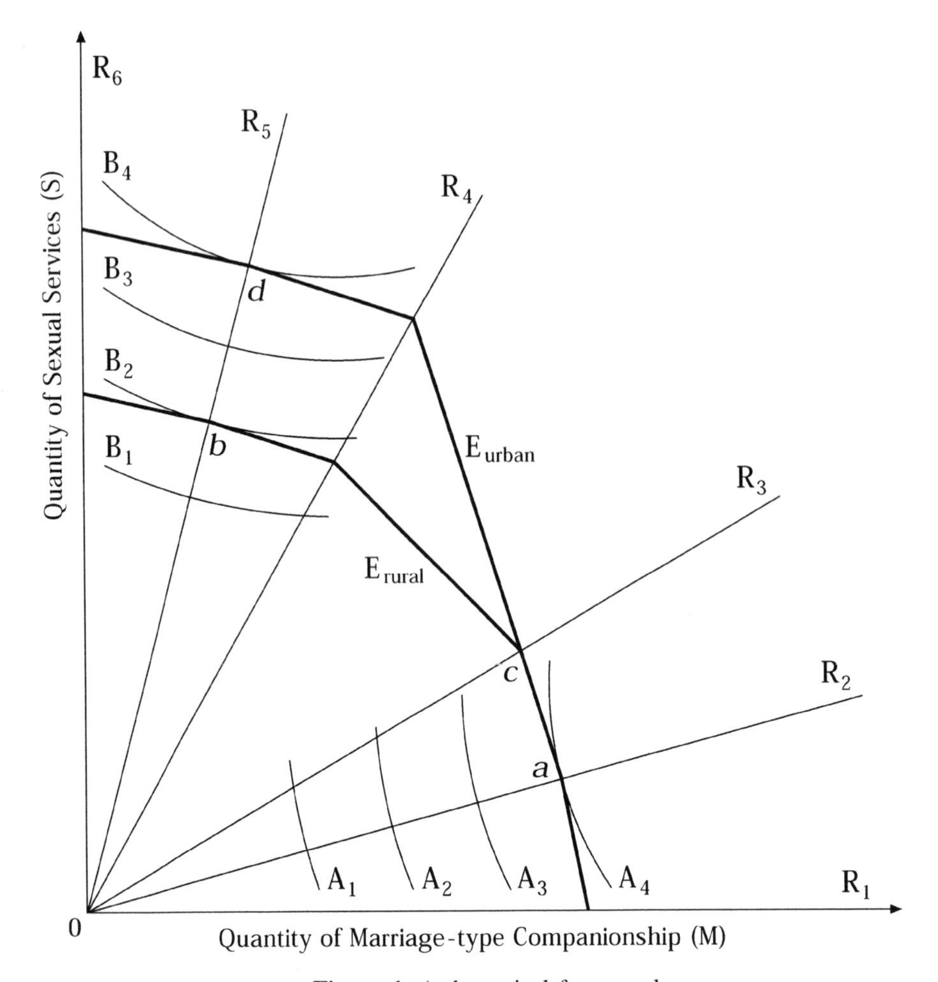

Figure 1. A theoretical framework.

economic framework, higher levels of utility must be preferred to lower levels. The curves labelled A_1 to A_4 depict the indifference curves for individual A. The curves labelled B_1 to B_4 depict the indifference curves for individual B.

The rays R_1 through to R_6 represent possible relationship types (or services) featuring various proportions of **S** and **M**. The ray R_1 which overlays the horizontal axis denotes a purely platonic relationship whereas R_6 which overlays the vertical axis, depicts a purely sexual relationship type containing no **M** element—as could be expected, for example, from the consumption of standard prostitution services. R_2 denotes a relationship type characterised by high levels of **M** and very little **S**, whereas R_5 denotes a relationship type characterised by very little **M** and a high proportion of **S**. Such relationship preferences as depicted in R_5 may also be met through consumption of commercial prostitution services or possibly sexual consumption characterised by 'one night stands' and frequent sexual partner turnover. In the case sketched in Figure 1, individual A who is strongly **M**-orientated, may obtain maximum utility at point *a* (on ray R_2) in both rural and urban settings. Individual B, who is much more **S**-orientated, would only achieve a maximum level of utility depicted by point *b* in a rural setting. In an urban setting, individual B could achieve a substantially higher level of maximum utility, denoted by point *d* in Figure 1.

Given such a substantial difference in

maximum attainable utility or satisfaction levels in these two settings, residency in an urban location is likely to be optimal or strongly preferred by individual B. This is not to say that this individual's tastes may not change, such that, for example, as they age, a greater **M** orientation may be developed (perhaps with child rearing in mind). Allowing for such changes does not necessarily pose a consistency problem in rational choice economics. Some economists (Becker and Stigler, 1977) would contend that we can hold well established preferences or desires for what we want over the entire course of our lifetime, but we plan for these to vary at different key points in our lifespan—for example, during child rearing or approaching retirement.

The concepts and terminology introduced in this framework colour and texture the following overview commentary. The framework presented above also offers some very basic insights to help account for the varieties of linkages between urban living and sexual desire. The wide range of papers in this Special Issue address such linkages from primarily economic, historical, geographical or sociological perspectives. Some papers advance or examine broad theoretical contentions. A number of papers present detailed case study material or original qualitative survey-based evidence. In the next section, each key theme within which these papers are grouped is subject to some brief prefacing commentary.

Key Themes

While edited volumes inevitably involve choices, the themes around which this Special Issue is structured identify key issues in academic and policy worlds. The papers within each highlighted theme do not all reflect mutually supportive positions, but they do principally address particular matters of sexual consumption and lifestyles that have contemporary topical resonance to urban residents and policy-makers. Each is considered in turn.

Commercial Sexual Spaces, Markets and Community Responses

The various papers bound together under this banner address the broad range of commercial sexual experience, ranging from urban spaces offering no-contact pornographic media, to the physical contact derived from street and off-street prostitution services. The paper by Cameron offers a reasoned and detailed typology to help provide some broad insights into how paid sex markets evolve and mature. For Cameron, the emergence of paid sex markets is a product of various locational economies. He also suggests that the physical clustering of traded sex commodities can enhance the progression of the consumer's accumulation and development of experiences (laddering) of different commercial sexual services by heightening the stimulus to enter new markets for the first time.

The importance of such physical clustering seems also to dovetail neatly with notions of urban locations offering time cost saving, or at least minimising time cost uncertainty when searching for commercially traded sexual services (to satisfy more **S**-orientated individuals in the model terms). Minimising time cost uncertainty may be of particular relevance if one is trying to conceal extra-relationship sexual activity, especially from a cohabiting partner. This idea is also picked up in the paper by Ryder who provides a high resolution case study of the 'adult entertainment zone' found around Times Square in New York. Locational economies and clustering are again recurring themes, but the notion of moral spaces and their delineation and regulation enters his econo-geographical lexicon more explicitly. Counterpointing the more organic economistic reasoning in the work of Ryder, and doing so with a wholly different geographical focus (London and Paris), the paper by Hubbard also considers moral spaces and their regulation. For him, the recent expression of 'zero tolerance' initiatives to displace sex workers from red-light districts is seen in the context of a principally revanchist envisioning of the behaviour and operation of policy-makers. He interprets their actions as consistent with a desire to assert their central oversight of moral order in the city. The underlying motivation for this, he would contend, is to permit reappropriation, for capital accumulation, via more family-orientated (or **M**-orientated in the model terms) urban redevelopment. A somewhat conflicting account of London's Soho redevelopment, that questions the validity of this revanchist explanation, is set out in the paper by Collins, grouped in the following sub-theme, since its primary focus is to investigate the development of urban 'gay villages' in England.

The paper by Cameron also provides general theoretical reasoning to explain how risk conditions the nature and development of sex markets in various locations from both demand-side and supply-side perspectives. A higher-resolution urban sociological analysis of the management of such supply-side risks, as faced by street-based sex workers, is presented in the paper by Sanders. Based on extensive ethnographic fieldwork in Birmingham, England, she articulates in extensive detail the personal strategies and coping mechanisms employed by these sex workers 'on the ground' and how they connect with the nature and terrain of their urban spaces. That some commercial (hetero)sexual spaces may co-develop with, or come to be dominated by, more discernibly gay urban spaces is an observation that is picked up within various papers in the next section theme.

The Evolution and Development of Urban Gay Spaces

Quite obviously, in terms of absolute numbers, there are fewer homosexuals than heterosexuals, so clustering is a probable long-run equilibrium feature if partner search and gay lifestyles are to be sustainable. Some slight revision to the efficiency frontier locations would also be necessary in the model context to reflect that both cross-gender **M**-type characteristics as well as gay and lesbian **M**-type characteristics are more likely, or more easily found in urban settings. This can be explained in terms of the more liberal social climate one might expect in the city as compared with a deeply rural setting and the higher density of gay and lesbian people already in the city. These explanations have been given recent popular cultural expression in the 2004 UK British Broadcasting Corporation comedy series *Little Britain*, where the rural Welsh character of Dafydd repeatedly bemoans the difficulties of being "the only gay in the village".

Accordingly, the milieu by which the clustering of homosexuals has long been a discernible feature (at least since the classical era of ancient Greece) is the city environment. A broad historical and systematic global sweep of the connections and contexts by which homosexuality and the city are intertwined, across various continents, is presented in the paper by Aldrich. This task is effected by a panoramic review of an extensive range of literary, historical and social science book-length English language sources. These help to locate firmly the concept of the city in the broad gay imaginary. In so doing, he also alludes to the functionality and purposes of city locations around the world in supporting the conduct of gay lifestyles, partner search and commerce.

Moving on from this analytical *tour d'horizon* but retaining a strong historical perspective, the paper by Sibalis offers a high-resolution account of the evolution of a particular gay space in a single city—the Marais in Paris. Drawing on a wealth of historical documents, as well as introducing more contemporary sources as required by the unfolding chronology, he traces its developmental path and status in Paris and France. Sibalis draws attention to a number of themes, such as the enabling role of enterprise and the increasing assimilation of the area into the mainstream social landscape. Many of these themes are also echoed in the general developmental model of urban gay villages in England, advanced in a later paper by Collins.

The papers by Collins and Bell and Binnie both draw attention to the fact that the presence of gay communities now features in the competitive strategy toolkit of entrepreneurial city governance. The latter paper draws particular attention to the implications they perceive this has had in terms of tightening regulation in some sexualised urban spaces. For Bell and Binnie, this challenges the authenticity of such spaces as they unfold or are transformed into acceptable expressions of a 'new homonormativity'. These they seem to consider as engineering or inviting assimilation into the urban social mainstream. For Collins, stochastic processes, *ad hoc* development and gay community entrepreneurial effort are also key drivers shaping urban gay space. Further, the assimilation process is considered more as an inevitable outcome of an evolutionary development process, although it is contended that this need not challenge the existence or sustainability of a commercial gay presence in the city.

Turning attention to a more contemporary Asian focus, the paper by Lim investigates the socio-spatial expression of homosexuality in the island city-state of Singapore, a society where homophobia remains rife and where Singaporeans are not formally permitted to speak publicly about homosexual-related issues. Yet despite these restrictive socio-political constraints, he finds that homosexual expressions are becoming increasingly visible and their spatial manifestations are challenging the course of heteronormativity. This, he argues, has led to discontinuities in the social production of heterosexual space and has helped to advance the inclusion and broader cohesion of urban gay lifestyles.

Urban Sexuality

In the conceptual model previously outlined, the scope for changing relationship preferences over the course of one's lifespan was raised in the simplified model context of adjustment from a more **S**-orientated set of preferences, towards a greater **M** orientation, the latter possibly being associated more in a heterosexual context with the prospect of child rearing. This adjustment process is given some voice and a degree of validation in the paper by Brewis. She charts how work and life experiences as well as family and peer expectations have triggered and shaped trajectory changes, over a wide range of preferences, for a group of working British women in their 30s. These changes relate to career aspirations, desired relationship outcomes and altered reasoning regarding urban, suburban and rural residential and work location. Drawing on a detailed analysis of focus group transcripts, she also implicitly questions how comfortably these changes and experiences sit with the popular cultural representations of urban sexuality that are presented for such an age cohort of working women (for example, the US television drama series *Ally McBeal* and *Sex and the City*).

Drawing also on focus group evidence as well as interviews, the paper by Skeggs, Moran, Tyrer and Binnie examines urban sexuality from a principally gay male perspective. By further contrast, it develops a much more explicit analysis of the contribution of popular cultural representations to examining the nature and authenticity of a gay urban imaginary. Its geographical focus is Manchester, England, and its key cultural reference is the 1999 UK Channel Four television drama series *Queer as Folk*. The paper explores questions of authenticity and sexual politics. It also examines the sense of place provided by viewing through the cultural lens of this production's *dramatis personae*, controversial plotlines and scene locations. Additionally, the authors point to the significance of some snappy reactive entrepreneurial vision in drawing upon and effectively constituting this visual imaginary of gay Manchester. By these means, they helped to refresh the tourism potential of the 'rather *passé*' gay village by making it trendy once again on an international as well as a national level.

Concluding Remarks

This collection provides a useful platform for those engaged in social and economic explorations of urban sexual themes. It was not from the outset intended to offer comprehensive coverage of such themes. Indeed, while this collection of papers has offered contributions from a broad range of disciplinary approaches on the interplay of sexual desire and urban living, there are undeniably some clear topics of merit that have been underplayed or wholly neglected. Further, even for those themes accorded considerable research attention, the geographical focus has arguably been particularly directed towards developed economy contexts. There remains still considerable work to be done beyond the involvement of social anthropologists, in providing more systematic analysis of the economic evolution, spatial development and regulation of sexualised spaces in various developing country contexts. Of related concern are the spillover effects that sex-related commerce may have on other urban and rural centres—for example, in labour market terms—and also in terms of the linkages with overlapping geographies of borders, organised crime and human trafficking.

Another issue warranting attention within this broad research agenda pertains to the assessment of the urban and regional economic effects of sex-related commerce. Newspapers and various web media sources have, it seems, frequently turned to unusual academic sources for quotable headline numbers regarding the money flows associated with sex-related enterprise. Many feminist studies, sociology and criminology academics have, it seems, not been slow to offer or reinforce crude estimates of the revenues, profits, turnovers and local/regional economic significance (in money numeraires) of sex-related commerce (see, for example, Hughes, 2000, 2002; and McCaughey, 2004), as well as the monetary scale of sexual consumption.[2] This is perhaps somewhat surprising given that researchers in these disciplines are not generally noted for their interests or skills in financial analysis and regional economic modelling. Nor do they tend to offer such economic and financial projections in other policy contexts. Such estimates are typically deeply flawed by implausible assumptions and a lack of caveats regarding data limitations and probable confidence intervals. That said, in part, these crude figures may merely have arisen to fill a relative void of formal economic modelling work in these market contexts and, accordingly, the exceedingly rare appearances in appropriate peer-reviewed journals.

Another potentially rich vein of research endeavour relates to relatively few detailed economic, spatial and social impact assessments of various specific planning and urban policy initiatives on sex commerce regulation. In terms of the accumulation of knowledge to support the development of evidence-based public policy, there seem to be few attempts to systematically gather such data for 'before and after' and comparative studies. What could Edinburgh really learn from Hamburg—for example, in terms of the design and management of various gradations of 'sex for sale' toleration zones? Do they actually minimise land use and user conflicts and other related negative externalities, or merely simply modify their spatial margins? Other initiatives worthy of similar formal impact assessment could involve attempts by some cities to induce or incentivise relocation of sex business away from central business districts or residential areas, towards industrial warehouse districts, or other low value and low density land areas. The use of out-of-city locations for brothel residences features in a number of US states such as Nevada. Yet, in the context of mature city-based sex markets featuring legal brothels (for example, in various cities in Greece, the Netherlands, Belgium, France and Australia), other approaches that have not yet been subject to widely reported formal economic and social assessment include regulation by increasing or decreasing the supply of brothel licences. In the context of Athens, Greece, this is a subject of considerable topical public interest and controversy, as the city authorities have planned to increase the number of brothel licences in time for the 2004 Olympic Games—presumably with a view to boosting tourist spend in the city economy.

Yet despite all these research gaps and

as well as preventing them from clustering in central business districts. The extreme case of this comes from the Progressive Policy Institute's publications on the 'new economy' (see their website at http://www.neweconomyindex.org/metro/part6.html) which seeks to give guidance to American urban development, suggest demolition of such areas as desirable to eradicate prostitution, although one wonders where they think the activity is going to move to. This 'shifting of sex zones' is a relatively recent focus of attention in urban policy as it has emerged in the past 100 years (Frances, 1994). Daniels (1984) observed, of English-speaking countries that

> The growth of an urban middle class which accompanied the industrial expansion of the 19th century created a class of leisured wives and daughters who sought to use urban space in new ways, most notably by shopping and promenading in the central business districts. A variant of this was the fashionable Melbourne pastime of 'doing the Block', or promenading around the Collins, Swanston, Bourke and Elizabeth Streets block of shops. With more 'respectable' women using the streets, the presence of what they regarded as 'nuisances' had to be minimised and preferably eliminated. Hawkers, beggars and drunks were all targets of this campaign, but prostitutes were especially targeted. The reason for this is obvious: while it might be annoying for a bourgeois woman to be accosted by beggars and so on, it was extremely embarrassing for her to have to encounter 'fallen women' and, worse still, to be mistaken for one of their number and propositioned by men (Daniels, 1984).

Where the sex economy is regarded as a deviant and only reluctantly tolerated sector of the city in the North American and British economies and to a lesser extent in some European, Asian and Latin American cases, its contribution is downplayed and there is no coherent policy towards it. Returning to the question of the Progressive Policy Institute's suggestion to demolish sex enclaves, clearly, in the short-run the volume of trading may fall in absolute terms. There will be some displacement to more fragmented trading within the jurisdictional area and there may be some migration of demanders and suppliers to nearby regions. The latter case would in some circumstances create an externality problem of exporting 'social pollution' across urban centres where there is an excess amount of spending due to lack of co-ordination between different policy jurisdictions that are competing with each other in the export of 'undesirable sex markets'.

This kind of problem has been highlighted in general economic models of crime as a form of negative externality (Weicher, 1971; Hellman and Alper, 1997). In the longer run, one would expect the dedicated enclave simply to re-emerge in another run-down area as it is unlikely that we could have perfectly balanced growth in a city where some locales do not become economic graveyards.

The notion of demolition or razing the sex market enclaves of cities is more a ritualistic posture of cleansing a disease from the social body of the community than a rational economic development policy. It also confuses cause and effect. That is, how can we attribute urban blight to the presence of sex markets when there has been a prevalent policy stance of forcing sex trade into areas that are already blighted.

5. Typology of Sex Market Maturity

Having looked at the idea of a city and its sex market as a club good, we now turn to the issue of how markets are able to develop under the limitation that sex trading is circumscribed by policies, reflecting greater or lesser degrees of tolerance, as described above. The term maturity is used here in the sense that something which is 'mature' has had the opportunity to reach its full potential. In the present context, we are talking about economic maturity so that complete maturity would be where sex trade is able to establish a level of expansion commensurate with what would be obtained if it were treated like more non-contentious everyday consumer goods. I do not enter into any debate about what is the morally correct strategy for the regulation of a city sex economy.

5.1 The Sporadic Sex Economy

In the sporadic sex economy, fringe criminal or poorly equipped entrepreneurs run isolated, usually poor-quality, establishments (or in the extreme case sell their bodies on the streets) in low-value economic spaces with little expansion potential. There is little continuity in such markets as traders are forced to move due to episodic enforcement of the law or restrictive application of ordinances in letting and licensing. This creates spaces which lack feelings of safety for consumers and street sex sellers, thus further curbing the economic development of such spaces. Sporadic markets have a self-perpetuating nature as, being isolated enclaves, they also lack the property of providing excuses for being in the area for other reasons and thus are not good 'opportunity spaces'. There will be some concentration of sellers in the market due to the Hotelling model type of reasons but this will not be sufficient to push the sex economy into a more dynamic phase and hence we would not term it clustering. These weak clusters of the sex economy may even be located on the outer ring of a city's structure, but they remain sporadic because of lack of linkages with more tolerated activities in the inner rings. The area around the King's Cross railway station in London is a good example of a sporadic sex economy. There are other examples to be found in Britain's cities, a notable example being Belfast where traditional religious fervour meant that there were no sex shops at all for a long time: currently there is one, Anne Summers high street lingerie shop (as in most UK towns of any size), and a very small collection of porn/sex toy/etc. shops on the less traversed edges of the Smithfield market area which is in the shadows of the major high street retailing enclosed shopping centre.[1]

5.2 The Partially Clustered Sex Economy

The current idea of a market cluster is usually defined with respect to the ideas of Porter (1998) who considered it to be a group of industries connected by specialised buyer–seller relationships or related by technologies or skills. There is a degree of hair-splitting in deciding whether a prostitute or lap dancer or booth stripper are in the same industry or different segments of the same industry, but the general Porterian concept seems to apply to sex markets. For example, there are cognate skills or technologies employed throughout the industry. One skill that is common to these employments is putting new and apprehensive, or morally inhibited customers, at ease.

In this case, we have groups of sellers of the same products in specific locations which may also tend to be assisted by the presence of other types of trade in the locations to provide increased opportunity to visit the area. The clustering will be due to more than the Hotelling type reasons—that is, clustering provides variety in the offer of products. Partial clustering can take place in low-value spaces if they are sufficiently easily accessed in a safe manner. The low-value spaces may also be more likely to be tolerated by authorities if they are located with respect to higher-value spaces in such a way that they do not threaten them or indeed may enhance them. The classic model for the partially clustered sex economy can be seen in various phases of the late 20th-century development in Soho, London, of sex book/video shops and strip clubs and in the window prostitution of Amsterdam.

5.3 The Partially Laddered Sex Economy

The term 'laddering' is being here applied to some degree of directional clustering from the view-point of the consumer. That is, they may move up or down between products which offer them different degrees of safety from various costs. Most of what was said in the last category also applies to this one. We may of course have combinations of the degree of clustering and laddering in any given sex market but laddering is conceptually a higher stage of development than clustering. The laddered sex economy provides entry points for those who feel more comfortable 'cruising' an area with an eye to becoming a regular consumer of sex products. Whilst cruising, they get the opportunity to view other goods and services they may not have previously been aware that they wanted. The existence of such demand complementarity linkages creates an incentive for firms to locate near each other in order to reap the benefits. Nevertheless, ambiguous partial tolerance and zonal restriction prevent the emergence of full-scale laddering and clustering.

5.4 The Mature Sex Economy

The mature urban sex economy displays a high level of both clustering and laddering. The extreme case would be where a whole city is geared up to the provision of sex goods and services. This is difficult to find in the real world. For example, the sex economy of Amsterdam is far from being the totality of the city's trading. It does, like the sex economy of Vienna, exhibit pronounced clustering and some degree of laddering but these are not fully developed due to limits on the nature of franchising and licensing. The general point to be made here is that legalisation of trade does not necessarily guarantee a fully mature sex economy unless the legalisation is across the board. For example (Chapkiss, 1997, ch. 7), the 1971 Nevada leglislation provided legal brothels but only in smaller counties, thus partitioning the gambling sin economy and the sex sin economy, by allowing legal full-scale gambling in the urban areas.

The Amsterdam and Viennese situations display the existence of a full sex economy rather than just isolated forms of trade. There is overt prostitution plus 'cut-down'/partial prostitution (that is, in Vienna the '*kabinsex*' provides 'hole-in-the wall' relief for men from an unseen participant) and nearby other mild forms of titillation in clubs and bars and porn vendors of print and filmed product. This is a pedestrianised model where the consumers circulate through the levels of provision with wide scope for search over the product aspects.

The concentric sex market model is one where we have the most socially unacceptable (partially and reluctantly tolerated) in the outer 'ring' of the conurbation which may literally be on a ring road which facilitates entry into the circles of the sex markets. That is, trade will be away from high-value residences and retail and largely out of sight of those most likely to experience discomfort from its presence. Total maturity would involve increasing levels of tolerance as we move successively into the inner rings. At the innermost ring, in the central business district, we have the 'sanitised' sex economy of hotels with pay-for-view porn, erotic lingerie shops alongside the standard high street stores, and erotic sections within mainstream retailers of print and digital media. This inner ring may contain the hostess club, lap dancing club, escort service providers, etc. if these can be sited and presented in a way that is conformable with prevailing social norms. This requires that sex trading by such specialists be normalised as a business practice and thus is free from associations with drug dealing, extortion, etc. which will require it to be pushed into an outer ring.

A good example of a move to maturity in the UK can be found in the movement of the 'Spearmint Rhino' lap dancing chain into Birmingham, one of the largest cities in the UK. Here, Spearmint Rhino have an out-of-town motorway-situated club, but also a large venue on the edge of the centre of town in an area of potential regeneration. Birmingham is in a process of regeneration having demolished its notorious 'Bull Ring' inner-city shopping centre. The new inner-city Spearmint Rhino has a large ostentatious façade in the style of a mock Edwardian public house or music hall. This is in marked contrast to the semi- or fully legal *bordellos* in some fairly tolerant and mature sex markets in continental Europe which use fairly anonymous frontages (i.e. explicit absence of descriptions of the purposes of the premises). The style of the inner-city Birmingham Spearmint Rhino is significant as it sends the iconic message of 'harmless fun' of a sexual nature outside a designated 'out-of-the-way' market tolerance zone. Not all Spearmint Rhino outlets are of this type as there are some which attempt to convey the 'anonymity' message by appearing more like a fairly middle-market health club. This kind of tailored niche marketing is a considerable factor in creating a drive to maturity in city paid sex markets.

A mature sex economy clearly can be beneficial to the economy of a city if fully developed and, in principle, social welfare can be higher with specialisation of certain cities in the sex economy while others specialise in other goods and services. The mature sex market creates multiplier effects of spending from the trade generated by inner-circle integration in the life of the city. This may also provide a laddering effect as individuals progress into other sex market goods located in the further-out circles of the city. The location of an outermost circle on a ring road with easy motorway access creates an additional revenue stream from the 'exporting' of consumers of the least-tolerated goods/services from other areas. In the mature sex economy, we expect the degree of tolerance to be reflected in licensing over where a seller can be in the ring. That is, there will not be raids, arrests or disruptions to the business practices in the outer ring even if they are strictly speaking illegal. Their illegality will simply be ignored or the cover used to disguise the establishment will not be too deeply probed. In a fully mature sex economy, street prostitution of the most literal type would tend to erode as the vendors are likely to be integrated into a licensing system of the Amsterdam 'window prostitute' type or esconced in the legally fronted quasi-*bordellos* of massage parlours, etc. Even if all forms of prostitution are legal, it would be more efficient for the protection of consumers and suppliers if they are building-based rather than street-based.

Any vestigial trace of street prostitution will reflect either excess supply which may be due to poverty—that is, those who are destitute or addicted but unable to gain employment in sexual trading may attempt to sell themselves by undercutting the market price—or violating normal safety practices which are upheld by vendors in the normal building-based market. As I am here discussing the maturity of sexual markets from an economic point of view, I have not said anything about issues of whether regulators will use health checks, curfews, etc. on prostitutes. However, if we are to have integration of sex markets into the general framework of markets, then ultimately they have to be subject to health and safety regulations, worker rights and product liability in the same way as all other goods. These issues are coming to maturity in some cases as sex workers are now joining trade unions in the service sector. In the less-than-fully-mature sex economy, there is some *ad hoc* provision for the welfare of employees. Depending on the city in question, this takes the form, in the UK at any rate, of multiagency approaches from differing charities, health organisations, other NGOs or temporarily funded project bodies.

6. The Internet and Sex Market Maturity

The influence of the Internet on sex markets is indisputable. Anyone with a credit card and modest abilities in using a search engine can access porn images and buy sex toys in a very short space of time. Even if they are not trying to do these things, they are highly likely to receive such information by accident whilst searching for other things, through pop-up advertising or even overt 'page-jacking' by intrusive software. Although we have commented on its influence at certain points above, it is probably useful to provide some specific commentary on how it may influence the functioning of cities as sex club economies.

The rapid growth of the Internet is a dynamic element which disrupts the established immature and semi-mature urban sex markets. There are a number of different *nega-*

tive effects which the internet has on the prosperity of urban sex centres

(1) The capacity to render the product in digital form would seem to undermine factors leading to local specialisation in the sex market. That is, one can obtain audio-visual products down a modem or ISDN connection without having actually to visit a retail outlet in person or have a product delivered by mail order. This would erode the need for clusters of video stores or porn cinemas.
(2) The growing feeling of security over on-line credit card transactions in terms of both guarantee of service and escape from detection, by those who may disapprove, will intensify desires to use the Internet instead of an on-site purchasing method.
(3) Whilst it might be felt that there are sex services for which the Internet cannot provide adequate substitutes, one could argue that it may provide more intimacy for consumers of overt sex acts. The live webcam broadcast is a fairly close substitute for the 'peep show' type of club performance, but it may provide more intimacy for the consumer than they can obtain in the live performance situation. Live, and static, sex websites seem to exhibit a quasi-celebrity 'fan' relationship between suppliers and consumers. This is reflected in them often following the generic model of a website for an actor or musician—that is, sections for a guestbook, requests, news, updates, etc. This grants a certain ownership and influence to the consumer, at a 'safe distance', which has traditionally been excluded from most live paid sex transaction situations.

Given the above factors, it would seem that Internet sales can remove the advantages of a clustered sex market as it does not matter where the product is sold from and Internet sales may have certain advantages for consumers. However, there may be a number of spillovers from increased Internet sales to increased commercial sex in the city due to various factors. An Internet presence may enhance the sales of established brands in the sex market such as—for example, *Playboy* magazine. Internet advertising may alert people to the presence of traders in their own region of which they were previously unaware. Internet information dissemination may also attract additional sex tourists to established mature urban sex markets. This can span a number of dimensions such as reassurances on risk and product quality from previous consumers.

7. Concluding Remarks

This paper has looked at the total collectivity of sex market activity as an analytical category rather than delving into individual case studies or the statistics for the individual product categories. The factors behind the development of specialist urban centres of sex market activity have been identified and a typology of the maturity of sexual markets has been proposed. Entrepreneurs in these markets have been circumscribed by risk and opportunity due to the unpredictable and ambiguous legislation under which they have operated. This does generate high profits for some entrepreneurs, but it curtails the overall expansion of the market even in the cases where the market is at a relatively mature growth state. The continued expansion of the Internet presents a new element in the evolution of paid sex markets. At first sight it may appear to undermine the stability and structure of paid sex markets in cities, but it may be that it will ultimately facilitate the transition to mature sex economies.

The above remarks relate solely to the issue of the best self-interested policy by urban planners in one location. It may, of course, be the case that formal or informal competition over the sex trade could be counter-productive to some extent due to lack of co-ordination between urban centres. That is, a degree of specialisation in trade in nearby cities may be welfare-improving. For example, a city which is in economic decline near to one which is growing, could in effect 'hive off' the less tolerated transactions and become the hub or periphery of the growth pole city in the region. The main obstacle to such a policy is the difficulty of ensuring enforceable side payments (or transfers) between the gainers and losers from the policy.

Note

1. This information is based on the author's personal experiences of visiting the city in December 2003 and on many earlier occasions.

References

BERLANT, L. and WARNER, M. (1998) Sex in public, *Critical Inquiry*, 24(Winter), pp. 547–566.

BERNSTEIN, E. (2001) The meaning of the purchase: desire, demand and the commerce of sex, *Ethnography*, 2(3), pp. 389–402.

BINNIE, J. (1995) Trading places: consumption, sexuality and the production of queer space, in: D. BELL and G. VALENTINE (Eds) *Mapping Desire*, pp. 182–199. New York: Routledge.

BUCHANAN, J. M. (1965) An economic theory of clubs, *Economica*, 32(1), pp. 1–14.

CAMERON, S. (2002) *The Economics of Sin: Rational Choice or No Choice At All?* Cheltenham: Edward Elgar.

CAMERON, S. and COLLINS, A. (2000) *Playing the Love Market: Dating, Romance and the Real World*. London: Free Association Books.

CAMERON, S. and COLLINS, A. (2003) Estimates of a model of male participation in the market for female heterosexual prostitution services, *European Journal of Law and Economics*, 16(3), pp. 271–288.

CHAPKISS, W. (1997) *Live Sex Acts: Women Performing Erotic Labour*. London: Cassell.

CHRISTALLER, W. (1933/1966) *Central Place in Southern Germany*, trans. by C. W. BASKIN. Englewood Cliffs, NJ: Prentice-Hall.

DANIELS, K. (1984) Prostitution in Tasmania during the transition from penal settlement to civilised society and 'St Kilda voices', in: K. DANIELS (Ed.) *So Much Hard Work*. Sydney: Fontana (quoted in: http://www.hartford-hwp.com/archives/24/230.html).

FELS, R. (1980) The price of sin, in: H. TOWNSEND (Ed.) *Price Theory: Selected Readings*, pp. 302–303. Harmondsworth: Penguin.

FRANCES, R. (1994) The history of female prostitution in Australia, in: R. PERKINS, G. PRESTAGE, R. SHARP and F. LOVEJOY (Eds) *Sex Work and Sex Workers in Australia*, pp. 27–52. Sydney: University of New South Wales Press.

HELLMAN, D. A. and ALPER, N. (1997) *Economics of Crime: Theory and Practice*. Needham Heights, MA: Simon and Schuster.

HIRSCH, F. (1976) *The Social Limits to Growth*. Cambridge, MA: Harvard University Press.

HOTELLING, H. (1929) Stability in competition, *Economic Journal*, 39, pp. 41–57.

HUBBARD P. (1998) Community action and displacement of street prostitution: evidence from British cities, *Geoforum*, 29(3), pp. 269–286.

JACOBS, J. (1970) *The Economy of Cities*. New York: Vintage Books.

KNOPP, L. (1995) Sexuality and urban space: a framework for analysis, in: D. BELL and G. VALENTINE (Eds) *Mapping Desire*, pp. 149–161. New York: Routledge.

LATTIN, J. M. and MCALISTER, L. (1985) Using a variety-seeking model to identify substitute and complementary relationships among competing products, *Journal of Marketing Research*, 22, pp. 330–339.

LINDELL, J. (1996) Public space for public sex, in: E. G. COULTER *ET AL.* (Eds) *Policing Public Sex: Queer Politics and the Future of Aids Activism*, pp. 73–80. Boston, MA: South End Press.

LÖSCH, A. (1939/1954) *The Economics of Location*. New Haven, CT: Yale University Press.

LYOTARD, J. (1993) *Libidinal Economy*, trans. by I. H. GRANT. Bloomington, IN: Indiana University Press.

MANSKI, C. (2000) Economic analysis of social interactions, *Journal of Economic Perspectives*, 14(3), pp. 115–136.

MARCUSE, J. (1955) *Eros and Civilization*. Boston, MA: Beacon Press.

MCCOY'S GUIDES (1998) *McCoy's British Massage Parlour Guide No. 4*. Stafford: McCoy's Guides.

MCMILLEN, S. (1998) Adult use and the First Amendment: zoning and non-zoning controls on the use of land for business. Land Use Law Center, Pace Law School (http://www.pace.edu/lawschool/landuse/adult.html).

MOFFATT, R. (2004) Prostitution, in S. BOWMAKER (Ed.) *Economics Uncut: A Complete Guide to Life, Death, and Misadventure*. Cheltenham: Edward Elgar (forthcoming).

MORGAN, S. and TRIVEDI, T. (1996) The order of the brand choice process revisited: some new perspectives on measurement and data issues, *Journal of Business and Economic Statistics*, 14, pp. 221–229.

O'TOOLE, L. (1998) *Pornocopia*. London: Serpent's Tail.

PORTER, M. E. (1998) *The Competitive Advantage of Nations*. New York: Free Press.

REYNOLDS, H. (1981) *Cops and Dollars: The Economics of Criminal Law and Justice*. Springfield, IL: C. C. Thomas.

SANDLER, T. and TSCHIRHART, J. (1980) The economic theory of clubs: an evaluative survey,

Journal of Economic Literature, 18(4), pp. 1481–1521.

SCHLOSSER, E. (2003) *Reefer Madness: And Other Tales from the American Underground*. London: Allen Lane.

SCITOVSKY, T. (1986) *Human Desire and Economic Satisfaction*. Chichester: Wheatsheaf Books.

TOWN OF HYDE PARK (1996) *Adult use study of the Town of Hyde Park*. February.

TUCKER, D. M. (1997) Preventing the secondary effects of adult entertainment establishments: is zoning the solution, *Journal of Land Use and Environment Law*, 12(2), pp. 385–431.

WEICHER, J. C. (1971) The allocation of police protection by income class, *Urban Studies*, 8(2), pp. 207–220.

WILSON, E. O. (1975) *Sociobiology: The New Synthesis*. Cambridge, MA: Harvard University Press.

WRIGHT, R. (1994) *The Moral Animal: The New Science of Evolutionary Psychology*. New York: Vintage.

The Changing Nature of Adult Entertainment Districts: Between a Rock and a Hard Place or Going from Strength to Strength?

Andrew Ryder

Introduction

In recent years, particularly in US cities, there has seemingly been a spectacular decline of traditional adult entertainment districts.[1] This is often attributed to increased policing, new land use controls, better enforcement of existing regulations and a general 'get tough' attitude towards such activities. Many point to New York City under Mayor Giuliani as an example of what can be done through 'zero tolerance' policing, new land use ordinances, new adult entertainment laws and stricter enforcement. However, the alleged success of zero tolerance ignores profound changes which have occurred in the field of adult entertainment and in the underlying moral climate since the 1960s. The article argues that it is these which have caused a decline in the number of outlets and led to their spatial deconcentration. These changes parallel those in the entertainment sector as a whole. It seems likely that zero tolerance itself is not the cause of decline and change, but rather that it has coincided with changes in the adult entertainment industry. In addition, the apparent success of zero tolerance ignores a shift in sales outlets for many adult products, from specialist stores to newsagents and mainstream stores.

This article examines the location of adult

Andrew Ryder is in the Department of Geography, University of Portsmouth, Buckingham Building, Lion Terrace, Portsmouth, PO1 3HE, UK. Fax: 023 9284 2512. E-mail: andrew.ryder@port.ac.uk. *The author wishes to thank three anonymous referees for their useful comments.*

entertainment districts and different adult entertainment outlets in cities, looking particularly at New York City and the Times Square area. It focuses on peep shows, book stores, video outlets, clubs and cinemas, but not prostitution which, one might suggest, is far more footloose.

It begins by discussing the nature and location of adult entertainment zones in cities in a general sense. It reviews their history and the literature about them, and attempts to predict future trends. It discusses the location of adult entertainment districts in cities, enumerating locational criteria and arguing that accessibility is a key factor. It proposes a model of the location of such districts and the activities in them, then examines a range of 'command-and-control' techniques which have been used to control the development and spread of adult entertainment outlets and districts. It proceeds to examine New York City in greater detail, looking particularly at the Times Square area. It evaluates the impact of 'zero tolerance' policies on the area, looking at underlying reasons for its apparent success. The paper also briefly examines the rise of specialist districts, suggesting that these have a long but unrecognised pedigree. It suggests that, in many cases, specialist districts constitute a different kind of adult entertainment district which is organised on the basis of different locational constraints.

The paper argues that adult entertainment districts are part of an industry which has always been in flux and continues to evolve and change, reflecting changes in the broader entertainment industry, as well as constraints which are specific to the adult sector. These include zoning, legal constraints and efforts to license, restrict or control the growth and content of the sector. Until recently, straightforward economic location factors, such as access to customers and ability to pay rents, only partly determined the location of such districts, since all adult clusters were constrained by a layer of government regulation and moral stigma. Recently, the redefining of 'adult' suggests a relaxation of moral standards, although this seems, paradoxically to have been accompanied by attempts to increase control over the sector. A key argument in this paper is that until recently the traditional adult entertainment district was largely based on clustering due to a lack of information among customers about the location and nature of retail outlets and their merchandise. The growth of published information, information on the Internet regarding stores and services, and the sale of products by mail and over the Internet have caused traditional districts to decline and have inhibited their rise elsewhere. More significantly, this paper also argues that the apparent decline in adult outlets is because 'adult' entertainment has been redefined to exclude a wide range of activities which formerly came under the classification. All this, rather than the use of command-and-control techniques such as zoning or licensing, has led to the decline of adult entertainment districts and a shift in the distribution of outlets.

Adult Entertainment Zones in the City

Since the 19th century if not before, adult entertainment districts, often referred to as vice districts, have been features of cities. Sociologists and their precursors have written about such districts in cities since the mid 19th century (Booth, 1889; McCabe, 1872, pp.186–193 and 579–617; McCabe, 1882, pp. 154; 250–257 and 640–643). In 1925, Park and Burgess included a vice zone in their description of Chicago and, in 1926, Reckless wrote about the distribution of vice in the city from a sociological perspective. More recently, Ashworth *et al.* (1988) have written about red-light districts in the west European city.

Adult entertainment districts are often distinctive, well known and persistent, and include the Bowery and Times Square area in New York City; the Reeperbahn in Hamburg; Pigalle, Rue St Denis, St Lazaire station in Paris; London's Soho; Zeedijk in Amsterdam; Gare du Nord, Brussels; and East Colfax in Denver. In San Francisco, an adult entertainment or 'vice' district—the Tenderloin—has been present since the city's founding and, far from being marginalised, appears to have been closely linked with the city's growth (Shumsky and Springer, 1981). In many older cities, adult entertainment districts are found in one or more of just a few locations: alongside the business core, at gateways to the city, such as large railway stations or bus terminals, or alongside waterfront or port districts. However, particularly in urban systems characterised by large numbers of local governments within metropolitan areas, adult districts may move to more friendly jurisdictions. In Hollywood and West Hollywood, Kenney (2001, pp. 38–39) argued that 'gay' activities moved into West Hollywood because it was outside the city boundary of Los Angeles and more friendly to non-sanctioned commercial activities. In some cities, such as Amsterdam, Boston, New York (to some degree) and London, adult entertainment areas have been concentrated in one or two districts, sometimes called 'combat zones', but in others, such as Paris (and increasingly New York), they have been scattered throughout the city. In Paris, Amsterdam and other cities, different districts contain different types of entertainment, aimed at different markets. Ashworth *et al.* (1988) noted that in Amsterdam the Rembrandtsplein and Leidseplein contained more 'respectable' and 'legitimate' forms of entertainment, whereas the Zeedijk was more 'risqué'.

Areas such as Times Square are exceptional, generally found in older cities, including Boston, Massachusetts; London, England; Paris; Amsterdam; and Hamburg. Zones in these cities are fairly central, often close to large transport centres or alongside the central business district. In younger cities, such as Portland, Oregon, clusters of adult entertainment outlets are found outside the urban core, located along main arterial roads (City of Portland Police Bureau, 1994, 1997; see Figure 1). In addition, particularly in American cities, adult entertainment outlets are scattered throughout the suburbs, although they avoid some jurisdictions in favour of others.

As long ago as 1925, Burgess argued that adult entertainment zones were found in areas of high population turnover where the social order had broken down and, since that time, such zones have generally been associated with moral decay and crime. Although the empirical evidence for this is ambiguous (New York City, Department of City Planning, 1994; Lasker, 2002), many observers have taken this argument for granted. Despite abandoning the view that adult entertainment is 'vice', recent work has continued to describe adult entertainment districts as part of a range of activities whose marginalisation in space reflects marginalisation in society, conflating social marginalisation with spatial marginalisation, both confusing correlation with cause and effect and anthropomorphising a process which may be far more random and accidental than it appears. Several researchers have written about adult entertainment districts in terms of geographies of difference and exclusion, arguing that adult entertainment zones are relegated to marginal areas due to social stigma, even conflating such activities with criminals and drug addicts (Papayanis, 2000; Winchester and White, 1988) (although this is not to say that those activities do not sometimes overlap). However, 'marginalisation' is an imprecise description of location. The 'marginalisation' in space of adult entertainment activities is no different from the 'exclusion' of other services to sub-central districts. Adult entertainment districts are no more marginalised than antiques districts, charity shops, industrial districts or used car lots. Describing adult entertainment districts as marginalised does not explain why they end up where they do. In most urban areas, there are many 'marginal' areas into which adult entertainment can go. In New York City, these include the Bowery, the far west side, the Times Square area, or the outer boroughs—so why Times Square in particular? Some would argue that proximity to theatres and live entertainment was the main factor and there appears to be a correlation with the development of the Times Square area as a theatre district and the growth of adult-oriented activities in the area (Eliot, 2002; New York State Urban Development Corporation, 1981). As Krugman (1996) has noted, chance

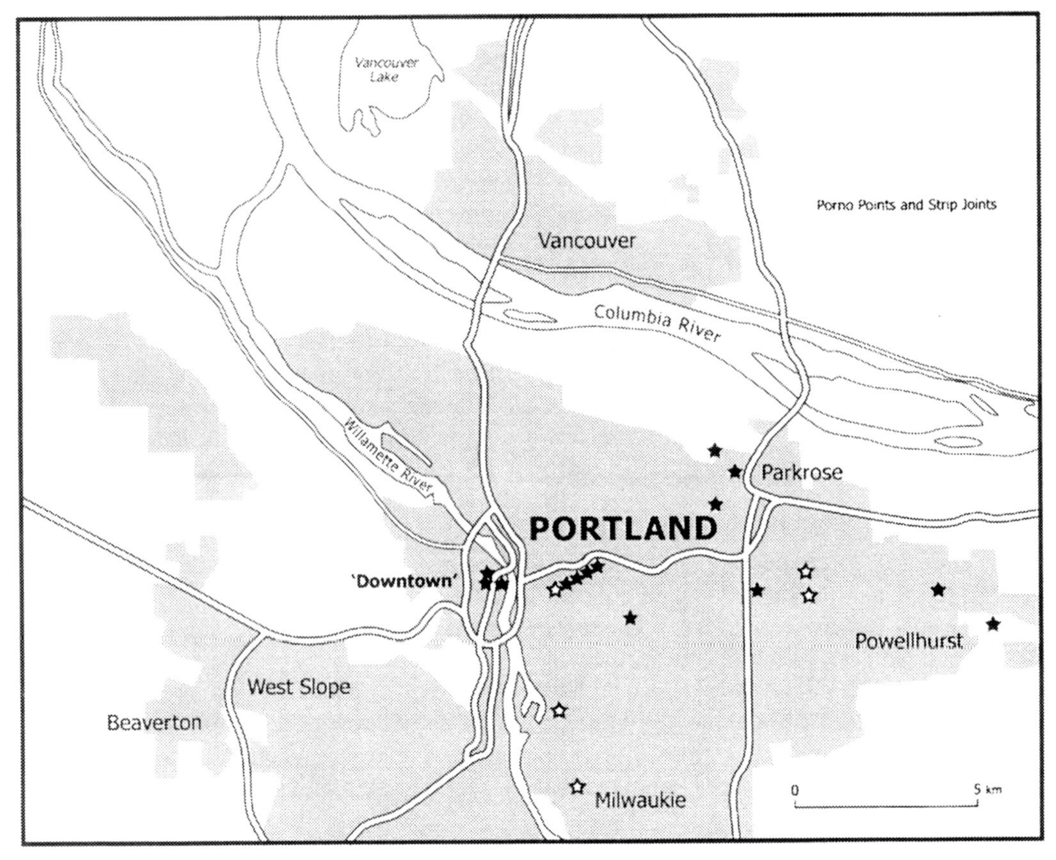

Figure 1. Adult entertainment outlets in Portland, Oregon, 1997. *Key*: White stars: existing in 1994, not 1997; black stars, existing in 1994 and 1997. *Source*: City of Portland Police Bureau (1994, 1997).

may also play an important role. For example, a book cluster developed on lower Fifth Avenue in Manhattan in the 1960s, when the popularity of the Barnes and Noble bookstore attracted a French and Spanish book store (a branch of a store in Rockefeller Center), a Russian book store, a Chinese book store and a Barnes and Nobles annex. The large numbers of people visiting the area attracted other activities. With the exception of the original Barnes and Noble, the book stores closed or migrated, and the street is now lined with fashionable clothing shops.

Moreover, marginalisation is not an accurate description of what is happening. In New York City, the Department of City Planning (1994, p. 27) suggested that adult outlets clustered together because of the 'copycat factor'. In other cases, outlets may be attracted to and in turn enhance a mix of alternative retailing and services. Writing about the 'red-light district' in Denver in 1983, McNee described East Colfax Avenue as having more variety than the "rather sterile, tightly zoned parts of Denver" and having "small shops … some with charming craft items and 'off beat' merchandise" (McNee, 1984, p. 23). Adult entertainment is also increasingly a mainstream business (Williams, 2002) and includes international and national chains such as Private, Ann Summers and Beate Uhse (Datamonitor, 2002).

In many cities, particularly in older, inner-city areas, adult entertainment districts have persisted in the same location for exceptionally long periods. In New York City, the Bowery had a reputation as a vice area from the mid 19th century onwards and, even today, is a kind of 'skid row'. By the start of the 20th century, New York's 'vice' district had migrated to the Times Square Area, gradually moving up Manhattan's west side, from 14th Street, through 23rd Street, to 39th–40th streets, between 6th and 9th Avenues, to what was then 'the north western fringe of developed Manhattan' (Eliot, 2002; New York State Urban Development Corporation, 1981). By the turn of the century, Times Square had 76 burlesque and strip theatres but, by the 1930s, there were just 33 and, in 1942, the last one closed. This was said to be due to the efforts of the then New York City mayor, Fiorello LaGuardia, who followed a policy of 'zero tolerance' towards adult entertainment, although some have suggested that his "destruction of the shady side of 42nd street" merely drove "the strip shows, the gambling, the bootlegging and the prostitution … ever further underground" (Eliot, 2002). Moreover, by the start of the 1970s, Times Square had 120 adult entertainment outlets (Times Square BID, 2000) and by 1976 it had 151 (New York City, Department of City Planning, 1994, p. 19).

In Boston, the area around Scollay Square was the traditional 'vice' zone. Attempts were made to bulldoze it out of existence in the early 1970s as part of an urban renewal scheme—it was even renamed Government Center—but the adult entertainment district remained in the area. However, book stores, peep shows, cinemas and related activities did not become widespread until the 1950s, after the easing of censorship and pornography laws. Ward (1975) noted that these outlets often appeared in skid rows. The benefits of clustering were obvious for consumers and shop-keepers: consumers knew where to go, in the absence of advertising or listings, and shops could capitalise on concentrated demand. In many cases, such districts became self-reinforcing, both encouraging similar activities and repelling dissimilar ones. As Firey (1945, p. 140) noted, "ecological processes which apparently cannot be embraced in a strictly economic analysis" appear to have been at work in many areas. "Locational activities are not only economizing agents but may also bear sentiments which can significantly influence the locational process" (p. 140). Although Firey was explaining the persistence of high-status residential districts in central areas, this may be true in a negative sense. Planners, local governments and real estate developers have argued that the reputation of a district as an adult entertainment zone may discourage new development: people fear crime, 'contamination', stigmatisation from being associated with adult entertainment by proximity and loss of property values and discouragement of custom (New York City, Department of City Planning, 1994, pp. 40–42, 53–53; Lasker, 2002, pp. 1157–1170; Los Angeles, Department of City Planning, 1977). Ward (1975) argued that such districts are more tolerant towards adult entertainment and become what Park called 'moral regions'

> Every neighborhood, under the influences which tend to distribute and segregate city populations, may assume the character of a 'moral region'. Such, for example, are the vice districts which are found in most cities. A moral region is not necessarily a place of abode. It may be a mere rendezvous, a place of resort (Park, 1952, p. 49).

Current attempts to control the spread of such outlets are generally based on command-and-control systems based on some form of licensing or land use zoning. It is currently argued that such controls have led to the elimination of outlets, or to their severe reduction. However, attempts to suppress such outlets in one area have often led to their appearance elsewhere. Equally, the disappearance of dedicated adult entertainment outlets has led to the distribution of adult material through new outlets. In New York City and elsewhere, the closure of book and magazine stores, and the changeover in others from printed material to videos, has actually resulted in more outlets selling adult material than before: newsagents open to all now sell a variety of adult magazines which were once only available in adult shops.

Therefore, despite the apparent decline in the number of adult shops in the city, outlets selling adult material have sharply increased. In the UK, changes in obscenity and censorship laws have effectively narrowed the definition of a sex shop. Since 2001, the licence is mainly based on the sale of adult videos depicting sex acts—R-18 (Farrow, 2001), while shops selling other sexually oriented products no longer require licences.

Where Is Adult Entertainment?

As long ago as 1926, Reckless (1926) suggested that vice is located in zones of transition, alongside the central business district. He argued that vice was found in slums ("The slum as the habitat of the brothel", Reckless, p. 168), but in slums of a particular kind, characterised by high population turnover where "social life tends to be unregulated and disorderly" (Reckless, p. 172). He further argued that such neighbourhoods were often characterised by population decline, adding that there was generally a correlation of high land values and low rents, in areas which were in "direct line of business expansion" (Reckless, p. 175) and therefore allowed to deteriorate—that is to say, in zones of acquisition rather than zones of discard. However, he noted that not all slums were adult entertainment districts and also noted that some activities were migrating away from city centres and were accessible only by car or taxi. Interestingly, in 1994, the New York City Department of City Planning argued similarly, writing that

> Adult entertainment businesses tend to be transitional, and locate in areas that are 'moving upwards'; they are rarely found in poorer neighborhoods. One statement submitted ... maintained that some major real estate developments owe their existence to the ability of landlords to warehouse property by renting space to adult businesses that are willing to accept high rents and short leases during the period when a major assemblage is underway (New York City, Department of City Planning, 1994, p. 26).

On the other hand, Rowley (1978) found that bordellos were a feature of the 'skid row' in Winnipeg, which was a zone of discard as the central business district migrated southwards.

Thus, in some cities, the adult entertainment district has been the same as 'skid row', in a former business district, along a waterfront, or in a semi-industrial area. However, adult entertainment districts are often separate from skid rows, although sometimes they over lap. Skid rows are characterised by low rents, but as Ward (1975) noted, are often highly accessible by public transport or car. Ward's description of the development of skid row is not dissimilar to explanations for the existence of neighbourhoods like the Times Square district. Skid rows develop in a spatially restricted city where most commercial activities cling to central areas. They remain where they are because of inertia; they develop in response to a heavy demand for single men's accommodation and warehouses are easily converted into 'hotels' to cater for the influx. The major passenger arrival points are (or were) located close to the CBD and were the initial entry points for early skid row residents, so that services competing for their custom located in the area. Ward wrote that skid rows in several cities were being renovated or becoming respectable and noted that this often led to displacement. For example, in Sacramento, former skid row residents moved a few blocks to the east of the old area. Ward also noted that in some cities, such as New York, Baltimore, and Washington, DC, they were losing population. He described Baltimore's skid row as an urban desert of vacant buildings. Bogue, writing in 1963, observed that most of the skid rows in some 41 cities were losing population. Like Park (1952), Ward described the skid row area as a moral district in which residents felt territorially secure, adding that

> further tolerance of this spatially selective tolerance is manifested in the presence of commercial activities such as adult books stores and girlie shows that frequently

> cluster on the edges of skid row districts. They do not cluster there in response to a demand for such services from the skid row inhabitants, but rather in response to the high tolerance threshold (Ward, 1975, p. 293).

Unlike skid rows, adult entertainment districts are often characterised by a high throughput of people and high anonymity. They are often near transport terminals, in areas where there is a high concentration of hotels. This is true of zones in Paris, London Kings Cross, Brussels, Amsterdam and other cities, and is particularly true of New York City. By the 1920s, Times Square had become a major interchange on the New York City subway system and the intersection of several streetcar and bus routes. It was within 10 minutes' walk of Penn Station and Grand Central Station, the two main rail gateways to New York, about 10 minutes from the terminus of the Trans-Hudson tubes and not far from several long-distance bus terminals. The opening of the Sixth and Eighth Avenue subways in the 1930s and of the Port Authority Bus Terminal after the Second World War enhanced its position as a transport node. By the 1960s, some 150 000 people were passing through the bus terminal every working day and several hundred thousand were passing through the district on the way to and from work. The area was also an important hotel district and the city's main theatre district. In its 1999 business report—for example, the Times Square Business Improvement District (Times Square BID, 1999) noted that 15 subway lines and 15 city bus lines pass through the Times Square area and that every year over 2 million buses pass through the Port Authority Bus terminal which adjoins the district, carrying about 57 million passengers. By the end of 2001, the local employee count was expected to reach almost 250 000. It was estimated that about 1.5 million people passed through the square every day. Of those living in the Times Square area, although the average income was slightly lower than that for the rest of the city, the same survey showed that 42 per cent were college educated and 76 per cent were high school graduates. In addition, the square was listed as one of the top 10 tourist attractions in the US, attracting about 27 million visitors annually. A 1998 survey showed that 61 per cent were under 35 years old, 44 per cent had a post secondary degree, the median income was $64 500 and only 32 per cent came from New York City. Thus, visitors to Times Square are young, single, male, educated and wealthy (Times Square BID, 1999, p. 21). Interestingly, this demographic group is very similar to that found using the Internet to buy sexually oriented products: a Nielson survey in 2000 found that only 6 per cent of adult content site visitors had annual incomes of under $25 000 (and only 35 per cent had incomes below $50 000), 58 per cent were under 35 years old and 44 per cent had a post secondary degree (2-year associate arts degree or higher). An additional 25 per cent had completed some university course (Datamonitor, 2002, pp. 33–34).

Anonymity is an important feature of many districts. The unsavoury reputation deters the casual visitor or those not interested, ensuring that customers will remain relatively unobserved. To some extent, the anonymity is a function of the high throughput. As Milgram (1970) observed, so-called illicit activities can flourish in areas with a high density of visitors. In 1970, he noted that 220 000 people were located within 10 minutes' walk of Times Square and argued that areas of such high density and throughput led to "overload", which "characteristically deforms daily life on several levels, impinging on role performance, the performance of social norms, cognitive functioning and the use of facilities" (Milgram, 1970, p. 1462). In such a situation, the creation of coping mechanisms and institutions "simultaneously protects and estranges the individual from his social environment" (p. 1462). He suggested that in such a situation, "moral and social involvement with individuals is necessarily restricted" (p. 1462). An unsavoury reputation can further deter casual visitors or, in the case of an area like Times Square, can deter lingering as people pass through quickly on their way to somewhere else, enhancing the anonymous nature of the area. He also suggested that such districts can even become invisible to the beholder, writing that

> Principles of selectivity are formulated such that investment of time and energy are reserved for carefully defined inputs (the urbanite disregards the drunk sick on the street as he purposefully navigates through the crowd) (Milgram, 1970, p. 1462).

In many adult entertainment districts, rents for shop space are relatively cheap. Although outlets selling sexually explicit material often pay high rents for store space, shops selling other adult-oriented products, particularly small start-ups, cannot. Thus, while often seeking centrality, they are forced to move outwards along the bid-rent curve in search of suitable locations. Here again, new shops may depend on propinquity to other similar establishments in order to capture custom. In Berlin—for example, after the fall of the Berlin Wall, the area of East Berlin around the Oranienburg Strasse and along the Schoenhauser Allee became an adult entertainment district (Waldau, 1990; Lothar, 2001). Shop space is a requirement, not only for adult entertainment districts, but for all entertainment activities. In New York City, by 1981, 8 per cent of all shop space in the immediate Times Square area was vacant and 29 per cent of office space above street level was vacant (New York State Urban Development Corporation, 1981).

On-going urban decline, particularly in the US and the UK throughout the 1960s and 1970s, may have been another reason for the growth of adult entertainment districts. In the UK, state policy aimed to reduce population in inner cities and restrict the expansion of offices and industry. In the US, it was due to a lack of new migrants to replace out-migrants. For example, New York's population fell from a post-1945 peak of 7 894 862 in 1970, to around 7.1 million in 1980—a loss of over 10 per cent (New York City, Department of City Planning, 2001). London's population fell from 7.977 million in 1961 to 6.770 million in 1988 (Greater London Authority, 2002). During this period, rents and property values in many cities collapsed and vacant space increased sharply. In the US, abandonment of buildings became widespread. Between just 1976 and 1979, New York City lost 150 000 apartments, mainly to arson (Quindlen, 1979) and, in 1978, the US General Accounting Office stated that abandonment and arson were a major threat to US cities (Roper, 1978). New York, Cleveland, Camden, New Jersey, St Louis and Toledo were said to have major problems and Chicago, Columbus Ohio, Detroit, Philadelphia and other cities were said to have substantial problems. The adult entertainment industry expanded to fill empty spaces. This is particularly true of cities like Detroit, where office workers remained concentrated in the city centre, while residential population declined.

Single occupancy rooms nearby are also a feature of adult entertainment districts in many cities. As Ward (1975) notes, these are characteristic of 'skid rows' but are also found in inner cities with older hotels which are no longer considered tourist class. New York's Times Square area is a good example of this. The area was overbuilt with hotels in the 1920s and early 1930s, many of which were converted to single room occupancy, let to non-tourists or let on a short-term basis in the 1960s and 1970s. In the 1980s and 1990s, many hotels moved up-market, tourism increased, occupancy rates grew and new hotels were built (Times Square BID, 2000). Moreover, today's car orientation means that places of assignation are no longer always alongside the location of the actual transaction.

Empty theatres or empty shops are not sufficient for the formation of a successful adult entertainment district. In New York City, in 1959, a local guide (Honan, 1959) noted that 14th Street east of 3rd Avenue in Manhattan had two movie theatres, one showing Spanish films. Nearby, on Second Avenue and 10th Street was a live theatre showing what it claimed was burlesque, and another theatre on 3rd Avenue and 13th

Street showed a mixture of older films and adult films. From at least 1960, there was an adult book store on the corner of 14th Street and 3rd Avenue. The area was a few minutes' walk from Union Square, location of two major department stores and a major movie theatre, The Academy of Music, which later became the Palladium. By the late 1960s, the Spanish cinema had converted to sex films (in English) before closing down. Another cinema across the street shifted from English films by the start of the 1970s and, by the mid 1970s, was showing Spanish-language sex films, before also closing. By then the department stores had closed and many stores in the area were empty. Despite a relatively central location, near to several subway lines, several bus lines and the PATH railway to New Jersey, despite empty shops and despite empty theatres, the area never took off.

A customer orientation is important in the development of adult entertainment districts. In an era when there was little public information about specialist stores and outlets, clustering allowed customers to find readily what they were looking for. Even in 1993, New York City found that "the basic locational criterion for adult entertainment businesses is to be 'where the customers are' "(New York City, Department of City Planning, 1994, p. 26). In Manhattan, this was where the tourists were. Elsewhere, it meant where there was "easy access to public transit, main arterials, and plenty of parking for local residents and customers passing by on the way home" (p. 26). In a statement which brings to mind Hotelling-like clustering (1929), it was also suggested that concentrations of outlets were "due to the 'copycat factor' and the tendency of patrons to want to 'barhop';" (New York City, Department of City Planning, 1994, p. 27). McNee made a similar point, writing that "the concentration of prostitution in a particular part of the city makes it handy for locals and adult male visitors", adding that "the attraction of 'comparison shopping' has long been a factor in retail clusterings" (McNee, 1984, p. 22).

Today, in contrast to the past, a host of specialist guides exist which allow customers to target a particular shop or address—for example, *Sexy New York City*, embracing all forms of adult entertainment, the *Damron Men's Travel Guide* (published since 1964 and oriented towards homosexual activity) and the *Damron Women's Traveller* (oriented towards lesbians) (Gatta, 2002a, 2002b). However, location still matters in many cities. Although a trawl through the Internet reveals a lot of information about gay entertainment, there is relatively little about heterosexual entertainment districts and shops (although see Taylor, 2003).

Models of Location: The Traditional District

Adult entertainment districts could be viewed as being in the same class as so-called 'Bohemian' districts, such as those described in McCabe (1872, 1882), but not all Bohemian districts become adult entertainment districts. Like Bohemian districts and skid rows, one might argue that adult entertainment areas are 'moral regions', like those described by Park and others. In this sense, they resemble skid rows. Regarding skid rows, Ward wrote that

> further evidence of this spatially selective tolerance is manifested in the presence of commercial activities such as adult books stores and girlie shows that frequently cluster on the edges of skid row districts. They do not cluster there in response to a demand for such services from the skid row inhabitants, but rather in response to the high tolerance threshold (Ward, 1975, p. 239).

Rowley (1978) suggested that skid rows developed because of 'moral symbolism' which resulted in the area being considered a special urban region. This is virtually the same as Park's description of adult entertainment districts and other areas of cities as 'moral communities':

> We must accept these moral regions and

> the more or less eccentric and exceptional people who inhabit them, in a sense, at least, as part of the natural, if not the normal, life of a city (Park and Burgess, 1925/1984, p. 45).

> It is not necessary to understand by the expression 'moral region' a place or society that is necessarily criminal or abnormal. It is intended rather to apply to regions in which a divergent moral code prevails, because it is a region in which the people who inhabit it are dominated, as people are ordinarily not dominated, by a taste, or by a passion, or by some interest which has its roots directly in the original nature of the individual. It may be an art, like music, or a sport, like horse-racing (Park and Burgess, 1925/1984, p. 151).

However, in many cities, adult entertainment outlets, which include shops selling not just videos or publications depicting sex, but also sexually oriented products, including clothing and accessories, are located in neighbourhoods which might be described as incubation districts: relatively central, but underdeveloped, with relatively cheap rents and small manufacturing and retailing spaces. Such districts often attract a mix of businesses, like that described in Denver by McNee (1984). Many will fail, but some may become successful. More recently, Florida and Lee (2001) have suggested that such neighbourhoods are both a product of and foster diversity which leads to innovation and growth. The implication is that the different moral setting in the district can itself stimulate innovation and that moral districts can be assets as well as liabilities.

Thus, if one were to summarise the characteristics of a 'traditional' adult entertainment district, it would be characterised by centrality—as close to the business core as possible. It would be central in terms of the overall urban area, alongside the central business district, often located between the main business district and major gateways or transport terminals serving that district. However, located on the periphery of that district, it could appear to be peripheral to the social and economic life of the city. It would be characterised by a high throughput of people, high anonymity and a large number of single occupancy rooms. It might be located in areas with a relatively low permanent residential population, although a high population of transients, including visitors and commuters. It could be in a zone of discard, but equally could be in a zone of acquisition, into which the central business district is likely to expand, or is expected to expand at some uncertain time in the future—in what is effectively a holding zone, in which landowners and speculators are reluctant to make major investments in renewal since they expect to sell out or undertake substantial redevelopment in the future.

One might also argue that although the initial concentration may be an accident, like types of activities attract similar ones, in a pattern similar to that described by Hotelling (1929). Retailing depends on information: customers and clients need to know what is on offer and where. This is particularly true in the case of adult entertainment. In contrast to most forms of retailing, the adult entertainment sector is characterised by an information deficit. Until recently, merchants could not (or would not) advertise or publicise their wares or location. Therefore, they had to locate where customers were, rather than calling or attracting customers to them. Although word of mouth might have attracted some customers, for the most part, customers went to those places in which they knew shops were located—in turn forcing adult entertainment outlets to locate in those places to which customers would come. The clustering of such outlets could not only attract imitators, but could repel other non-conforming businesses by creating a moral atmosphere which discouraged other kinds of retailing and entertainment and by driving up rents for stores, effectively pushing up the bid-rent curve. Moreover, turnover among shops is high. In New York City, in the early 1990s, it was found that about 15 per cent of all outlets in the city disappeared over a 6-month period. In a sector in which turnover in shops and businesses was high, it makes additional sense for shops to cluster: although individual stores may disappear, the overall orientation of the district will remain the same. To the extent that the lack of information leads to clustering, one might hypothesise that increased information about store location and about what stores and services are on offer would reduce clustering and lead to the dispersal of adult entertainment activities. In New York City, this appears to be the case.

Controlling Adult Entertainment Outlets through the Zoning Process

Throughout the 19th and 20th centuries, constant attempts have been made to control vice and the spread of vice, mainly through laws and legal restrictions. As long ago as 1910, Chicago police regulations "prescribed that no house of ill-fame shall be permitted outside certain restricted districts, or to be established within two blocks of any school, church, hospital, or public institution, or upon any street car line" (Haller, 1970, p. 631). In the 1930s and 1940s, the mayor of New York City, instituted a policy of 'zero tolerance' towards adult entertainment

> The legendary LaGuardia crackdown on 42nd street was intended to make an example of those whose moral breakdown had helped to depress the city economically. One by one he personally padlocked the streets notorious burlesque houses, strip joints, game parlors, and houses of prostitution. ... LaGuardia's grandstand destruction of the shady side of 42nd street resulted in his accomplishing little more than driving the strip shows, the gambling, the bootlegging and the prostitution ... ever further underground (Eliot, 2002).

More recently, reformers and town planners have attempted to eliminate vice through a mixture of legislation and zoning, using a command-and-control system based on zoning and licensing to reduce its impact and to contain its spread. Comand and control techniques often involve greater than normal supervision of the financial affairs of adult entertainment outlets and the vetting of their owners to ensure that they have no connection to organised crime. In some cases, they involve detailed measurements of floor space to ensure that zoning requirements are upheld. Under some zoning laws, stores selling adult videos come under a separate licensing and zoning regime if more than a fixed percentage of the floor area is devoted to x-rated products. In the 1970s, Boston, Massachusetts, attempted to limit adult entertainments by creating what came to be called the 'combat zone' akin to the 'zones of toleration' in Utrecht described in Hubbard (1997). The city established a two-block adult entertainment area in which 90 per cent of the adult entertainment outlets in the city were already located (New York City, Department of City Planning, 1994, p. 11). The Boston legislation was aimed at "adult entertainments and bookstores which are characterized as such because they exclude minors by reason of age" (New York City, Department of City Planning, 1994, p. 11). Similar laws were passed in Seattle, Washington and Camden, New Jersey. In other cities, zoning laws were passed to ensure that no adult establishment was near a school and that no clustering could take place. Detroit—for example, passed the first major law of this type in 1972 (although it was not upheld by the US Supreme Court until 1976) which said that no more than 2 adult businesses could be located within 1000 feet of each other or within 500 feet of a residential area. Similar laws were passed in Atlanta, Georgia, Kansas City, Missouri, and Los Angeles, California. In the case of Chicago, which passed a similar law, all adult use businesses were required to obtain a licence which was designed to prevent the involvement of organised crime. (Detroit, Boston, Camden, Atlanta, Chicago and Kansas City all lost population in the 1970s.) Many adult retailers may support such dispersal laws which guarantee a spatial monopoly, eliminating what Gruen and Smith (1960, pp. 45, 105) termed 'pirating' by other stores. A growing number of suburban American communities also passed zoning laws restricting the spread or location of adult entertainment outlets. In 1980, Islip, New York, a dispersed, low-density, suburban community, passed a law restricting adult uses to specified industrial areas (New York City, Department of City Planning, 1994, pp. 3–14; Lasker, 2002; Philips, 2002).

In America, where the use of these kinds of planning controls has been most widely applied, not all vice is 'illegal'. Adult book stores, video shops, cinemas and live entertainment are, in the eyes of the law, difficult to distinguish from general book shops, video shops, general cinemas and general entertainment and nightclubs. This makes it difficult if not impossible to zone them out of existence. In addition, the term 'adult use' is defined differently from community to community. In some, it is based on the content of materials, the nature of activity shown, the depiction or presentation of specific sexual acts, or specific parts of the anatomy. In others, it is defined by the exclusion of minors by reason of age. However, the definition of 'pornography' continues to change, as does the nature of material sold. As Tait (2003) noted, writing about the UK, "more and more mainstream shops were now selling lingerie, sex toys and accessories—for example, Selfridges, which has promoted a 'window of love' display with a vibrator taking pride of place". More recently, authorities at Harvard approved the publication of a student magazine, *H Bomb*, featuring nudity and erotica, announcing that the editors "will not be permitted to conduct nude photo shoots inside university buildings". According to the editors, the magazine will not focus on pornography, but on creating a forum for a "discussion of sex on campus" (Goldenberg, 2004).

Times Square and New York City

For many, the Times Square area, shown in Figure 2, epitomises the adult entertainment district. It typifies recent trends in the location and nature of adult entertainment and undermines the myths of marginalisation and zero tolerance. As early as the 1870s, New York's adult entertainment districts were featured in guidebooks (McCabe, 1872, pp. 186–193 and 579–617; 1882, pp. 154, 250–257 and 640–643). Although it was not recognised at the time, by the 1870s the city had become a magnet for sex tourists (McCabe, 1872), a role it retained throughout the 20th century. Some have gone so far as to suggest that the city's reputation in this realm transformed the nature of sex in America (New York Museum of Sex, 2002).

By the end of the 19th century, the sex trade was centred on the Times Square area. In 1901, as well as being the location of the Metropolitan Opera House, the area was a focal point for prostitution, which preceded the development of offices and hotels in the area (Sagalyn, 2001, p. 43; Gilfoyle, 1991). The area was also a centre of the burlesque industry, which although not entirely based on sex, had a strong sexual component. Far from being marginalised, it was suggested in 1906 that, "the city is being rebuilt, and vice moves ahead of business" (Gilfoyle, 1991, p. 299). Before moving to Times Square, the adult entertainment district had moved from the Bowery and lower Broadway, slowly progressing up-town along Broadway and gradually extending tentacles westwards. By the 1890s, an area known as the Tenderloin, stretching from 23rd Street North to about 40th Street, and from Sixth Avenue to the port area along the Hudson River had become the city's main adult entertainment district (Eliot, 2002).

Located at the intersection of Broadway, Seventh Avenue and Forty-second Street, Times Square was already a major transport intersection by the 1880s. New York's theatre district moved to the area in the 1890s (New York State Urban Development Corporation, 1981). The development of the elevated railroads in the 1870s, the electrification of trams in the 1890s and the opening of the first subway line in 1904 cemented the area's centrality. The opening of the Port Authority Bus Terminal in 1950 further enhanced it. Suburbanisation and the general increase in mobility, particularly after 1945, increased the throughput of people

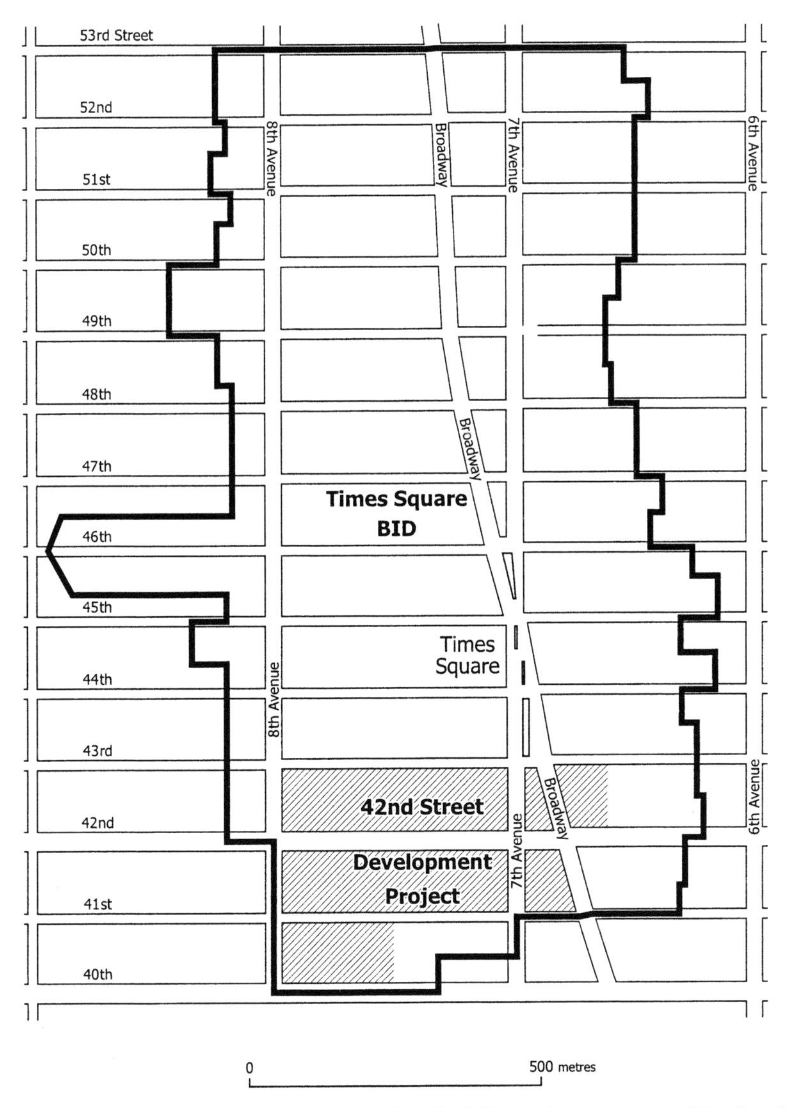

Figure 2. The Times Square area, also showing the 42nd Street Improvement District. *Source*: New York City, Department of City Planning (1994).

in the area. By 1950, although it was alongside rather than within the mid-town office district, it was definitely central in relation to Manhattan and the New York City region, intersected by two 42nd Street subways, the Broadway, 6th, 7th and 8th Avenue subway lines and the main bus gateway to New York City, and being less than one half-mile from both Grand Central Station and Pennsylvania Station, the two main rail gateways to the city.

Like San Francisco's Tenderloin, the vice district in New York was located between the growing office core of the city and the residential districts to the north (Shumsky and Springer, 1981; Reckless, 1926). What is interesting about the Times Square area and adult entertainment is that after 1900 adult entertainment became anchored on the Square—or, more accurately, on the area within one or two streets of the Square. Until 1900, it had continually moved northwards, in advance of the business and retail district, pushed by rising rents (or drawn by lower rents), but retaining its location between the main residential districts and the office areas, as was described by Park (1925/1984). It may have been trapped in the Times Square area because areas further up-town were rapidly developed for housing after the construction of the elevated railroads and the subway. Another factor may have been the passage of the city's first land use zoning law in 1916 which, along with subsequent amendments, restricted the spread of specialist districts and services. By 1904, the area was serving local residents, tourists and a growing number of office workers in the mid-town area who commuted by ferry from New Jersey, by train through Grand Central Station, a short distance away, and by subway and elevated railroad. The opening of the Pennsylvania Station at 32nd Street and Seventh Avenue in 1910, the opening of the Trans-Hudson tubes around the same time and the construction of new subway lines through the district made it more accessible and, more importantly, made it one of the busiest locations in New York.

The advent of Prohibition in 1919 substantially altered the area. Many 'legitimate' theatres had depended on cabarets, restaurants and nightclubs, using theatrical performances as a kind of 'loss leader'. Prohibition made them unprofitable and, in the first of several waves in which owners of obsolete theatres or venues turned to adult entertainment to remain profitable, new businesses were attracted to the area, mainly arcades, vaudeville, burlesque and movies, which were open 24 hours a day, making the area unique in the city (New York State Urban Development Corporation, 1981). During the 1920s, the area developed further as an adult entertainment district. As the first fingers of offices and lofts extended westwards and northwards from mid-town, the area was increasingly squeezed by new developments and appeared set to become an office district. Economic depression from 1929 onwards halted new construction and transformed the area from a zone of acquisition to a zone of discard, or perhaps more accurately, a 'zone of anticipation', although the movement of the garment district from lower Manhattan to Seventh Avenue in the 1930s further squeezed the adult entertainment area. By the late 1930s, Times Square was well known both locally and nationally as a centre of vice. Mayor Fiorello LaGuardia, elected in the 1930s, claimed to have eliminated 'vice' from the area by 1942 (Eliot, 2002) and live sex shows and vaudeville disappeared. However, the area remained characterised by low maintenance, high turnover uses, consisting of arcades, dance halls, cut rate shops, adult bookstores and 24-hour movie theatres (New York State Urban Development Corporation, 1981).

After 1945, there was a global explosion of adult entertainment. This was partly due to changing values, reflected in declining censorship of films and publications. It may have been due to increasing living standards, dropping costs of publishing, particularly colour printing and the rise of the super-8 home movie and the video. All of these substantially transformed the adult entertainment industry. In addition, just as the rise of the cinema led to the decline of live theatres which turned to burlesque and vaudeville to stay in business, television furthered the decline of cinemas, which had started in the 1930s (Pautz, 2002). In 1930, 65 per cent of the US population went to the movies at least once a week. By 1954, the figure had fallen to 30 per cent, and, by 1966, to 10 per cent. The result was that many theatre owners were unable to attract viewers to legitimate films and turned to adult films instead, particularly in relatively tolerant cities like New York. In the 1960s, relaxed censorship led to an explosion of publications and to the opening of new clubs and bars featuring sex. By 1969, many viewed the film *Midnight Cowboy* as an accurate depiction of the Times Square area.

For decades, local governments, New York City included, attempted to restrict the opening of adult entertainment outlets and the spread of districts. By 1972, an explosion of outlets led many US cities to pass zoning laws to restrict their location. The explosion of adult shops was not restricted to New York, where the number of outlets went from 9 in 1965 to 151 in 1974. In Detroit, where the population fell from 1 849 568 to 1 514 063 between 1950 and 1970, the number of adult entertainment outlets grew from 2 adult movie theatres, 2 adult bookstores and 2 topless bars in the early 1960s to over 100 by 1972 (Lasker, 2002). By 1977, Hollywood was described as

> a slum ... massage parlour girls flaunting their wares in doorways and windows. Dirty book stores. Clean book stores. More dirty book stores. Magazine stands, mostly dirty. Trolling homosexuals ... Jockers in leather and chains. Hustling black pimps. Listless whores ... Paddy hustlers, pigeon droppers, pursepicks, muggers (Wambaugh, 1978, pp. 75–76).

Initially challenged on the grounds that they infringed freedom of speech, the zoning laws were ultimately upheld, although the courts stated that an ordnance must be content neutral (Phillips, 2002, p. 321) because a content-based ordnance violates the first amendment of the Constitution protecting freedom of speech. As the courts noted, "Entertainment is constitutionally protected speech, and nudity alone will 'not place this otherwise protected material outside the mantle of the First Amendment' " (Phillips, 2002, p. 320). This has been taken to mean that the ordnance must serve a public interest and that it should not unreasonably limit alternative avenues of communication. Despite this, not all cities passed zoning ordnances immediately. New York City—for example, attempted to do so in 1977, but failed and did not pass a law until October 1995. That law was not upheld by the courts until January 1998 and did not become effective until June that year. The law established minimum distances between outlets and proscribed their location with fixed distances from churches and schools. It applied to all adult establishments with an area of over 10 000 square feet, those in which 40 per cent or more of floor space was given over to adult products, or those in which a substantial amount of floor space was devoted to adult products (Times Square BID, 2000).

Since the 1950s, the redevelopment of the Times Square area has been viewed by New York's planners as being imminent. Formal efforts to foster and control the development of the Times Square area go back to the late 1960s.[2] The initial goal was to restrict office development in the area, but at the same time encourage new construction, preserve live theatres and maintain the residential population (Barnett, 1974). Eliminating adult use activities was almost incidental—it was believed that development was imminent and that the main problem was to control and restrict growth rather than encourage it. Policies aimed at preserving theatres while encouraging the area's reconstruction were maintained through the early 1980s (New York City, City Planning Commission, 1982). From the 1970s onwards, the governments of the city and state of New York announced plans to redevelop the area (Quindlen, 1979), bringing new offices, hotels and housing to the district, through the auspices of the New York State Development Corporation. However, the Corporation went bankrupt and the plans were sidelined.

The Times Square area was also an important source of revenue for the city. Although characterised by high crime—the police precinct in which the district was located had the highest number of reported crimes in the city (New York State Urban Development Corporation, 1981)—it also generated jobs and taxes. In 1994, it was estimated that topless clubs in New York City employed 1500 dancers and grossed $50 million annually, and an analysis of 2 clubs showed that they had 218 employees, an annual payroll of $1 302 627 and grossed over $600 000 monthly from credit card payments alone (New York City, Department of City Planning, 1994, pp. 17 and 26). This continued a tradition of sex tourism dating back to the 1870s.

According to official figures, by 1965, adult entertainment in the Times Square district was a shadow of its former self (New York City, Department of City Planning, 1994, p. 19). However, from the 1960s onwards, the number of adult entertainment establishments in New York City as a whole skyrocketed, from just 9 in 1965, to 151 in 1976, 131 in 1984 and 177 in 1993. Of these, about 97 were in mid-town Manhattan, which includes the Times Square area, in 1976, but just 49 in 1984, 45 of which were in Community District 5 where Times Square is located. By 1993, the number in district 5 had risen to 53, of which 36 were in the Times Square area or along 8th Avenue just north of it.

In this sense, perhaps adult entertainment stabilised the Times Square area and even preserved it from the arson and decay which were found in other areas of New York throughout the 1970s and early 1980s. Between 1976 and 1979 alone, it was estimated that the city lost 150 000 apartments due to arson and abandonment (Quindlen, 1979) and the 1970s saw a massive loss of jobs and population (New York City, Department of City Planning, 2001). One could suggest that, rather than being pushed into the area due to sanctions, adult entertainment activities repelled other activities, less because of the negative social reputation of adult entertainment, than because of its ability to pay high rents. Due to high profits, adult entertainment outlets were able to pay a higher rent for space than would have been paid by non-adult uses, which made it possible (if not profitable) for landlords to retain and maintain existing structures, including theatres, even when they were unable to let office space on upper floors. Without these uses, buildings in the area might have been abandoned and even demolished, as happened along 8th Avenue between 41st and 42nd Streets. Instead, theatres which once showed sex films have become the showpiece of the area's redevelopment. In effect, adult entertainment outlets paid for centrality and, at the same time, kept an obsolete and otherwise unmarketable district intact. Other adult entertainment clusters existed (and continue to exist) alongside the Wall Street office district and in Greenwich Village. The distribution in New York City 1993 is shown in Figure 3. The distribution in Manhattan is shown in Figure 4.

The adult entertainment industry has continued to evolve. Cinema attendance has declined further, leading to the closure of large numbers of movie theatres. Simultaneously, the rise of the adult video, predicted as early as the start of the 1970s (Deighton, 1974, p. 227), substantially altered the nature of the adult entertainment industry. In 1984, New York had no adult video shops, but by 1993 it had 64 (New York City, Department of City Planning, 1994, p. 22)! During this same period, the number of bookstores and peepshows declined from 29 to 22 and the number of cinemas and live theatres fell from 23 to 11, a phenomenon replicated in other cities (Needham, 1990). By contrast, topless bars and nude bars increased from 54 to 67 between 1984 and 1993 (after increasing from 23 in 1976), in line with a nation-wide trend (New York City, Department of City Planning, 1994, p. 17). Overall, in New York City, the total number of adult entertainment outlets increased from 131 in 1984 to 177 in 1993—the increase being among video stores and topless entertainment.

More recently, the adult entertainment industry has undergone another shift. In the US, between 1999 and 2002 alone, Datamonitor (2002, p. 14) estimated that CD-ROM sales increased by 93.8 per cent and on-line sales by 89.1 per cent, whereas magazine sales fell by 6 per cent. During this period, the combined sales of CD ROMs and on-line material rose from 12.2 per cent of the total spend on adult entertainment to 19.1 per cent of an estimated $13 240.4 million and the share of magazine sales, strip clubs and DVD/video sales (most of which are sold through fixed outlets) fell from 70.6 per cent to 63.9 per cent. This shift coincides with a drop in outlets in New York City. Often it is attributed to zero tolerance policies, but it continues a pattern of rapid expansion of outlets and then a gradual decline in numbers seen with bookstores and peepshows a generation ago. It is not dissimilar to patterns of growth and decline in the broader field of entertainment, such as the increase and then die-back in numbers of multiplex cinemas.

What is most marked about the Times

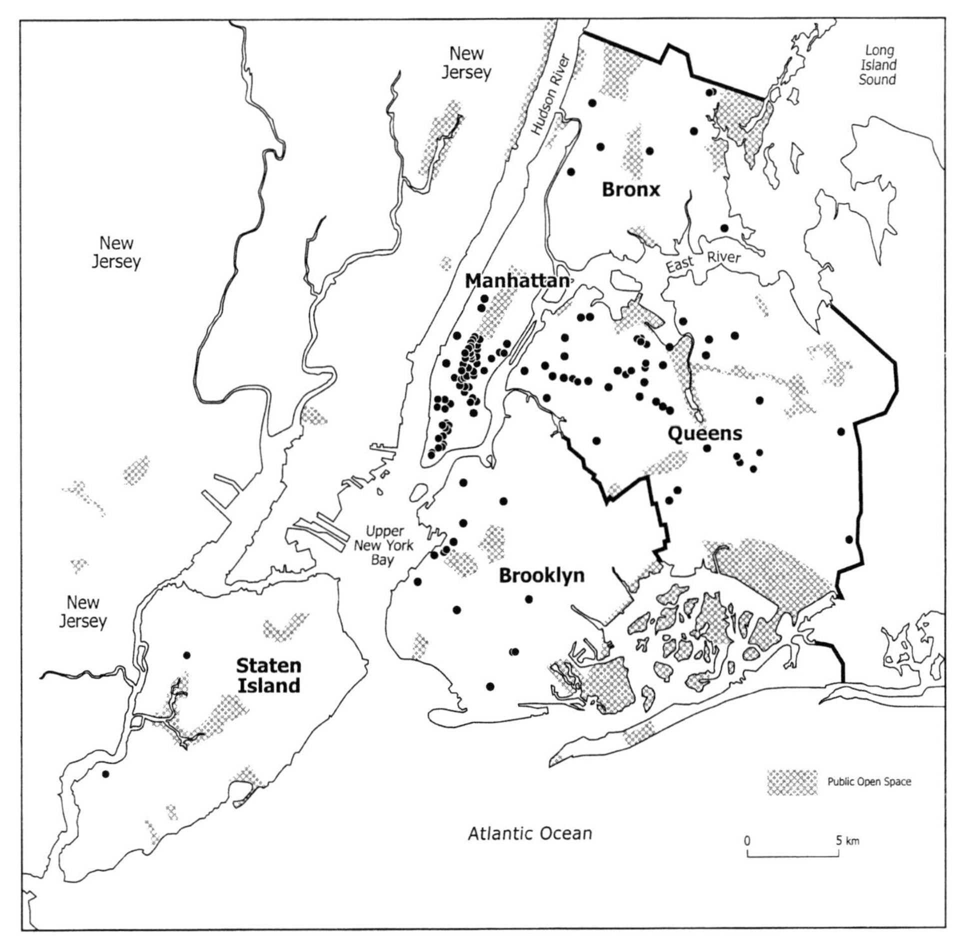

Figure 3. The distribution of adult entertainment outlets in New York City, 1993. *Source*: New York City, Department of City Planning (1994).

Square area is its persistence as a centre of adult entertainment. From the 1930s onwards, although the New York City government attempted to remove adult activities from the district, they remained rooted to the area. Vice sweeps, crackdowns on prostitution, massage parlour closings, anti-pornography zoning and strict licensing laws failed to close down or eliminate the district, although at times they made it less visible. Sagalyn (2001, p. 45) suggested that the causes of inertia were three-fold: small parcels of land which made assembly of land for new development both expensive and difficult; the high cost of land, due to high rents and income from advertising; and the area's reputation for pornography crime. She also noted that

> the configuration of parcels in this area looked no different than that in any other commercially undeveloped—meaning low density—area of mid-town where as many as 80 parcels might make up a single block (Sagalyn, 2001, p. 46).

She suggested that property ownership in the area was deliberately obscured, so that owners and landlords could distance themselves from activities going on in their buildings. She also noted that 80 per cent of the property in the immediate Times Square area was controlled by just 3 corporations with major holdings of theatres and small properties.

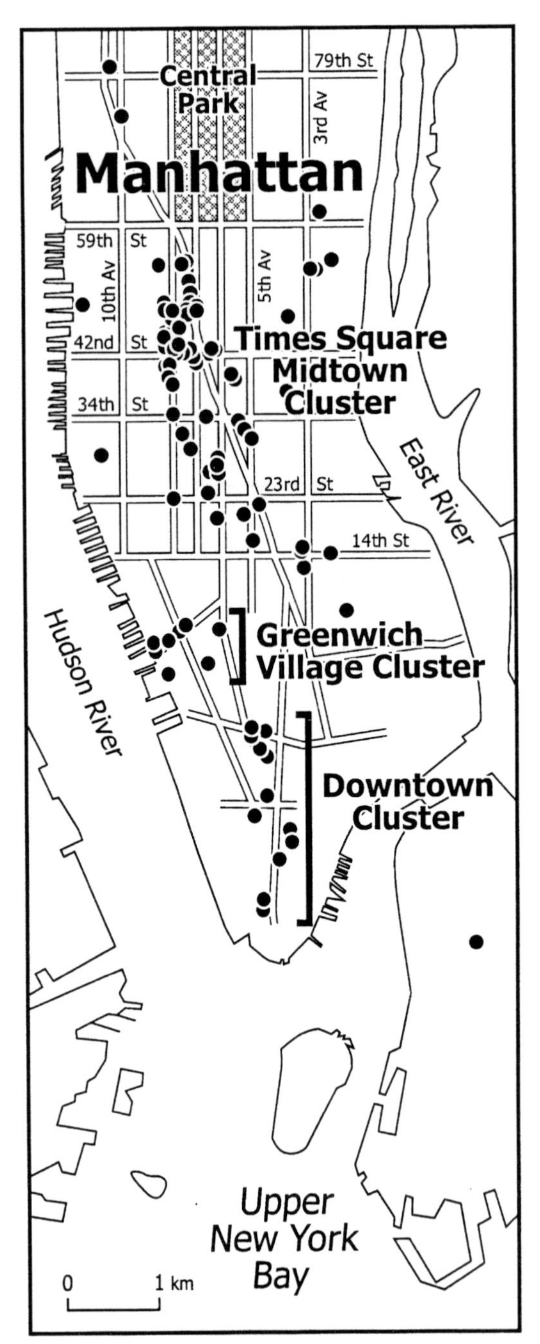

Figure 4. The distribution of adult entertainment outlets in Manhattan, 1993. *Source*: New York City, Department of City Planning (1994).

Along 42nd Street, there were under 30 landlords (Sagalyn, 2001, p. 48), although the New York State Urban Development Corporation (1981) claimed that there were at least 73 parcels of land, which made land assembly difficult.

One could also argue that the neighbourhood retained its basic characteristics because there was no demand to change it. From the 1920s onwards, Manhattan lost population, a process which spread to all of New York City during the late 1960s and 1970s. Consequently, demand for new construction declined and property values went down. In effect, outside a few core areas, the entire city became a zone of discard as more people left than moved in. In 1981, in an 'assessment of blight' (required to initiate a redevelopment process), the New York State Urban Development Corporation (1981) found that the area immediately along 42nd Street between 7th and 8th Avenues was characterised by an unusually high percentage of vacant land—16 per cent—and another 5 per cent of the area was occupied by one- or two-storey buildings. Thirty per cent of the block bounded by 7th and 8th Avenues and 41st and 42nd Streets was vacant. Although the area had been zoned for high-density office development, it contained only 2.4 million square feet, or 32 per cent of the allowable floor space, as opposed to the 'east side' which was overcrowded. In addition, although the vacancy rate for offices in the mid-town area was only 1.75 per cent, the rate in the Times Square area was 25 per cent: 29 per cent on upper floors, and 8 per cent of shop space on ground floors. Building floor plates were small, 70 per cent being under 5000 square feet, and many buildings were defined as being technologically obsolete, lacking fireproofing and other amenities. Despite this, rents were high, reflecting the high throughput of people and the high rents adult entertainment outlets were willing to pay. Sagalyn (2001, pp. 47–48) notes that, before redevelopment, rents paid by adult entertainment outlets on 42nd Street per front foot in the Times Square area were twice as high as those paid by other retailers, and writes about a small bookshop with a narrow street frontage paying a then-high rent of \$32 000 per year. She cites a City University of New York study which estimated that peep shows in the area grossed between $76 000 and $106 000 weekly, or over $5 million per year.

Sagalyn also recognised that, by the early 1970s, the adult entertainment industry had identified office workers as its best customer group. Times Square and the surrounding area were well placed to intercept this customer group travelling to and from work. Throughout the 1950s, the number of office workers in the mid-town area to the east of Times Square expanded, and it became the city's main business centre, supplanting the area around Wall Street. Around 1970, Times Square had a head count of about 49 000 during the morning rush hour, versus about 12 000 at Rockefeller Center. More recently, in 1999, the Times Square Business Improvement District (2000) found that the site in front of the Virgin Megastore at Broadway and 44th Street was passed by 54 500 people on a weekday and 62 400 on a Saturday. In 1981, the New York State Urban Development Corporation noted that the Times Square area was crossed by 16 subway lines (serving 3 stations), 9 city bus lines and the Port Authority Bus Terminal, served by 7200 suburban and long-distance buses daily. In 2000, the Times Square BID similarly observed that 188 000 people per day passed through the Port Authority Bus Terminal and that the area was intersected by 11 subway lines and 9 city bus lines. In addition, Times Square was and remains a centre for tourism. In 1996, it was estimated that some 20 million tourists visited the area and, in 1999, the number was estimated at 29.4 million—80 per cent of the 36.7 million visitors to New York City. In 1996, the area contained 24 hotels with 12 417 rooms, 24 per cent of Manhattan's total and, by 1999, it contained 30 hotels with 13 503 rooms, 23.5 per cent of the rooms in Manhattan (Times Square BID, 1996b, 2000).

In the 1990s, the number of adult entertainment outlets in the Times Square area declined substantially, from an estimated 36 in 1994, to 25 in 1995 (Times Square BID, 1996b); 21 in 1996 and 1997 (Times Square BID, 1997); and 17 in 1998, 1999 and 2000 (Times Square BID, 1997, 2000). This decline is not directly related to the redevelopment of 42nd Street, since the Urban Development Corporation (later Empire State Development) and its agent, 42nd Street Development Corporation, took possession of most of the buildings along the street in 1990. Nor does it seem to be due to the city's zoning legislation, since that legislation did not become effective until January 1998 and was not enforced until June that year. Since 1998, the number of outlets has remained static at 17, perhaps because outlet owners had shut down in anticipation of the change. The decline in numbers in the Times Square area can also be viewed as the continuation of a trend which goes back to 1976, when there were 97 outlets in the mid-town area. Given the high number of stores and the competition among them for clients, it is also important to note that the adult entertainment industry, like all retail industries, is characterised by a high failure rate and high turnover. In 1994, the Department of City Planning (1994, p. 27) noted that in the 6 or so months between the completion of a census of adult outlets and the mailing of letters to them to invite them to a hearing on adult entertainment in the city, 27 out of 177 letters were returned, mainly because the shops were no longer operating at that address. This suggests a failure rate of 15 per cent per 6 months, or about 30 per cent per year. For this reason, the figure of 36 outlets in 1994 for the Times Square area may be an overestimate. If the 15 per cent failure rate is applied uniformly across the city, the number in the Times Square area may have been as low as 31 by 1994, suggesting that the number of outlets in the area fell by less than half after the new zoning regulations went into effect. Volatility of numbers is evident in the figures provided by the Times Square Business Improvement District (Times Square BID, 1996b, 1997): in the 1996 report, 15 adult uses were said to be in the area in 1996, but in the 1997 report, 21 were listed.

Although some outlets may have disappeared relatively quickly (as is the case with all retail outlets), a city-wide comparison with the listings of clubs, strip clubs, restaurants and book-store-peep-show-video outlets in Manhattan in 1994 and 2000 suggests that many outlets have considerable staying power. The New York City Department of City Planning (1994) listed 107 adult outlets in Manhattan in 1994 (although, again, assuming a 15 per cent fall in numbers, it could have been as low as 94). In 2000, 27 were still listed in *Sexy New York City 2000*. Despite the new zoning ordnance, in Manhattan alone there were still 36 video outlets, 10 strip clubs, 3 other clubs and 1 male 'burlesque' (although in several cases, the trading names were different). This ignores new types of businesses listed in the guide, including 5 'swing clubs' and 12 dungeons, 'addresses available on request'. Nor does it include clothing stores, novelty stores, specialist book stores and restaurants.

In addition, while the Times Square area has seen its outlets decline, other areas in New York City have seen them grow. In 1994, at hearings on the adult entertainment industry, residents of the 'Chelsea' neighbourhood on Manhattan's west side between 14th and 31st Streets were said to have "routed four out of nine" establishments in the neighbourhood (New York City, Department of City Planning, 1994, pp. 43 and 62) and were singled out as taking a stand against adult entertainment in the area. However, by 2003, the area along 8th Avenue in the 1920s, which had no adult outlets in 1994, was described as "a predominantly gay strip ... [with] no fewer than 6 shops carrying pornographic movies ... [and] ... bawdy accoutrements" (Lee, 2003). The growth of this district highlights several trends in the industry, including the rise of specialist districts ('predominantly gay'); a dispersal of sex shops away from central business district locations and from leading employment areas; and a move towards residential districts, suggesting that some adult entertainment clusters are not just oriented to tourists or people on their way to and from work. Instead, watching adult videos is a leisure time activity, often no different from watching any other kind of video or cable television film. This again parallels changes in the broader entertainment industry. In 2003, film receipts from the sale of DVDs in the US were some $22.5 billion, versus just $9.2 billion in box office sales (*The Economist*, 7 February, 2004, p. 63) and over 50 per cent of receipts came from video and DVD sales.

The decline in outlets may also reflect changing patterns of consumption and retailing. In January 2003, field research and interviews with the managers of adult entertainment outlets in Manhattan, and field research in Berlin in 2002 and early 2003, suggested that books and magazines had virtually disappeared from adult 'book stores' in those cities—a view later substantiated by interviews with adult shop managers in Exeter and London in the summer of 2003.[3] Field surveys in Berlin and New York City showed that what was traditionally thought of as pornography has been mainstreamed and can be bought at newsagents in both cities—in Berlin, they are even sold at newsagents on the U-bahn. In New York City, although some blamed the shift away from publications on Mayor Giuliani, store managers in Manhattan repeatedly emphasised that videos and DVDs are more profitable than other products, as they did in Exeter and Berlin. In New York City, managers pointed out that adult magazines were on sale in newsagents. Observations in New York, Berlin and elsewhere confirmed this (although it was not the case in London). In many cases, shop customers are repeat visitors. In Exeter, the manager of one store noted that it was part of a national chain of retailers and that all the magazines and videos sold in the store were produced in-house. He added that much of the material on sale was available over the Internet, from the same chain, but argued that the store continued to attract customers because it offered price reductions on multiple purchases of videos, which were already sold for less than the Internet prices, and also because the store bought back and exchanged videos which had been purchased there.

Adult entertainment outlets have also

fallen back on videos because they have lost their monopoly on published erotica and sex-related goods. Magazines are sold in newsagents; clothes and 'sex toys' in clothing stores; and books in mainstream book stores under erotica or specialist fiction as well as in specialist book shops aimed at particular markets, such as gay or lesbian book stores. Nor is this limited to large cities in western Europe and North America. In Poznan, Poland, in September 2003, newsagents were selling a range of adult-oriented publications, both heterosexual and homosexual. If one included all the clothing stores, book stores and newsagents selling sexually oriented material in a census of 'sex shops' or adult entertainment outlets, the number of stores in New York City (as well as elsewhere) would not only be far greater than is publicly recognised, but be several fold greater than the number in the 1980s and 1990s.

Today, the industry is undergoing another evolutionary shift. Since the early 1990s, video stores, book stores and peep shows have been supplanted by the Internet and computers, and the global decline in telephone prices has led to new telephone chat lines (Datamonitor, 2002). World-wide sales of adult entertainment products, including DVD/videos, escort services, magazines, strip clubs, phone sex, cable and satellite pay television, on-line sex products and CD-ROMs grew by 16 per cent between 1999 and 2002, but on-line sales grew by 82 per cent and CD-ROM sales by 116 per cent (Datamonitor, 2002). Although the absolute amount sold has increased, the share of goods sold through shops and retail outlets, including magazines, videos/DVDs and strip clubs declined from 68 per cent of total sales in 1999 to 65 per cent in 2002; the share of the Internet grew from 4.4 per cent to 7 per cent of total sales of $41 501.24 million. Publications suffered most, actually declining in absolute sales terms by 6 per cent, leading some firms to abandon publishing all together (Rees and Tomkinson, 2002; Walsh, 2002).

The Rise of Specialist Districts

As well as general adult entertainment districts like Times Square, many cities have specialist districts. In 1872 and 1882, McCabe described different 'vice' districts in New York City, mentioning Sixth Avenue, the Bowery, lower Broadway and Bleecker Street in Greenwich Village as catering to markedly different kinds of moral communities and sheltering different kinds of vice aimed at different customers and income-groups.

As early as the 1870s, Greenwich Village was described in guide books as a 'bohemian' district, a reputation which persisted in the 1930s (Ware, 1935/1994, pp. 249–263). In 1872, McCabe described the Village as a

> a suspicious neighbourhood, and he who frequents it must be prepared for the gossip and surmises of his friends. No one but its denizens ... knows anything with certainty about its doings or its mode of life (McCabe, 1872, p. 387).

In 1882, he went further, writing that, Bleecker Street

> cannot be said to be bad or even disreputable, but it is at best a doubtful neighbourhood, which people with reputations to lose avoid. Life here is free from most of the restraints imposed elsewhere, and so long as the denizens of the neighbourhood do not actually violate the law, they may do as they please... [Bleecker Street] is emphatically a street in which no questions are asked (McCabe, 1882, p. 276).

From the late 1800s, Greenwich Village was an immigrant neighbourhood, but between 1910 and 1930, the population fell from 67 719 to 38 045 (Ware, 1935/1994, p. 462). By the 1940s, if not earlier, it had become one of several centres of 'gay' life in New York, along with Times Square around 42nd Street and the east 50s in Manhattan. This last area gives a good idea of how districts form and spread. Kaiser writes that, after the opening of the 8th Avenue subway in the 1930s, there was a direct link from Times

Square to a former speak-easy district in the east 50s. By the 1940s, the speak-easies had become nightclubs and, later, gay bars (Kaiser, 1997, pp. 106-107). By the 1960s, Christopher Street was the centre of the 'gay' village, lined with bars, bookstores and clothing stores, and remains a focal point of gay-oriented adult entertainment outlets which have slowly extended northwards along 7th and 8th Avenues.

Whereas specialist districts like Greenwich Village were once rare, today they are found in most major cities, although they often overlap with heterosexually oriented activities. In the US and Canada, one or more recognisable districts exist in Denver (East Colfax), Columbus, Ohio ('the Short North'), San Francisco, West Hollywood, Phoenix, Miami Beach, Chicago (North Halstead), Philadelphia—south-east of City Hall, San Diego, Washington, DC (Dupont Circle), Atlanta, Toronto and Montreal. Gay resorts exist at Fort Lauderdale, Palm Springs, Provincetown, Fire Island, Laguna Beach, Key West and New Hope, Pennsylvania. (Columbia Fun Maps). As Collins (2004) has noted, a similar district has developed in Soho in central London and identifiable districts exist in Hamburg (the Reeperbahn and the area behind the Main Station); Zurich (across the river from the main station), Hanover (around 'Am Marstall'), Paris, Cologne, Manchester (Canal Street) and other cities. In Toronto, the 'gay' district is located in part of what was skid row, just east of Yonge Street, which it had colonised by 1970. It had become a noticeable cluster by 1989 and had developed further by 2002 (David James Press, Ltd, 1989; Columbia Fun Maps, 1998, 2002a). In Montreal, it is located in an eastern inner-city area between the Berri UQAM and Papineau Metro stations, and even has its own tourist office (Columbia Fun Maps, 2002f). This last is a good example of how an adult entertainment district can colonise a declining urban region. The Montreal city government views the area as a declining one, characterised by population loss and underdevelopment (interview, June 2001, with representatives of CED-Q and city authorities). Berlin provides an example of a recent development of an adult cluster. The traditional adult entertainment district was located along and near the Kurfurstendam in the western part of the city, not far from the Bahnhof Zoo. The 'gay' district was located in and around the Nollendorf Platz. After the fall of the Berlin Wall, the Nollendorf Platz expanded as a specialist area, and a new area catering to both heterosexual and homosexual orientations formed in the eastern part of the city, from the Alexander Platz northwards along the Schoenhauser Allee. The first signs of this new district appeared as early as 1990 and had become pronounced by 2000 (Waldau, 1990; Lothar, 2001). Such districts are not restricted to 'gay' activities. For example, the London Fetish Map (Whatsyours Publishing, 2002) identifies a cluster of outlets in Camden Town, as well as in Soho. In some cases, specialist districts aimed at a particular submarket or sub-group overlap (or even overlay) older or non-specialist adult entertainment districts, such as Soho in London, but in many cases, such as Greenwich Village and Chelsea in New York, they are in separate locations.

Conclusion

Given the amount of interest adult entertainment districts have attracted in the popular press, and among planners, law enforcement officials and moral crusaders, research on adult entertainment districts has been limited. Until recently, adult entertainment districts have generally been located in central areas and, far from being marginalised in space, have been at the heart of things. Until recently, they appear to have been attracted to locations characterised by centrality, high throughput of people, ready shop space and, sometimes, hotel rooms and anonymity. Certainly, the Times Square area has all these features. This does not explain the original choice, nor does it explain the way in which a few outlets may result in the creation of a new centrality with regard to adult retailing and services. Park and Burgess explained their location in terms of moral regions in a zone of transition alongside a central business district, but failed to explain why they chose a particular point as opposed to any other point along the 'ring' of transition surrounding the CBD. Nor did their model explain the persistence of such districts, since each ring was supposed to be simultaneously moving outwards and becoming broader. Perhaps such activities moved into tolerance zones or moral districts (although that fails to explain why zones of tolerance develop where they do).

Describing an activity as marginalised fails to explain its location at a given time in a given place. Describing an adult entertainment district as a moral region, or in terms of Firey's 'sentiment and symbolism' does not explain the 'why' of the reason for its location there, although it may explain its persistence. However, marginal activities can end up in any number of areas with similar characteristics. In Manhattan, why did Times Square and 42nd Street become the main focus of adult entertainment, and not Union Square and 14th Street, or the Bowery? Why did Greenwich Village become a 'bohemian district', and not Brooklyn Heights?

Landownership may play a role, as in the case of Soho where Paul Raymond, described as "the most powerful man in British porn" (Walsh, 2002), is said to own 60 acres of freehold property. The answer may be more akin to the process of chance described by Krugman (1997) than to any particular sequence of cause and effect. Much as Barnes and Noble on lower Fifth Avenue in New York created a book-store district which lasted for almost 20 years before evolving into a district of up-market clothing stores, the initial stimulus may be due to accident or chance, one successful outlet attracting others.

In a motorised age, 'centrality' is an ambiguous term. This can be seen in newer or younger urban areas, such as Portland, Oregon, or Los Angeles. Centrality to a hinterland need not mean central in the sense of being near a core district, as can be seen in the case of Hollywood and West Hollywood. This is evident in suburban Islip's passage of an adult entertainment zoning law in 1980 and is also evident in the location of specialist adult outlets in suburbs on Long Island and in New Jersey (Columbia Fun Maps, 2002c, 2002d). In geographical terms, accessibility and centrality need not coincide. In some cases, adult outlets have moved into incubator districts, which are homes to new businesses and start-up activities, characterised by transience and change. In other cases, they have moved out towards the edge of the urban fringe, following a pattern observed in retailing since the 1950s (Gruen and Smith, 1960) and pursuing a location strategy followed by Walmart since the 1970s.

Changing morals have redefined adult entertainment, effectively narrowing the definition to just a few kinds of products such as sexually explicit videos. There are now two kinds of adult entertainment outlets: official and unofficial. The former include those which are subject to licensing and zoning constraints, and the latter consist of those which are not. Changing morals have also reduced constraints on information. Adult entertainment outlets no longer need to cluster to find customers. If lack of information among consumers originally led adult entertainment outlets to cluster in search of custom, that is no longer true. Today, information is available from published sources and on the Internet regarding location, goods on offer and even the quality of goods and service (Taylor, 2003). Freed from the constraint of centrality, and freed from the need to locate near each other, in many cases, they are moving to areas of lower rent. Although this may be due to stricter enforcement in former high-profile areas, and may be due to ordinances prohibiting excessive concentration, it may also be due to other decentralising influences, including high rents for shops in more central locations and the redevelopment of central areas due to the recent regrowth of cities.

More recently, some activities, particularly clothing sales, have gone mainstream, as have sales of sexually oriented publications. In many cases, retailers have deconcentrated, leaving traditional adult entertainment areas and moving to 'new' locations in cities. These include shops run by Beate Uhse in Germany and elsewhere, Ann Summers in the UK, Private Media, based in Barcelona, Goalie Entertainment, based in Denver, Colorado, with 60 US stores and Adultshop.com, based in Western Australia (Datamonitor, 2002; Johnson, 2003). Many mainstream book stores have erotic sections, adult sections, or gay and lesbian sections, blurring the boundary between them and 'adult' outlets. An even greater threat to the traditional adult outlet and the adult entertainment district is the rise of mail order, encouraged by the growth in credit card use from the 1960s onwards. The growth of the Internet, which makes it possible to order goods on-line and to download images and printed material, has hastened the trend towards aspatial retailing and aspatial adult entertainment, also characterised by 'telephone-sex' chat lines—although to some extent one can view this as a continuation of a trend which began with the development of the telephone, which allowed people to arrange appointments with prostitutes and order goods by mail. Adult entertainment is increasingly characterised by 'floating activities': special interest-groups or activities which meet at different places on different nights (for example, Detroit Adult Webmasters, 2003), in a manner analogous to weekly street markets, not dissimilar to those described by Skinner (1964/65). These are specialist groups which cannot sustain a regular daily or weekly venue or outlet, and are often findable only through the Internet or word-of-mouth. For example, the *London Fetish Map* lists two 'fetish fairs' or markets which meet at different venues: one on the first Sunday of each month in Islington, to the north of central London, and one near the South bank of the Thames, which meets on the last Sunday of each month (Whatsyours Publishing, 2003).

At the same time, suburbanisation and the deconcentration of consumers have occurred throughout the motorised world. In the past, it was argued that adult entertainment districts had three main sources of customers: tourists, ship personnel and office workers. Ships have gone and office workers are increasingly dispersed across urban areas (Mills and Hamilton, 1994, p. 83; Yeates, 1998, p. 320). Adult entertainment has been slow to suburbanise, but has started to do so, as can be seen in the *Fun Maps* of New Jersey and Long Island (Columbia Fun Maps, 2002c, 2002d; City of Portland Police Bureau, 1994, 1997) (see also Figure 1).

This raises another issue. Much of the work on adult entertainment districts treats them as fixed features in the urban fabric, ignoring the fact that there was an explosion of retailing in the sector between the 1960s and 1980s, reflecting changing morals, values and economies of particular times and places. Perhaps the number of shops and outlets in the recent past was exceptional. It was due to a series of continuous changes not just in views about what constituted adult entertainment, but also changes in the industry itself, particularly cheap colour printing, home movies and videos, relaxed censorship, de-urbanisation, higher living standards and, more recently, the CD ROM and the Internet. Like the mainstream entertainment industry, adult entertainment is characterised by successive trends which lead to explosions in particular kinds of retailing and then see a die-back in outlets, such as that which occurred in the video rental and retailing sector in the 1980s and 1990s or multiplexes in the late 1990s and early 2000s. In fact, in New York City, many video stores found themselves unable to compete with successive new entrants into the field and turned to pornographic videos as a result (New York City, Department of City Planning, 1994). Perhaps the large number of adult entertainment outlets was never sustainable in the medium to long term. As the novelty wore off and new products and new methods of delivery developed, most of the shops were likely to close anyway. In New York City and elsewhere, the passage of zoning legislation restricting the location of so-called adult entertainment outlets coincides with the rise of new forms of entertainment, including chat lines, the Internet, and CD-ROMs.

Moreover, the increase in adult entertainment outlets in the 1970s and 1980s may have been due to the collapse of land values and rents in many older city centres, and motivated by a desire on the part of landlords to maintain some kind of rental income in the face of declining demand for space.

It is tempting to suggest that 'zero tolerance' has closed down adult entertainment districts, but there is little hard evidence of this. Hubbard (2004) writes that, in Westminster, authorities announced that the borough would have no more than 16 licensed 'sex shops', but a few years earlier, it was announced that the number would be increased from 10. In 2000, Smith (2000) estimated that Soho already had 16. If anything, the growing number of apparent attempts to close down adult establishments reflects their spread to ever-more locations and, as UK censors have noted, failure to permit outlets to open may merely drive adult entertainment underground (Brooks, 2000; Smith, 2000). Even in the UK, Smith estimated that there were no more than 65 licensed shops in the entire UK in 2000, but in 2001, Farrow (2001) suggested that there were 120. More recently, an attempt has been made to create an Internet list (Taylor, 2003). In the UK, the number of outlets selling sexually oriented adult material has grown but, thanks to a redefinition of the terms, local governments can claim to have reduced their number.

By the start of the 1990s, the traditional adult entertainment district was already in decline in many cities, as 'sex' and 'adult entertainment' became legitimate and as new forms of media began to compete with older ones. The spread of adult entertainment—particularly explicitly sexual entertainment—is the tip of an iceberg. Despite appearances, adult entertainment is no longer a mom-and-pop cottage industry. It has swollen to immense proportions, as figures from Datamonitor (2002) show. A growing number of firms not only operate shops, but have on-line ordering services, run Internet sites with downloadable videos, produce films and videos, maintain cable and satellite broadcasting services, and operate telephone sex lines (Datamonitor, 2002) and, for such firms, retail outlets are only one, increasingly minor, part of their total sales strategy.

As Tom Lehrer wrote, "When directly viewed, everything is lewd" (Lehrer, 1965). Adult entertainment and pornography have been substantially redefined since 1945. As various guides and maps show, adult entertainment includes not just activities of an explicitly sexual nature, but also clothing, toys, baked goods, tourism, networks, interest-groups and more. This is reflected in the growth of specialist zones—or moral regions—in which adult entertainment may be present to a considerable degree, but in which it no longer plays a central role. Manchester's Gay Village is one example, as are the area around the Pompidou Centre in Paris, the eastern part of central Montreal and West Hollywood. Some cities have encouraged such districts as a means of fostering economic growth (Collins, 2004). In addition, the advent of the Internet has led to the dispersal of adult entertainment to every home and every PC. Far from being marginalised or forced onto the Internet, adult entertainment has been sucked into it and, as experience in the Middle East shows, 'scandalous' behaviour by Internet users in search of adult material is widespread (Whitaker, 2003). Perhaps the reasons for centrality, for clustering, or for a zone of tolerance are disappearing and the future form of adult entertainment districts will be very different from that in the past.

Notes

1. In this paper, the term adult entertainment is used to describe those retail and service activities which have a sexual orientation, including the sale of x-rated or 'adult' videos, clothing, publications, so-called sex toys and other similar items, as well as cinemas which show sexually explicit movies, topless bars, sex shows and other related forms of entertainment. Although not everyone considers such activities 'adult', this is the current term used to describe them.
2. It is interesting to note that efforts in other cities to change the moral characteristics of urban districts have also often failed, or

taken a very long time to show results. For example, in the case of le Marais in Paris, efforts to transform the district through gentrification go back to the 1960s (Kennet, 1972, pp. 55–61).

3. In New York City, interviews were conducted with all the managers of shops listed in *Sexy New York 2002* located between 45th Street and 23rd Street and between 6th and 8th Avenues in Manhattan. In Berlin, they were conducted in April 2002 and May 2003 in shops located in the vicinity of the Kurfurstendam near the Bahnhof Zoo. In London, they were conducted in July 2003, in all the stores shown in the London Fetish Map in central London; in Exeter, they were conducted in July 2003 in two stores in Fore Street.

References

ASHWORTH, G. J., WHITE, P. E. and WINCHESTER, H. P. M. (1988) The red light district in the west European city: a neglected aspect of the urban landscape, *Geoforum*, 19, pp. 201–211.

BARNETT, J. (1974) *Urban Design as Public Policy*. New York: Architectural Record.

BOGUE, D. J. (1963) *Skid Row in American Cities*. Community and Family Study Centre, University of Chicago, IL.

BOOTH, C. (1889) *Life and Labour of the People of London*, Vol. 1. London: Williams and Norgate.

BROOKS, R. (2000) Film censor wants sex shop in every town. Based upon an article in: *The Sunday Times*, 5 November 2000. (www.melonfarmers.co.uk/brmore.htm; accessed on 1 September 2003).

BURGESS, E. W. (1925/1967) The growth of the city, in: R. E. PARK, E. W. BURGESS and R. D. MCKENZIE (Eds) *The City*, 4th edn 1997, pp. 47–62. Chicago, IL: University of Chicago Press.

CALIFORNIA ECONOMIC DEVELOPMENT DEPARTMENT (1992) *Los Angeles County Business Patterns*. Sacramento, CA: California Economic Development Department.

CHICAGO, DEPARTMENT OF ZONING (2003) *Chicago Zoning Ordnance: Adult Use Ordnance*, art E, chapter 16-16 (chapter 194C*) (www.ci.chi.il.us/zoning/ordnance/adult/html).

CITY OF PORTLAND POLICE BUREAU (1994) *Portland rape crimes with porno points, 1994* (carltown@Worldstar.com).

CITY OF PORTLAND POLICE BUREAU (1997) *Portland Metro Area 1997 number of non-violent crimes with porno points and strip joints*. Oregon Professional Microsystems.

COLLINS, A. (2004) Sexual dissidence, enterprise and assimilation: bedfellows in urban regeneration, *Urban Studies*, 41(9), pp. 1789–1806.

Columbia Fun Maps (1998) *Toronto*. New York City: Alan H. Beck.

Columbia Fun Maps (2002a) *Toronto*. Maplewood, NJ: Alan H. Beck (http://www.funmaps.com/).

Columbia Fun Maps (2002b) *Denver*, 2002 edn. Maplewood, NJ: Alan H. Beck.

Columbia Fun Maps (2002c) *New Jersey and Philadelphia*, 2001–2002 edn. Maplewood, NJ: Alan H. Beck.

Columbia Fun Maps (2002d) *Northeast Resorts: Fire Island, Long Island, New Hope, Asbury Park, and Country Inns*, 2002 edn. Maplewood, NJ: Alan H. Beck.

Columbia Fun Maps (2002e) *Los Angeles, Long Beach and Laguna Beach*, 2002 edn. Maplewood, NJ: Alan H. Beck (http://www.funmaps.com/).

Columbia Fun Maps (2002f) *Montreal and Quebec City*. 2002 edn. Maplewood NJ: Alan H. Beck (http://www.funmaps.com/).

DATAMONITOR, PLC (2002) *Online Adult Entertainment*. London: Reuters.

DAVID JAMES PRESS, LTD (1989) *Gay and Lesbian Map and Guide of Toronto*. 1989 edn. New York: David James Press.

DEIGHTON, L. (1974) *Close-up*. London: Pan Books.

DETROIT ADULT WEBMASTERS (2003) *Welcome to Detroit Adult Webmasters* (www.detroitadult webmasters.com/index2.php; accessed on 28 August 2003).

The Economist (2004) Romancing the disc, 7 January, pp. 63–64.

ELIOT, M. (2002) *Down 42nd Street: Sex, Money, Culture, and Politics at the Crossroads of the World*. New York: Warner Books (www.newyorkhistory.info/42nd-Street/; accessed on 18 August, 2002).

FARROW, B. (2001) Hello sex, goodbye dirty mac, *The Observer*, 3 June.

FIREY, W. (1945) Sentiment and symbolism as ecological variables, *American Sociological Review*, 10, pp. 140–148.

FLORIDA, R. and LEE, S. Y. (2001) *Inovation, human capital, and diversity*, Carnegie Mellon University, Pittsburgh (http://www.heinz.cmu.edu/~florida/pages/pub/working_papers/APPAM_paper_final.pdf; accessed on 20 August 2003).

FRIED, J. (1977) Arson, a devastating big city crime, *The New York Times*, 14 August section 4.

Friends: The Gaymap (2002) Asperg, Germany: Friends Meien (http://www.gaymap.info/download.html).

GATTA, G. M. (Ed.) (2002a) *Damron Men's Travel Guide 2003*. San Francisco, CA: Bob Damron.

GATTA, G. M. (Ed.) (2002b) *Damron Women's Traveller 2003*. San Francisco, CA: Bob Damron.

GILFOYLE, T. J. (1991) Policing sexuality, in: W. R. TAYLOR (Ed.) *Inventing Times Square: Commerce and Culture at the Crossroads of the World*, pp. 297–314. New York: Russell Sage Foundation.

GOLDENBERG, S. (2004) Harvard women deny new illustrated sex magazine is pornography, *The Guardian*, 13 February.

GREATER LONDON AUTHORITY (2002) *Planning for London's Growth*. London: GLA..

GRUEN, V. and SMITH, L. (1960) *Shopping Towns USA*. New York: Reinhold Publishing Corporation.

HALLER, M. H. (1970) Urban crime and criminal justice: the Chicago case, *Journal of American History*, 57, pp. 619–635.

HONAN, J. (Ed.) (1959) *Greenwich Village Guide*. New York: The Bryan Press.

HOTELLING, H. (1929) Stability in competition, *Economic Journal*, 39, pp. 41–57.

HUBBARD, P. (1997) Red light districts and toleration zones: geographies of female street prostitution in England and Wales, *Area*, 29(2), pp. 129–140.

HUBBARD, P. (2004) Cleansing the metropolis? Sex work and the politics of zero tolerance, *Urban Studies*, 41(9), pp. 1687–1702.

JOHNSON, L. (2003) Do blue movies make blue chips?, *Sunday Telegraph*, 2 March.

KAISER, C. (1997) *Gay Metropolis*. New York: Houghton Mifflin.

KENNET, W. (1972) *Preservation*. London: Temple Smith.

KENNEY, M. R. (2001) *Mapping Gay LA*. Philadelphia, PA: Temple University Press

KNIGHT, M. (1978) Scholars in new rift over 'white flight', *New York Times*, 11 June.

KRUGMAN, P. (1996) How the economy organises itself in space: a survey of the new economic geography, in: B. W. ARTHUR, S. N. DURLAUF and D. A. LANE (Eds) *The Economy as a Complex Evolving System II*, pp. 223–237. Santa Fe, NM: Santa Fe Studies in the Sciences of Complexity.

LASKER, S. (2002) Sex and the city: zoning 'pornography peddlers and live nude shows', *UCLA Law Review*, 49(4), pp. 1139–1185.

LEE, D. (2003) It's not the porn, it's the color scheme, *New York Times*, 15 June.

LEHRER, T. (1965) *Smut That was the Year that Was*. Los Angeles, CA: Warner Brothers

LOS ANGELES, DEPARTMENT OF CITY PLANNING (1977) *A Study of the Effects of the Concentration of Adult Entertainment Establishments in the City of Los Angeles*. Los Angeles, CA: DCP.

LOTHAR, A. (2001) *Berlin von Hinten*. Berlin: Bruno Gmunder Verlag.

MCCABE, J. D. JR (1872) *Lights and Shadows of New York Life or Sights and Sensations of the Great City*. Philadelphia, PA: National Publishing Company.

MCCABE, J. D. (1882) *New York by Sunlight and Gaslight*. Philadelphia, PA: Hubbard Brothers.

MCNEE, B. (1984) If you are squeamish, *East Lakes Geographer*, 19, pp. 16–27.

MILGRAM, S. (1970) The Experience of Living in Cities, *Science*, 167, pp. 1461–1468.

MILLS, E. S. and HAMILTON, B. W. (1994) *Urban Economics*. New York: Harper Collins.

NEEDHAM, J. (1990) Gone with the sin: closure of adult theater in Santa Ana reflects trend credited to—or blamed on—the videocassette revolution, *Los Angeles Times*, 14 August, p. E-1.

NETZER, D. (1978) The worm in the apple, *New York Affairs*, 4(Spring), pp. 42–48.

NEW YORK CITY, CITY PLANNING COMMISSION (1982) *Mid-town Zoning*. New York: Department of City Planning.

NEW YORK CITY, DEPARTMENT OF CITY PLANNING (1994) *Adult Entertainment Study*. New York: Department of City Planning.

NEW YORK CITY, DEPARTMENT OF CITY PLANNING (2001) *NYC 2000 Population Growth and Race/Hispanic Composition*. New York: Department of City Planning.

NEW YORK MUSEUM OF SEX (2002) *NYC sex: how New York city transformed sex in America* (Folder).

NEW YORK STATE URBAN DEVELOPMENT CORPORATION (1981) *Basis for blight finding 1981*. (xerox).

PAPAYANIS, M. A. (2000) Sex and the revanchist city: zoning out pornography in New York, *Environment and Planning D*, 18, pp. 341–353.

PARK, R. E. (1952) *Human Communities: The City and Human Ecology (The Collected Papers of Robert Ezra Park, Vol. 2)*. Glencoe, IL: The Free Press.

PARK, R. and BURGESS, E. W. (1925/1984) *The City*, reprinted 1984. Chicago, IL: University of Chicago Press.

PAUTZ, M. (2002) The decline in average weekly cinema attendance 1930–2000, *Issues in Political Economy*, 11 (http://www.elon.edu/ipe/pautz2.pdf).

PHILLIPS, A. C. (2002) Comments: a matter of arithmetic: using supply and demand to determine the constitutionality of adult entertainment zoning ordnances, *Emory Law Journal*, 51, pp. 319–353.

QUINDLEN, A. (1979) Must the city sacrifice the poor to lure back the middle class?, *New York Times*, 12 August, section 4.

RECKLESS, W. C. (1926) The distribution of commercial vice in the city: a sociological analysis, *Publications of the American Sociological Society*, 20, pp. 164–176.

REES, J. and TOMKINSON, M. (2002) Profits fall but

porn baron still nets GP £7 million, *Mail on Sunday*, 11 August.

ROPER, J. (1978) Abandoned housing is major threat to cities, *Newark Star–Ledger*, 20 August.

ROWLEY, G. (1978) 'Plus ça change …': a Canadian skid row, *The Canadian Geographer*, 22(3), pp. 211–224.

SAGALYN, L. B. (2001) *Times Square Roulette*. Cambridge, MA: MIT Press.

Sexy New York City 2000 (2000) New York: On Your Own Publications.

Sexy New York City 2002 (2002) New York: On Your Own Publications.

SHUMSKY, N. L. and SPRINGER, L. M. (1981) San Francisco's zone of prostitution 1880–1934, *Journal of Historical Geography*, 7(1), pp. 71–89.

SKINNER, G. W. (1964/65) Marketing and social structure in rural China, *The Journal of Asian Studies*, 24, No. 1 (November 1964), pp. 3–43; No. 2 (February 1965), pp. 195–228; and No. 3 (May 1965), pp. 363–399.

SMITH, A. W. (2000) Why every town in Britain needs its own sex shop, *The Independent*, 11 November.

TAIT, N. (2003) Ann Summers wins High Court battle over ban on recruitment adverts at job centres, *Financial Times*, 19 June, p. 6.

TAYLOR, D. (2003) *Sex shops directory* (www.melonfarmers.co.uk/r18sexsh.htm; accessed 28 August 2003).

TIMES SQUARE BID (BUSINESS IMPROVEMENT DISTRICT) (1996a) *Retail and market analysis, October, 1996*. New York: Times Square BID.

TIMES SQUARE BID (1996b) *Annual Report 1996*. New York: Times Square BID

TIMES SQUARE BID (1997) *Annual Report 1997*. New York: Times Square BID

TIMES SQUARE BID (1999) *Annual Report 1999*. New York: Times Square BID.

TIMES SQUARE BID (2000) *Annual Report 2000*. New York: Times Square BID.

WALDAU, R. (1990) *Berlin und DDR von Hinten*. Berlin: Bruno Gmunder Verlag.

WALSH, C. (2002) Media: videos kill magazine stars, *The Guardian*, 3 November.

WAMBAUGH, J. (1978) *The Black Marble*. London: Futura Books.

WARD, J. (1975) Skid row as a geographic entity, *Professional Geographer*, 27(3), pp. 286–296.

WARE, C. F. (1935/1994) *Greenwich Village, 1920–1930*, reprinted 1964. Berkeley, CA. University of California Press.

WHATSYOURS PUBLISHING (2002) *London Fetish Map*. London: Whatsyours.com.

WHATSYOURS PUBLISHING (2003) *London Fetish M|p*, 2. London: Whatsyours.com.

WHITAKER, B. (2003) Islam at the electronic frontier, *The Guardian*, 11 August.

WILLIAMS, H. C. 2002) Commission denies request for zoning, *The Augusta Chronicle*, 19 June.

WINCHESTER, H. P. M. and WHITE, P. E. (1988) The location of marginal groups in the inner city, *Environment and Planning D*, 8, pp. 37–54.

YEATES, M. (1998) *North American City*, 5th edn. New York: Longman-Addison Wesley Longman Inc.

Cleansing the Metropolis: Sex Work and the Politics of Zero Tolerance

Phil Hubbard

Introduction

On 8 March 2000, the English Collective of Prostitutes and the newly formed International Sex Workers' Union held a widely publicised protest on the streets of Soho, central London. Accompanied by a samba band, a 100-strong group of sex workers marched with placards proclaiming that they were, in effect, on strike. This unprecedented action was in response to continuing attempts by Westminster City Council to displace sex work from Soho through the compulsory purchase of properties that the authorities claimed were being used for 'immoral purposes'. This policy resulted in the eviction of a number of female sex workers, including some who had lived and worked in the area for 20 years:

> We organised a carnival in Soho, because not many people knew that there was going to be a prostitutes' strike. We wanted to make people aware of it and to think about the issues ... so we marched through the streets of Soho on the evening of International Women's Day, the day of the strike and it was brilliant, because it was good opportunity to celebrate what it is to be a sex worker. Although the action was firstly in support of those sex workers who were angry about being evicted ... it had a positive side. So it was a fun celebration, and that's very important—because it's important that sex workers feel pride in themselves, and pride in what they do (Lynn Clamen, spokesperson for the IUSW).[1]

Irrespective of this protest, which attracted the attention of the national media, Westminster Council persevered with the compulsory purchase of properties they suspected of being used for prostitution, handing the properties on to a housing association. In addition,

Phil Hubbard is in the Department of Geography, Loughborough University, Loughborough, LE11 3TU, UK. Fax: 01509 223 930. E-mail: P.J.Hubbard@lboro.ac.uk.

an eight-week crackdown by the Metropolitan Police 'vice squad' at the end of the year resulted in the arrest of 31 women in Soho, with officers claiming to have infiltrated a sex trafficking ring involving the illegal immigration of women from eastern Europe. Such claims were countered by the English Collective of Prostitutes, with spokesperson Sarah Walker suggesting that:

> It has nothing to do with illegal immigrants or the rights of women ... The police just want to gentrify the area ... It is to do with money and property values (quoted in Alleyne, 2001, p. 13).

In response, the ECP and ISWU organised another street carnival in Soho in 2001, timed to coincide with the May Day anti-globalisation protests.[2]

Mirroring the protests in London, on 5 November 2002, some 300 sex workers, mainly wearing white masks, occupied the space outside the French Sénat in Paris. In scenes reminiscent of the much-publicised church occupations of 1975 (which followed a series of murders of sex workers; see Jaget, 1980), these masked protesters held up signs proclaiming 'You sleep with us, yet you vote against us', 'End the war against the prostitutes' and, more pointedly, 'How much? 3750 Euros or two months in prison'. Later, they were to hold an impromptu street party in the Rue Vaugrigard, which, according to police estimates, involved upwards of 1000 participants (including delegations of sex workers from Lyon, Marseilles and Nantes, who held parallel protests in their own cities in the following weeks). Organised by the newly founded 'French Prostitution' movement, with support from ACT-UP Paris and the Green Party, this street demonstration was specifically designed to oppose articles in Interior Minister Nicholas Sarkozy's Internal Security Bill that proposed to make 'passive' soliciting a criminal offence punishable by two months' imprisonment or stringent fine. In addition to this attack on Sarkozy's Bill, the protesters also used the opportunity to draw attention to the deficiencies of alternatives to Sarkozy's proposal, including the reintroduction of official brothels mooted by some on the far right (including Francoise de Panafieu, the conservative mayor of Paris' 17th *arrondissement*). Others spoke to the press about the increasingly difficult working conditions they faced on the streets of Paris, alleging police brutality, confiscation of condoms and sexual blackmail (*Le Monde*, 2002a).

These tactical moves by sex worker groups are worthy of note, being part of an established tradition of street politics that exploits the visibility of the public sphere to make a powerful claim for rights and recognition (see McKay, 1996). Yet in this paper, I want to ignore questions concerning the effectiveness of such attempts to 'reclaim the streets' and instead to explore the context in which sex workers have resorted to these tactics. Specifically, I want to highlight recent changes in the regulation of vice in Paris and London, particularly attempts to displace sex workers through the adoption of (often-brutal) strategies of zero tolerance policing.[3] Bolstered by a rhetoric of spatial cleansing and purification—often shot through with xenophobic fears of Otherness—such policies enjoy a level of public and political support that means opposition to these swingeing new powers remains muted (the protests cited above being notable exceptions). Consequently, such policies are having major impacts on sex work in both cities, bequeathing sexual geographies that contrast markedly with the well-known patterns of sex work that characterised these two cities in the latter half of the 20th century.

In this paper, I thus engage with on-going debates surrounding the relations of sexuality and space by documenting recent attempts to remove (heterosexual) sex work from the public spaces of London and Paris. Specifically, I want to use this paper to argue that female sex workers are currently being identified as a threat to national values in an era when nebulous anxieties about difference and diversity are prompting the state to instigate 'public order' legislation which serves to criminalise specific groups (rather than particular offences). In turn, I seek to demonstrate that recent changes in the policing of commercial sex are connected to the rise of policies that regard prostitution and pornography as antithetical to the reinvention of city centres as safe, middle-class, family-oriented consumption spaces. In this regard, I suggest that there are in fact many similarities between punitive policies targeting commercial sex workers and those that challenge the presence of vagrants, buskers, gypsies, itinerant traders, teenagers and the homeless in city centres (see Sibley, 1995; Mitchell, 2001; Fyfe, 2004). Before I develop this argument, however, it is necessary to clarify the nature of the policies that are targeting sex workers with the avowed intent of making Paris and London safer, more commodious cities.

Criminalising Prostitution in Central Paris

Perhaps more than any other Western city, Paris has gained a reputation as a centre for commercial sexuality. Corbin claims it was Paris' 'extroverted nature' that encouraged the development of a highly sophisticated and varied sex trade in the 19th century, with "the bright artificial light of Paris by night stimulating the fantasies that sprang from that milieu" (Corbin, 1990, p. 205). While this reputation perhaps exaggerates the contemporary scale of sex work in the city, Paris has remained punctuated by a range of well-known areas notorious as spaces of elicit encounter, pornography and prostitution. These vary from the bright lights of Pigalle, a tourist-oriented and 'clearly formalised' district of sex shops and peep shows on the fringes of Montmartre (see Ashworth *et al.*, 1988, p. 204) to the more marginal Bois de Boulogne, which is notorious as '*l'allée des langues queues*' where transvestite prostitutes service a mixed clientele (including, over the years, many prominent political figures). Elsewhere, Rue St Denis in Beauburg offers a concentration of sex shops and street prostitution, well known for the West African sex workers who rent first-floor studio flats for upwards of 400 Euros per day (Montreyand, 1993). Beyond these notorious 'red-light' districts, mapped by Ashworth *et al.* (1988), street prostitution and off-street sex work in saunas and massage parlours are focused on the main boulevards around the *maraichaux* (the old ring road), with distinctive foci around the main rail stations and in the Bois de Vincennes.

This distinctive sexual geography is, in many respects, a reflection of French vice legislation as enshrined in the *Code Pénal.* While these codes have not made prostitution illegal, they amount to what is described as an *abolitionist* system (in stark contrast to the system abolished in 1946 that relied upon the enclosure of prostitution in state-sanctioned *maisons de tolerance*; see Corbin, 1990). The abolitionist system that replaced this stringent 'regulationism' insists that women and men should not be penalised for working in the sex industry because they are 'victims', yet argues that the sex industry should be discouraged by prosecuting those clients, pimps and brothel-owners who encourage prostitution. Hence, article 225-5 of the *Code Pénal* threatens those who procure sex workers with seven years' imprisonment and a 150 000 Euro fine, while article 623-8, introduced in 1994, advocates community service for kerb-crawlers. Yet if French vice laws proclaim a concern for the working conditions and safety of sex workers, the low number of prosecutions for living off immoral earnings (pimping) and kerb-crawling suggests that such concerns are currently given a low priority (Hubbard, 1999). Instead, it has been suggested that French vice legislation is principally (and pragmatically) designed to reduce the 'nuisance' experienced by people living in areas of street prostitution and to push sex workers away from whiter, more affluent areas (Montreyand, 1993). Indeed, most of the 350 annual arrests of prostitutes for soliciting in Paris occur when sex workers stray from recognised spaces of street prostitution into the more affluent *beaux quartiers*, where their presence often provokes moral approbation (Monteyrand, 1993). Thus, this abolitionist system generally tolerates street prostitution so long as soliciting is not active, with Article 225-10 stipulating that sexual exhibitionism and active soliciting are punishable, but passive soliciting is not.

The legal—as well as moral—ambiguities of French vice law have thus conspired to create a situation where street sex work is tolerated so long as it does not arouse public complaints about sexual exhibition or gross indecency. In the same manner, sex shops and 'peep shows' are tolerated under Article 227-24 so long as they are not within a hundred metres of a school and do not display obscene materials in their windows. This has allowed the development of well-known spaces of commercial sexuality in central Paris, typically away from wealthier *arrondissements* where the police might invoke Articles. Hence, the replacement of the French regulationist system by the abolitionist system has not been accompanied by a discernible decline in sex work, with contemporary estimates suggesting that there are around 15 000 female sex workers in Paris, constituting half the national total.[4] Yet in recent years, the continuing visibility of vice on the streets of Paris has prompted heated debates about the regulation of prostitution. For example, the leftist mayor of Paris, Bertrand Delanoe, has announced a plan to re-educate and retrain prostitutes to work in 'more acceptable' professions (*Le Monde*, 2002a), while some residents (for example, in the Chateau Rouge district in the 16th arrondissement) have taken it upon themselves to collect evidence to prove that active soliciting is occurring on their streets (*Le Monde*, 2000). Likewise, media concern has prompted a series of initiatives designed to 'clean up' the Bois de Boulogne (for example, by closing it to vehicular traffic at night and improving lighting throughout the park).

Simultaneously, the late 1990s saw increased allegations by prostitutes of police violence and harassment. Outreach workers distributing condoms and health advice among Paris' sex workers began to hear stories of the police demanding sexual favours from prostitutes threatened with arrest, while it was even reported that some sex workers had been required to clean cells and offices in police stations before the police would free them. Remarkably, it seems that possession of condoms was also being used a evidence of 'active soliciting' as some women reported regular arrest. The pressure group ACT-UP began to compile such evidence, presenting it to the Mayor's Office as evidence of the need for closer monitoring of police in the capital and the fostering of better relations between police and prostitutes.[5] However, in the context of a move towards 'order-maintenance policing' (inspired by the example of New York, where Zero Tolerance policing has been held up as a panacea for urban crime), new Agents of Security funded by the Mayor's Office have actually sought to take a tougher line with prostitutes as part of their mission to tackle so-called quality of life offences (Body-Gendrot, 2002). Inspired by declining crime rates in New York, and drawing on Wilson and Kelling's (1982) 'broken windows' theory, which holds to the idea that minor misdemeanours precipitate more serious crimes (see Fyfe, 2004), this witnessed Paris' reconstituted (and expanded) police force tackling signs of disorder in public places (for example, rounding up stray dogs, tackling drunkenness and removing peripatetic portrait painters from the streets). Street prostitution has thus joined this list of urban 'nuisances', targeted by the authorities as part of an attempt to diffuse concern about urban criminality. Although Parisians have evidently not accepted the logic of zero tolerance in all respects (for example, attempts to impose new speed limits on Paris' roads have largely been ignored), there is apparently much enthusiasm for this attempt to tackle 'street crime'—as well as increasing acceptance of the need for CCTV in public places (long anathema to French notions of *liberté*). Thus, the punitive policing of sex work currently enjoys a level of popular support, being regarded as part of a wider attempt to 'cleanse' public space.

In this context, the inclusion of amendments relating to prostitution in Nicholas Sarkozy's Interior Security Act (passed in February 2003, and enacted from 18 March

2003) can be viewed as the culmination of a series of policies aimed at restricting the visibility of sex work in Paris. Dubbed 'Monsieur Matraque' (truncheon), Sarkozy has quickly become one of the most feted of all French politicians, widening the appeal of the centre-right Chirac UMP party by pandering to those far-right politicians (such as Le Pen) who have argued for more stringent law and order policies while outwardly maintaining a concern for human rights (as manifest in his attempt to provide better facilities for the asylum seekers held at Sangatte, see Campbell, 2002). Espousing zero tolerance policing ('Repression is the best form of prevention' being one of his favoured slogans), Sarkozy pointed to an 8 per cent decline in crime over his first 6 months as Interior Minister (May–November 2002) as a vindication of his decision to invest in law and order and recruit 13 000 new police.

Although opposed by many civil liberties groups in France (Campbell, 2002), Sarkozy's Internal Security Act has resulted in over 75 major changes to the articles of the *Code Pénal*. Among these are new powers of imprisonment for aggressive begging (a maximum of 6 months), a fine of up to 30 000 Euros for swearing at the police and a maximum of 2 months imprisonment for youths who repeatedly loiter in stairwells or public areas of flats. Significantly, the new Act also makes passive soliciting an offence, with up to two months in jail for 'soliciting by any means, including dress, position or attitude'.[6] Equally controversially, once arrested for soliciting, any foreign nationals may have their Temporary Residence Permit withdrawn, with a Provisional Authorisation of Stay for three months granted only if the sex worker denounces their 'procurer or pimp' and agrees to move to protected lodgings and take up a new occupation. While Sarkozy has suggested that this law is 'balanced perfectly, as the prostitute can keep silent or contribute to the dismantling of prostitution networks' (*Le Monde*, 2002b; author's translation), those fighting sex trafficking do not unanimously share his enthusiasm. For example, Jean Phillippe Chauzy, a spokesperson for the International Organisation of Migration, claims that "deporting the victims of trafficking is the worst thing you can do, as without any reintegration assistance, there is a very high probability they will be trafficked again" (quoted in Richburg, 2002).

Given that as many as 70 per cent of Paris's street workers are east European or west African, including some from Sierra Leone and Albania, Sarkozy's reforms present many with a stark choice: return home to an uncertain fate or incur the wrath of potentially dangerous pimps. Sex work groups stress that there are existing laws which could be used to deal with pimps and sex trafficking, and that targeting prostitutes will merely worsen prostitutes' working conditions. According to a spokeswoman for the prostitute rights' group Cabiaria

> the consequence of this is that sex workers will go underground, in dark corners where nobody can come to their rescue if they are attacked. They are going to have to turn tricks in a rush. As they are in hurry, they will be not be able to check their clients or insist on the use of condoms (Cabiaria press release, 29 July 2002.[7]

In the eyes of many prostitute groups, the changes ushered in by Sarkozy amount to the adoption of a *prohibitionist* system, which renders all sex work criminal and condemns sex workers to work in clandestine (and inevitably precarious) conditions. Indeed, the draconian way the new powers were invoked in the first six months following the introduction of the Internal Security Act underlines these arguments: 569 prostitutes were arrested, and 22 deported, in that period alone—far in excess of the total for 2002 (*Libération*, 2003).

When coupled with changes to Articles regulating the operation of sex shops (with their window displays monitored on a weekly basis and a new super-tax introduced on the sale of X-rated videos), these new laws provide an effective series of mechanisms for police seeking to 'cleanse' the streets of French cities. Yet while Sarkozy's reforms apply to the whole of France, they emerge from anxieties about the visibility of commercial vice at the heart of the Paris rather than in the provinces (it is notable that Sarkozy is an elected representative for Neuilly, one of Paris' *beaux quartiers*).[8] Likewise, it appears that their impacts will be most sharply felt in the metropolis, where concerns about crime (and terrorism) are particularly pronounced. Indeed, in one of his regular media addresses, Sarkozy led bemused journalists around the city's *boulevards*, outlining his plan to deal with the 'problem' of vice.

> It is necessary that to be effective a reform must be total. Those who live in Paris want the explosion of prostitution halted ... To do so, it is necessary to cease the hypocrisy that persists in making a distinction between active and passive soliciting. Why is the Albanian girl put on the Parisian pavement by procurers? Because these modern-day slave traders risk nothing. By only penalising passive soliciting we do nothing about the network that exploits prostitutes (Sarkozy, quoted in *Le Monde*, 2002b, p. 22; author's translation).

Describing his reforms as necessary "to restore safety, human rights and basic freedoms",[9] Sarkozy's *Loi pour Sécurité Intérieure* has nonetheless been accused of criminalising certain groups, bracketing together prostitutes, aggressive beggars and squatters in an Act ostensibly designed to tackle broad issues of national insecurity. In this regard, the fact that Sarkozy's Act includes Articles relating to gun control, counter-terrorism measures and the construction of a database of DNA samples from all those suspected of paedophilia underlines that the endemic fear that characterises contemporary society is provoking *revenge* on a diverse range of targets, including "the working class, women ... gays and lesbians, immigrants" (Smith, 1996, pp. 44–45) as well as sex workers.

Cleansing the Streets of London

As in Paris, the link between urban disorder and the presence of sex workers has been a prominent theme in the recent policing of London's street-spaces. Previously, London's sexual geographies were shaped by a suite of laws that did not criminalise prostitution *per se* but had the twin aims of preventing "the serious nuisance to the public caused when prostitutes ply their trade in the street" and penalising the "pimps, brothel keepers and others who seek to encourage, control and exploit the prostitution of others" (Edwards, 1987, p. 928). The 1956 Sexual Offences Act thus targeted 'the procurer'—any pimp who knowingly lives wholly or in part off the earnings of prostitution—while the 1959 Street Offences Act made persistent soliciting an offence (Hubbard, 1999). The (hastily drawn-up) 1985 Sexual Offences Act provided the third plank of British vice legislation, criminalising those men who solicit women in addition to those women who solicit men.

Collectively, British vice laws created a paradoxical situation where, although prostitution may not be illegal, it was impossible for women to sell sex without breaking a number of laws. Thus, police 'vice squads' were able to use their powers of arrest in effect to concentrate prostitution in red-light districts, where unwritten rules of engagement between police, punters and prostitutes conspired to create *de facto* 'toleration zones'. This spatial concentration of vice, traditionally in inner-city areas with transient populations and low rates of owner-occupation, allowed the police to enact strategies of surveillance designed, in theory, to minimise incidents of violence against female sex workers. More cynical observers have suggested that concentrating vice in relatively compact red-light districts also allowed the police to enact periodic 'crackdowns' against vice when there was a need to improve monthly arrest figures (Edwards, 1987).

Researchers have accordingly expended much energy describing the way these attempts to police vice bequeathed post-war London a distinctive moral geography characterised by pockets of prostitution and pornography (see Thompson, 1994; Mort, 1998; Linnane, 2003). In his audit of sex work in the metropolis, Matthews (1997) documents the existence of such pockets in Paddington, Kings Cross, Stoke Newington, Mayfair, Streatham and Soho. He estimated that 600 street workers were active in these areas, a total that could be doubled if the number of those working from private residences were included. In addition, he estimated that 2000 women were working in massage parlours and saunas across the capital, as well as upwards of 500 working in hostess clubs in the West End. The majority of the latter are located in Soho, an area somewhat equivalent to Rue St Denis in Paris in offering an "apocryphal and irregular version of metropolitan life" (Mort, 1998, p. 893). Located just off the main thoroughfares of the West End (Oxford Street, Regent Street and Charing Cross Road), Soho has long been notorious as the centre of London's sex industry. In part, this reputation is a function of the state and law's desire to concentrate vice in areas of low owner-occupation and transient residence, with Soho becoming, in effect, London's 'tolerance zone'.

By the mid 1970s, the impacts of this toleration were apparent in the existence of a variety of sex premises in Soho advertised through characteristically gaudy neon signs, including 54 sex shops, 39 sex cinemas, 16 peep shows, 11 sex clubs and 12 massage parlours (Thompson, 1994). Ironically, the success of such outlets began to push up rents in the area, with the Soho Society arguing that other long-established businesses were being displaced by commercial sex. Finding allies in the Community Standards Agency, the Soho Society pushed for stricter controls on sex businesses. Consequently, the 1982 Local Government Act and 1986 Greater London Council General Powers Act contained clauses that allowed Westminster City Council to refuse licences for sex shops and cinemas on a variety of grounds. Thompson (1994) describes how Westminster City Council used these powers in a widely advertised 'clean-up' of Soho in the early 1980s, although he notes their limited success, with many sex cinemas simply converting to peep shows (which did not require such licences).

Despite increasing local opposition, Soho thus persisted as a tourist-oriented area of sex work into the 1990s, at which time new voices began to add their weight to the campaign to displace vice. These voices included those of members of the Labour Party, including shadow Home Secretary Jack Straw who, following trips to New York to witness zero tolerance policing first-hand, began to espouse new ways of "reclaiming the streets for the law-abiding citizen" (quoted in Bowling, 1999, p. 532). In 1997, Tony Blair (then shadow Leader of the House of Commons) threw his weight behind 'Operation Zero Tolerance', a high-profile policing initiative that involved 25 officers removing the homeless, 'squeegee merchants' and prostitutes from Kings Cross (a district that Blair described as 'frightening').[10] Given that the initiative was acclaimed as a success, with a 'dramatic fall in anti-social behaviour in the area', similar campaigns were enacted by the Metropolitan Police in tandem with the London Boroughs. For insistence, in an eight-week programme in 2000, Westminster City Council and the Metropolitan Police co-operated to tackle a series of 'nuisances' including prostitution, unlicensed taxi drivers and 'bogus' portrait artists (the latter being blamed for a spate of handbag thefts).

Bolstered by government ministers espousing zero tolerance tactics (including Jack Straw at the Home Office), Westminster City Council also inserted clauses in its Unitary Development Plan (1999) insisting that planning permission for 'sex-related' uses would not be granted except in exceptional circumstances. Further, applicants had to demonstrate that there would be no adverse effects on residential amenity, community facilities (places of worship, schools, community centres, etc.) and the function of the area (including parking and the free flow of traffic). Where planning permission was granted, the council imposed stringent conditions relating to opening hours and window displays "to protect the amenity of residents and the general environment": in the words of the Chairman of the Planning and Licensing Committee, these measures were part of an on-going campaign "against the West End sex barons ... a war we will win."[11] Such measures mean that only 16 licensed sex shops and clubs remain in Soho (with Westminster City Council stipulating that additional licences will not be granted in the future). Coupled with a series of vice squad raids on private residences suspected of being used for the purposes of prostitution (the immediate provocation to the sex work strike of 2000), the effect has been to reduce the visibility of commercial sex on the streets of Soho (with similar trends being apparent in Kings Cross, Mayfair and, to a lesser extent, Paddington and Streatham).

Matthews' (1997, p. 7) exhaustive audit of prostitution in London thus noted a decline in street trade over the previous decade, "partly due to pressure from the local community, police intervention and a growing awareness of the dangers of working on the streets": this is a decline that has undoubtedly accelerated in the wake of new campaigns to clean up the streets. However, Matthews also noted a rise in the amount of off-street prostitution in the metropolis, a trend encouraged by the widespread availability of mobile phones, to the extent that calling-cards in public telephone boxes replaced street soliciting as the principal means of advertising to clients in the 1990s. Although data on this subject are conspicuously lacking, it is estimated that 13 million prostitutes' cards are distributed every year in central London. This implies there may be as many as 50 cards placed in each telephone box every day, with numbers even higher in some districts—particularly Soho. While these cards feature images that are largely innocuous (and, by the standards of existing British obscenity laws, perfectly legal), their increasing visibility provoked a number of complaints. Indeed, in one national survey of payphone users in 1996, complaints about prostitutes' cards represented the most frequent issue raised.[12] In response to such complaints, the owners of the majority of boxes (British Telecom) began to work with Westminster City Council to remove these cards on a regular basis, collecting over a million cards in one eight-week operation ('Operation Spotlight'). By the late 1990s, the task of removing cards is estimated to have cost Westminster City Council Street Enforcement Department around £250 000 per annum, with BT expending similar amounts barring incoming calls to advertised numbers (a policy that proved largely ineffectual given that only around 5 per cent of sex workers had BT accounts).

Hence, on 1 September 2001, Sections 46 and 47 of the Criminal Justice and Police Act came into force, making it an offence to place advertisements relating to prostitution in, or in the immediate vicinity of, a public telephone box.[13] Clarifying the scope of these powers, circular HOC 27/2001 stressed this applied to "any telephone in a place to which the public have access, whether on payment or otherwise" (Home Office, 2001, n.p.). Justifying this legislation, the Home Secretary David Blunkett suggested that:

> the primary intention of the measure is to deal with the nuisance and distress when prostitutes ply their trade in the streets and to penalize those who seek to encourage, control or exploit the prostitution of others (press release, 31 August 2002).[14]

The consequence of this is that the police have the power of arrest over anyone displaying advertisements that allude to sexual services in telephone boxes, with the offence punishable by a fine or up to six months' imprisonment. According to Westminster City Council Street Enforcement officers, this Act has resulted in a major reduction in the number of cards in Soho, although carders remain active in smaller numbers in Bayswater, Queensway, Paddington and areas north of Oxford Street.[15]

At the same time that those who distribute prostitute cards are being prosecuted under the terms of the Criminal Justice and Police

Act, the remaining cards are being used by undercover police to identify premises used for sex work. If the flat is rented, the owners are told by the police to evict any women working from that address (given that it is illegal for residential premises to be used for commercial purposes without appropriate planning permission). Coupled with the serving of Anti-social Behaviour Orders on street workers,[16] the net result of these powers is to make it increasingly difficult for sex workers to have any visible presence in the public spaces of the city, whether physically, through soliciting, or symbolically, through the calling-cards that proved an effective means for sex workers to advertise to clients. In effect, this means that prostitution in London is increasingly invisible, with punters forced to use adult contact magazines and websites to locate sex workers off-street, or to frequent the few remaining sites where sex workers furtively solicit for business (for example, Streatham Common).

Protecting the Nation: Prostitution and the Exclusionary Urge

While there are important differences between British and French law, this brief overview of recent attempts to regulate vice suggests that there are many similarities in the way sex work is being suppressed in London and Paris. Bolstered by a rhetoric of Zero Tolerance imported from the US (Body-Gendrot, 2002), both cities have introduced new laws designed to criminalise street work at the same time that police traditions of 'turning a blind eye' to vice have given way to more aggressive (and sometimes underhand) styles of policing. In both cities, such authoritarian strategies have been justified as attempts to protect both the public *and* prostitutes. Whether prostitutes benefit from the 'protection' offered is a moot point: while policy-makers persist with a rhetoric that suggests that *all* prostitutes are exploited, and potentially the victims of criminal cartels, they ignore the fact that many work independently (and successfully) on their own terms.

Yet the persistence of discourses that associate sex workers with criminal activity cements the idea that prostitution is a genuine threat to the quality of urban life (and, thus, unacceptable in the 21st-century metropolis). In this sense, the contemporary regulation of vice in London and Paris resonates with wider histories and geographies of prostitution, given that sex workers have often been scapegoated as the cause of social malaise and immorality. Indeed, in times of social crisis, the punitive treatment of prostitute women has often seemed a small price to pay for the restoration of social order (Hubbard, 1999). Here, there are clear parallels between sex workers and other marginal groups whose presence in Western nations has, over time, prompted periodic moral panic (for example, gypsies, Jews and the homeless). Sibley (1995) accordingly suggests that *spatial exclusion* is a recurring motif in the histories of such groups. For Sibley, such geopolitical strategies of exclusion are essentially concerned with the maintenance of social and spatial boundaries, with the exclusion of the disordered Other an attempt to physically and psychologically remove individuals labelled as different, deviant or dirty.

Significantly, Sibley (1995) suggests that the potential for exclusion is most pronounced when spatial, as well as social orders, are called into question in the midst of 'moral panics'. Such panics are characterised by the identification of a threat to national values, with a rapid escalation of media coverage serving to demonise this threat and propose ways the 'problem' can be solved. In a global era, such panics have typically concerned the (in)ability of the nation-state to designate, contain and protect its citizens: consequently, public anxiety has landed upon those groups who are seen to undermine national values, whether from within or without. For example, the stigmatisation of asylum-seekers and illegal immigrants in both the British and French media is indicative of the way that general anxieties about crime (as well as specific fears of terrorism) serve to identify specific groups as a national threat

(Sales, 2002). Likewise, the discursive portrayal of trafficking in women as a vast, widespread and growing global problem has lead to the demonisation of 'dark', 'Eastern' criminals who deal in "White flesh" (Berman, 2003, p. 54). Indeed, it is remarkable how much of the press coverage of sex work in London and Paris dwells on the role of Albanians, Serbs, Ukranians and Russians in instigating a 'flood' of immorality and criminality (for example, see *The Daily Telegraph*, 2003, on the 'sex empire' built up by an Albanian refugee; *Figaro*, 2003, on the Roma gypsies who allegedly ran a '100-strong vice ring'). As Berman (2003, p. 54) attests, the racial 'Otherness' of these criminals adds to the sense of fear and panic about criminal networks overrunning western Europe and destroying, with lawlessness and immorality, 'our' (White) way of life.

This conflation of sexual immorality and racial Otherness helps to explain why there has been a rising tide of concern about sex work in Paris and London over recent years. This anxiety is arguably related to two conflicting representations of those sex workers trafficked from East to West. In the first, sex workers are depicted as *victims* who are at the mercy of dangerous and violent pimps, seen as vulnerable individuals in need of protection. In the second, these sex workers are represented as a *threat*—individuals whose sexual immorality is seen to undermine national values. Apparently eschewing the traditional role that is allocated to women in Western societies—namely, that of the 'good wife and mother'—trafficked and migrant sex workers are not regarded as having the necessary self-control and sexual morality required to fulfil the requirements for citizenship. In this sense, the existence of French sex workers is eclipsed in the French media by talk of those eastern European prostitutes who threaten national security. Likewise, in Britain, the dominant representation of sex workers is one in which trafficked women are being routinely exploited and abused by organised criminal gangs, spreading immorality throughout the country.[17]

Regarded as being a site where many different forms of exploitation, criminality and migration come together, sex work is therefore widely depicted in both Britain and France as being in need of close surveillance lest it undermines the moral values at the core of the nation. In practical terms, this has meant that both sex workers and their pimps have been targeted by new legislation. Indeed, while recent Zero Tolerance strategies have been justified primarily with reference to the menace of male pimps and traffickers (witness—for example, the French Ministry of the Interior's claim that new laws will allow them to attack those 'pimps and go-betweens' who profit from prostitution), these policies have also served to criminalise female prostitutes. Indeed, while sex workers tactically adapt to new strategies of policing (Hubbard and Sanders, 2003), one immediate impact of zero tolerance policing is that the right of female sex workers to occupy public space has been severely curtailed. In this sense, while it is too early to provide definitive evidence of the impacts of zero tolerance policing on patterns and practices of sex work, the early signs are that sex workers are having to resort to more precarious forms of work as they become displaced from their traditional working environments. As is becoming apparent, this displacement takes different forms as, although the moral panic about sex work is arguably one about the security of *national* boundaries, it is also about the revalorisation and redevelopment of particular *local* spaces (and, conversely, the devaluation of others). It is to such questions of urban redevelopment and regeneration that I turn in the last section of this paper.

Sex in the Revanchist City

While recent policies determined to reduce the visibility of vice in Britain and France are connected to fears that the nation-state can no longer cohere and control its borders, it is crucially important to address the urban context in which such draconian policies have emerged. Indeed, it is highly significant that recent attempts to reduce the visibility of vice in the public spaces of central London and Paris correspond with the emergence of what Smith (1996) terms a 'new urbanism'. In his estimation, this is a form of urbanism based on neo-liberal policies of 'urban renaissance' led by governmental, corporate or governmental–corporate partnerships. For such entrepreneurial interests, gentrification has been a crucial urban strategy, the central plank of a 'new urban politics' that enables private capital to move into the vacuum left by the end of managerial, welfare-based urban policy (see Hall and Hubbard, 1996). This effectively allows urban real estate markets to become the prime vehicle of capital accumulation in the urban economy, with Smith (2002, p. 442) graphically detailing how obstacles to such gentrification policies—such as the presence of squatters, the homeless or street people in areas earmarked for redevelopment and improvement—have been ruthlessly dealt with through authoritarian tactics of repressive policing justified with reference to the 'scientific' doctrine of zero tolerance policing. Marked by a revengeful and reactionary viciousness, the revanchist (literally, *revenging*) city described by Smith is one where White upper-class men conspire against various populations whom they accuse of 'stealing' the city (Smith, 1996, p. xviii). For example, Smith alleges that, following the election of Mayor Guiliani and appointment of Police Commissioner William Bratton in New York in 1994, this urge to tame the disorder of the city was to trigger notorious police brutality against minorities, justified with reference to the need for improved quality of life, but actually intended to make the city safe for gentrification (and the associated invasion of White middle and upper-middle income-groups).

Smith's dystopian description of the contemporary city may be bleak, yet it is mirrored in a growing body of radical urban writing that stresses that Western city centres, traditionally characterised by diversity and difference, are giving way to single-minded spaces that exclude marginal groups (see—for example, Flusty, 2001; MacLeod and Ward, 2002). The concern for critical commentators is that the vitality and diversity that have traditionally typified the civil sphere have been lost as multifunctional public spaces make way for a predictable range of *nouvelle cuisine* restaurants, heritage centres, coffee shops, art galleries, shopping malls and multiplex cinemas. Hence, a common thread in critical urban studies is the idea that city centres are increasingly purified spaces, given over to the vicissitudes of the consumer society and what Sibley (1995, p. xi) terms "the white middle-class family ambience associated with international consumption style". Of course, it is easy to be nostalgic for the urban cultures that have been effaced by this recasting of city centres as a space of alluring consumption, and it is important to stress that not all exclusions are unjustified if one is to create a public sphere of mutual respect, rights and civility (see Merrifield, 2000). Nonetheless, many have sought to write the obituary of urban public space, lamenting the decline of a mode of metropolitan streetlife that was exciting, unpredictable and sometimes dangerous, yet which was open to difference:

> In the punitive city, the post-modern city, the revanchist city, diversity is no longer maintained by protecting and struggling to expand the rights of the most disadvantaged, but by pushing the disadvantaged out, making it clear that as broken windows rather than people, they simply have no right to the city (Mitchell, 2001, p. 71).

What seems to grate with many urban commentators is the fact that public space is designed to meet the wants and desires of affluent consumers while systematically excluding those adjudged unsuitable or threatening. As Flusty (2001) asserts, it is those who are Other to the real estate developers and their target markets who are effectively excluded from the new city centre. It is here that the duplicity of revanchism becomes apparent: all are apparently welcome in the spaces of the consumer city, "but only so long as they behave appropriately" (Flusty,

2001, p. 659). Nonetheless, Flusty insists that the consuming majority banally accept such exclusions, to the extent that exclusionary tactics have become taken-for-granted. In many ways, this banal acceptance of exclusionary spaces underlines Smith's argument that revanchist city politics, when coupled with strategies of gentrification, amounts to a successful recipe for capital accumulation. The visceral responses to the visibility of vice in the public spaces of metropolitan centres can only be understood when viewed in the context of such revanchism: sex workers are apparently seen as dangerous threats that need to be eliminated from the sight of the affluent (irrespective of the fact that many of the clients of sex workers are part of this group). In this sense, Smith's (1998, p. 1) assertion that "blaming the victim has been raised from a common political tactic to a matter of established policy" appears remarkably relevant in the context of post-September 11 anxieties about urban security. Evidently, the figure of the street prostitute has been reinvented as a cipher for a complex suite of anxieties about urban malaise—fears inevitably fuelled by xenophobic discourses that play on the imagined connections between illegal immigration, lawlessness and terrorist threat.

Accordingly, to understand why such Zero Tolerance approaches have been introduced in recent years, it is necessary to appreciate the convergence of national security fears and the increasing anxiety about urban crime and disorder. As Ellin (1997) argues, fear of urban living is now higher than ever before, intensified by sensationalist media reporting and a stream of environmental cues that symbolise the ever-present threats of criminality

> The fear factor has certainly grown, as indicated by the growth in locked car and house doors and security systems, the popularity of gated or secure communities for all age and income-groups, and the increasing surveillance of public space ... not to mention the unending reports of danger emitted by the mass media (Ellin, 1997, p. 26).

Against this backdrop of ambient fear, convincing investors and consumers that city centre spaces are 'safe' has become a key priority for those city governors and promoters keen to sponsor an 'urban renaissance'. Given that sex workers are depicted as part of a criminal class, reducing the visibility of sex work in the central city is an obvious way that policy-makers can send out a message that it is ripe for reinvestment. In this sense, the displacement of sex work can be viewed as an essential precursor to middle-class, family-oriented gentrification.

This putative link between state-sponsored gentrification and anti-vice policing is spelt out by Papayanis (2000) in her account of Guiliani's attempt to enhance 'quality of life' in Manhattan through the introduction of anti-porn legislation (partly at the bequest of the Disney Corporation, who set up operations on Times Square, a district long notorious as an illicit sexual marketplace; see also Boyer, 1996). At Giuliani's request, New York City Council approved amendments to their Zoning Resolution in October 1995 designed to "encourage the development of desirable residential, commercial and manufacturing areas with appropriate groupings of compatible and related uses and thus to promote and to protect public health, safety and general welfare" (quoted in *New York Times*, 23 February 1998). In effect, this resolution forced the closure of non-compatible land uses, including any business that has a "substantial portion of its stock-in-trade" in "materials which are characterized by an emphasis upon the depiction or description of specified sexual activities or specified anatomical areas". In this instance, such establishments are characterised as "objectionable non-conforming uses which are detrimental to the character of the districts in which (they) are located". Dictating that adult establishments should "be located at least 500 feet from a church, a school (or) a Residence District", this law stands as a remarkable attempt by legislators to reaffirm socio-spatial order by seeking to maintain distance between 'pornographic' and moral expressions of heterosexuality.

In Papayanis' account, the manner in which clandestine and underground expressions of desire in Times Square gave way to Disneyland emphasises the power of the state to eradicate sexual difference in the interests of cultivating an ambience of leisured consumption. Hence, we can interpret on-going attempts to remove prostitution from Paris and London both as strategies of *capital accumulation* (i.e. encouraging urban gentrification) as well as of *social reproduction* (i.e. marginalising those who threaten the moral values that underpin the reproduction of the nation-state). As Brown (2000, p. 83) insists, the production of sexual space cannot be understood solely by reference to the "the prevailing structures of heterosexuality" that dominate in a particular society, nor for that matter the "crises-resolution function of spatial fixes and their flexibility in the context of post-Fordism". On this basis, the identification of sex workers as a criminal Other appears an extremely effective strategy for displacing sex work from valued city centre sites and—*at the same time*—reasserting the moral values that lie at the heart of the nation-state.

However, given that sex work is also often highly profitable, the argument that it needs to be displaced to encourage capital accumulation in city centres seems somewhat contradictory. Brown (2000) works through such contradictions, reworking Lefebvre to suggest that the marginalisation of immoral sexuality in the cityscape serves as "a kind of spatial fix where sexual relations are commodified" (Brown, 2000, p. 85). In short, isolating sex work within marginal and liminal urban locations does not serve to devalue commercial sex work: rather, the opposite is true, with spatial marginalisation bringing sex work within the ambit of a *restricted economy* that hoards desire to commercial and capitalist ends (Bataille, 1993). Thus the segregation of family spaces from spaces of sex work serves to valorise both the family (as an idealised socio-sexual relation) and sex work (as an illicit but sought after socio-sexual relation characterised by high risks and high rewards). Policies designed to exclude sex work from the centre of London and Paris cannot therefore be viewed solely as attempts to repair the boundaries between eastern and western Europe, between family and non-family spaces, or between marginal and valued spaces: rather, they perform all of these roles simultaneously. In the final analysis, it is the congruence of those processes that favour gentrification and those that identify prostitution as a dangerous threat to national morality that is bequeathing new—and for many female sex workers, increasingly dangerous—geographies of sex work in London and Paris.

Conclusion

Beginning with a description of the punitive policing and public order acts that are serving to displace prostitutes from the central cities of Paris and London, this paper has sought to explore why prostitute women are currently being targeted as threatening Others. Alighting on debates concerning the logic of Zero Tolerance, it has been argued that sex workers stand at the intersection of a host of public fears about criminality, exploitation and disorder, and hence make convenient targets for policy-makers wishing to demonstrate their dedication to matters of law and order. For Sarkozy in France—as for Blair, Straw and Blunkett in Britain—introducing legislation designed to tackle low-level public disorder (apparently personified in the figure of the street prostitute) signals an intention to tackle crime and urban malaise. Critics suggest that this amounts to a policy of blaming the victim that does nothing to tackle the underlying causes of urban crime and poverty (an argument that holds particular weight given that some of the laws introduced to tackle the 'modern menace' of pimping and sex trafficking are impacting most severely on the sex workers themselves, not the pimps).

Although it is difficult to make a definitive statement about the ultimate impact of these policies, early evidence suggests that the net result will not be that sex trafficking or prostitution decreases. Instead, it appears that the outcome will be to remove vice from the central city. In terms of debates on public space, we should note that these policies may create spaces that many people find safer and more welcoming (encouraging gentrification and a 'urban renaissance'). But, as Flusty (2001) argues, this does not mean we should simply tolerate the effacement of difference that this involves: prostitution may well be deemed Other by the majority of the population, but does this imply that we should blithely accept the disappearance of vice from the streets of Paris and London? Certainly, some will always find the presence of sex workers in the public realm disquieting (in the same way, some will find it a source of excitement). But who has the right to deny prostitutes their claim to city space? And is it right that sex work is displaced to make way for sanitised public spaces that are as safe and sterile as shopping malls? These are difficult questions and there are no easy answers (see Merrifield, 2000). Yet in an era when sex workers (and other minority groups) find themselves increasingly marginalised because of the endemic fear and anxiety that grip society, urban researchers are surely duty-bound to explore such issues, to speak for (and with) such sexual minorities and to expose the processes that result in their marginalisation.

Notes

1. Interview with Lynn Clamen, spokesperson for the IUSW (http://www.sexworkerspride.org.uk/interview.html; posted April 2002).
2. The May Day protests in London 2001 comprised a series of mass protests, pickets and marches designed to oppose a number of targets, particularly businesses implicated in the perpetuation of the global capitalist economy.
3. Throughout, I use the term sex worker in preference to 'prostitute' as many sex workers do not identify as prostitutes. Additionally, the term 'sex worker' encompasses those who work in other segments of the sex industry, including those who work as lap dancers, striptease artistes, etc. On the definition of sex work, see especially Jeffreys (1998).
4. This figure does not include estimates of homosexual sex workers, who exhibit a very different geography (see Monteyrand, 1993).
5. ACT-UP open letter to Mayor's Office, dated 22 October 2002. ACT-UP are an activist organisation mobilising around issues concerning HIV/AIDS.
6. Amendment to Article 225-10, *Loi Pour Sécurité Intéreure* (2003) (author's translation).
7. Cabiaria press release (http://www.cabiaira.asso.fr; posted 29 July 2002).
8. In the first six months following the amendments to Code 225-10, the majority of the 569 arrests for soliciting were in the 16th and 17th arrondissements (see *Liberation*, 25 September 2003, p. 9).
9. Description from Ministry of Interior's website, posted 24 October 2002 (author's translation).
10. Blair's commitment to this concept was reiterated in 2002 when he launched a new Department of Environment, Food and Rural Affairs consultation document—*Living Places*—which suggests an overhaul of powers to deal with "beggars, rough sleepers, peddlers, buskers and others who are threatening or who engage in anti-social behaviour in public spaces" (quoted in Weaver, 2002).
11. Westminster City Council press release, Council wages war on porn barons (http://www.westminster.gov.uk/cex/wccnews/fa00234.htm; posted September 1999).
12. See note 3.
13. This Act also made kerb-crawling an arrestable offence, restricted drinking in certain public places and introduced fixed penalty notices (FPNs) for selected offences of criminal damage. The need for such powers to be enforced was reiterated in the governmental White Paper (2003) *Rights and Responsibility—Taking a Stand against Anti-Social Behaviour.*
14. Downing Street newsroom press release, New powers for crime fighters to tackle public disorder, (http://www.number10.gov.uk/news.asp/newsID = 2495; posted 31 August 2002).
15. Personal communication with Westminster City Council Enforcement Officers, 2002.
16. Anti-social Behaviour Orders were introduced in the Crime and Disorder Act 1998 to tackle low-level disorder and 'sub-criminal' activity. These orders prohibit an offender from specific anti-social acts or from entering a defined area for a period of two years.
17. On sex trafficking in London, see *Evening Standard* (2003); *The Guardian* (2003b); *The Times* (2002).

References

ALLEYNE, R. (2001) 31 held in Soho vice raids, *The Daily Telegraph*, 17 February, p. 13.

ASHWORTH, G., WHITE, P. and WINCHESTER, H. P. M. (1988) The red-light district in the west European city: a neglected aspect of the urban landscape, *Geoforum*, 19, pp. 201–212.

BATAILLE, G. (1993) *Symbolic Exchange and Death*. London: Sage.

BERMAN, J. (2003) (Un)popular strangers and crises (un)bounded: discourses of sex trafficking, *European Journal of International Relations*, 9, pp. 37–86.

BODY-GENDROT, S. (2002) *From zero tolerance to zero impunity: policing New York City and Paris*. Paper presented to the New Visions of the European City: Paris–New York, Congerence, New York University, April.

BOWLING, B. (1999) The rise and fall of New York murder: zero tolerance or crack's decline?, *British Journal of Criminology* 39, pp. 531–554.

BOYER, K. (1996) Twice-told stories: the double erasure of Times Square, in: I. BORDEN, A. KERR, A. PIVARO, and J. RENDELL (Eds) *Strangely familiar*, pp. 16–23. London: Routledge.

BROWN, M. (2000) *Closet Spaces*. London: Routledge.

CAMPBELL, M. (2002) Napoleon of crime-busting elevated to French pin-up, *The Sunday Times, 15 December, p.* 32.

CORBIN, A. (1990) *Women for Hire: Prostitution and Sexuality in France after 1850*. London: Harvard University Press.

EDWARDS, S. M. (1987) Prostitution, ponces and punters, policing and prosecution, *New Law Journal*, 137, pp. 928–930.

ELLIN, N. (1997) *Architecture of Fear*. New York: Princeton University Press.

Evening Standard (2003) Prostitution drive saves 120 women, 26 June, p. 6.

Figaro (2003) Roma de Ile St Denis avec cent femmes, 11 October, p. 10.

FLUSTY, S. (2001) The banality of interdiction: surveillance, control and the displacement of diversity, *International Journal of Urban and Regional Research*, 25, pp. 658–664.

FYFE, N. (2004) Zero tolerance, maximum surveillance: deviance, difference and crime control in the late modern city, in: L. LEES (Ed.) *The Emancipatory City*. London: Sage (forthcoming).

HALL, T. and HUBBARD, P. (1996) The entrepreneurial city: new urban politics, new urban geographies? *Progress in Human Geography*, 20, pp. 153–174.

HOME OFFICE (1999) *Consultation paper: new measures to control prostitutes cards in phone boxes*. London: HMSO.

HOME OFFICE (2001) *Home Office Circular 27/2001—Criminal Justice and Police Act 2001*. London: HMSO.

HUBBARD, P. (1999) *Sex and the City: Geographies of Prostitution in the Urban West*. London: Ashgate.

HUBBARD, P. (2000) Desire/disgust: mapping the moral contours of heterosexuality *Progress in Human Geography*, 24, pp. 191–217.

HUBBARD, P. and SANDERS, T. (2003) Making space for sex work, *International Journal of Urban and Regional Research*, 27, pp. 75–89.

JAGET, C. (1980) *Prostitutes: Our Life*. Bristol: Falling Water Press.

JEFFREYS, S. (1998) *The Idea of Prostitution*. Melbourne: Spinifex.

Le Monde (2000) A Paris, les habitants de quartier Chateau-Rouge refusent la logique du ghetto, 2 October, p. 5.

Le Monde (2002a) Les prostituées sous les fenêtres de Sénat, 6 November, p. 17.

Le Monde (2002b), Interview with Nicholas Sarkozy, 23 October, pp. 20–21.

Libération (2002) Le trottoir dans la rue, 6 November, pp 1–3.

Libération (2003) Nous allons doubler le ryhthm des interventions, 25 September 2003, p. 9.

LINNANE, F. (2003) *London—The Wicked City: A Thousand Years of Prostitution and Vice*. London: Robson.

MACLEOD, G. and WARD, K. (2002) Spaces of utopia and dystopia: landscaping the contemporary city, *Geografiska Annaler*, 84B, pp. 153–170.

MATTHEWS, R. (1997) *Prostitution in London: An Audit*. Centre for Criminology, Middlesex University.

MCKAY, G. (1996) *Senseless Acts of Beauty: Cultures of Resistance since the 1960s*. London: Verso.

MERRIFIELD, A. (2000) The dialectics of dystopia: disorder and zero tolerance in the city, *International Journal of Urban and Regional Research*, 24, pp. 473–489.

MITCHELL, D. (2001) Post-modern geographical praxis? Post-modern impulse and the war against homeless people in the post-justice city, in: C. MINCA (Ed.) *Postmodern Geography: Theory and Praxis*, pp. 57–92. Oxford: Blackwell.

MONTREYAND, F. (1993) *Amours á vendre: les dessous de la prostitution*. Paris: Editions Glenat.

MORT, F. (1998) Cityscapes: consumption, masculinities and the mapping of London since 1850, *Urban Studies*, 35, pp. 889–907.

PAPAYANIS, M. (2000) Sex and the revanchist city: zoning out pornography in New York, *Environment and Planning D*, 18, pp. 341–354.

RICHBURG, K. (2002) France may limit prostitution, *Washington Post, 11 August, p.* 20.

SALES, R. (2002) The deserving and the undeserving? Refugees, aslyum seekers and welfare in Britain, *Critical Social Policy,* 22, pp. 456–478.

SIBLEY, D. (1995) *Geographies of Exclusion: Society and Difference in the West.* London: Routledge.

SMITH, N. (1996) *The New Urban Frontier: Gentrification and the Revanchist City.* London: Routledge.

SMITH, N. (1998) Guiliani time: the revanchist 1990s, *Social Text,* 57, pp. 1–20.

SMITH, N. (2002) New globalism, new urbanism: gentrification as global urban strategy, *Antipode,* 34, pp. 434–457.

The Daily Telegraph (2003) Refugee jailed for trade in sex slaves, December, p. 9.

The Guardian (2002) Blair's war on dirty streets, 1 November, p. 3.

The Guardian (2003a) Swearing or loitering could be punished by jail in France, 15 January, p. 11.

The Guardian (2003b) Sex slaves to get safe houses, 10 March, p. 8.

The Independent (2001) Christians clear up prostitutes' cards from phone boxes, 9 April, p. 6.

The Times (2002) Albanian gangsters corner British sex trade. 6 July, p. 10.

THOMPSON, B. (1994) *Soft Core: Moral Crusades against Pornography in Britain and America.* London: Cassell.

WEAVER, M. (2002) Zero tolerance from urban summit, *The Guardian*, 31 October, p. 12.

WILSON, J. and KELLING, G. (1982) Broken windows, *The Atlantic Monthly,* 29 March, pp. 29–38.

The Risks of Street Prostitution: Punters, Police and Protesters

Teela Sanders

Introduction

The nature of street prostitution has been well documented in the UK and world-wide (Benson and Matthews, 1995; Hoigard and Finstad, 1992; May *et al.*, 1999; McKeganey and Barnard, 1996; Raphael and Shapiro, 2004; Williamson and Cluse-Tolar, 2002). These empirical findings have mainly concentrated on characteristics of the women involved in the street markets and visible harms such as violence, drug use and sexual health. The relationships between space, risk and prostitution have been made by scholars who have merged an analysis of geographical space with the dynamics of the social context in which women sell sex, in particular sociological discourses of sexuality and space (Duncan, 1996; Hubbard, 1999; Hubbard and Sanders, 2003; Larsen, 1992; Lowman, 2000; Porter and Bonilla, 2000; Sharpe, 1998).

This paper adds to the growing realisation that the space in which prostitution is advertised, negotiated and administered (Ashworth *et al.*, 1988) is an integral part of why and how prostitution happens in certain streets of cities and towns. The paper identifies how the geographical space that women rely on to make money is not a haphazard or neutral locale in the urban landscape. Using the case study of Birmingham, I identify how the place of street prostitution is often highly politicised by competing interests such as community protesters, services that advocate for sex workers, and law enforcement agencies and the sex workers. Sites of prostitution are the target of resources from public services, especially the growth of multiagency

Teela Sanders is in the School of Sociology and Social Policy, University of Leeds, Leeds, LS2 9JT, UK. Fax: 0113 233 4415. E-mail: t.l.m.sanders@leeds.ac.uk. *Special thanks to Martin Smith and Phil Hubbard for introducing the relevance of geography to my work. Alan Collins has shown great encouragement and the two reviewers from Urban Studies have been helpful with their critical direction. It goes without saying that the women who shared their lives are ultimately to be thanked. The outreach workers who relentlessly take services to the streets in politically hostile circumstances showed me how persistence and commitment can make a difference* partnerships and forums created to consider and act upon what has historically been considered a 'spoiled identity' (Pile, 1996), associated with 'drugs, diseases, dysfunctional families and danger' (Boynton, 1996).

Secondly, this paper clarifies the types of occupational hazards involved in street prostitution by dividing risks into three main categories: violence from male clients, issues relating to policing and community protesters. Thirdly, findings from the case study advance the literature on survival techniques or 'strategies of resistance' by illustrating how street sex workers manage these risks through calculated tactics that manipulate, manage and reinterpret space to their own advantage. Street workers constantly assess the levels of risk from punters, police and protesters in the urban market before engaging in commercial transactions. This paper makes a fourth contribution to the literature that is established by arguing that despite sex workers' active reactions to the risks they face evidence suggests that the policing and regulation of the sites of street prostitution increases the prevalence of danger for individual women. The urban site of prostitution is the target of punitive policing policies, intense community actions sanctioned by the state that victimise and criminalise street workers rather than address the issues that make commercial sex dangerous.

Risk, Relationships and the Realities of Sex Work

Understanding the relationship between sex workers' responses to occupational hazards and the site of prostitution in the urban landscape can be contextualised within wider theories of risk. Mary Douglas (1992), in her essay *Risk and Danger*, analyses voluntary risk-taking in a society where the reality of danger has been overlooked by the perception of risk. Douglas argues that taking risks is not irrational or a trait of a skewed personality but is understood as a character flaw because of the culturally biased model of risk perception in modern industrial society. Douglas (1992, p. 41) claims that risk-averse culture has vanquished the risk-seeking culture and that this bias should be corrected in order to understand risk as a choice in society. Douglas (1992, p. 102) proposes that the self is risk-taking or risk-averse according to the relationship between the person and others in the community.[1]

In prostitution, the consequences of risk are different across markets because "different categories of women have different risk profiles ... they have different degrees of control over their exposure to these risks" (Gysels *et al.*, 2002, p. 190). The risk of violence from pimps and dealers is not the same for those who work indoors as it is for street workers (May *et al.*, 1999). Douglas (1992) argues for a contextual approach to understanding risk from qualitative work that explores cultural, individual and interactional aspects. Responses to danger should not be quantitatively categorised or marked as 'rational' or 'irrational' according to our judgements, but we should try to make sense of different communities' dispositions towards authority and boundaries. Staying close to the sex workers' words and retelling their experiences avoids making judgements but tries to locate their dilemmas and decision-making in their own realities. If we are to understand how others interpret their social environments in deciding what is too risky and what is worth the risk, their reactions to the space in which they face the dilemma is an integral part of understanding risk in society. Individuals do not simply engage in risk-taking or risk-averse behaviour as a result of predisposed traits or irrational responses. Sex workers react to their surroundings and, through a complex process of assessing their own biography, skills and experience, decide whether to take or avoid risks.

The public nature of the street market makes this site of commercial sex significantly vulnerable and exposed. The commodification of an essentially private act in the public realm challenges acceptable mores about the expression of sexuality and the place of sexual behaviour. Sibley (1995) explains how spatial and social boundaries are part of the process that excludes and controls groups who do not conform to dominant ideologies and practices. It has been argued that isolating and separating those who are created as 'others' takes place in the urban city through policies and practices. Sexual minorities have been documented as a group that have been 'other-ed' through a spatial as well as social process of seclusion (Skeggs, 1999; Duncan, 1996). The historical case of street prostitution in Birmingham is an example of the continuation of 'othering' in the urban city where sex workers are depicted as a social threat to moral and familiar cohesion (Hubbard, 1998a). As a consequence, women are shifted from one location to another, in an attempt to remove them from the 'safe spaces' to the margins. The dislocation of sex workers from certain spaces interacts with the processes of risk-aversion and risk-taking. As Douglas states, if risk is specifically about the relationships individuals have with those in the community and not necessarily a reflection of their character, it is the interaction between social groups that must be the focus of any understanding of risk-taking. This paper examines the relationships between sex workers and both their physical geographical environment and the social environment and relationships which determine their position in the risk dilemma.

The Hazards of Prostitution in a Public Space

It has been well documented how street sex workers endure working conditions that stigmatise, criminalise and pose threats of physical, emotional and psychological harm (Hoigard and Finstad, 1992; O'Neill, 2001; Phoenix, 1999; Sterk, 2000). The occupational hazards open to street sex workers can be divided between the public and the private manifestations of risk. In public, women fear violence from clients and other people on the street, arrest from the police with the increasing possibility of imprisonment and further harassment from community protesters. In private, the stigmatisation and marginalisation as a result of working in prostitution are as equally stressful as women constantly fear that their friends, family or partner could discover their money-making activities. Although there is no time in this paper to consider fully the private dynamics of risk, the implications of the public risks are inevitably expressed in the private domain (see Sanders, forthcoming).

The prevalence of violence within prostitution has been well established in Britain (Barnard, 1993, Day and Ward 2001, Sanders, 2001) and world-wide.[2] The street market is increasingly vulnerable to violence and robbery. From a study across three cities in Britain, Church *et al.*, (2001) found that 81 per cent (93) of street workers had experienced violence from clients, compared with 48 per cent (60) who worked indoors. Benson (1998) found that 98 per cent (49) of street prostitutes experienced some form of violence at work, while Kinnell (1992) reported that 75 interviewees experienced 211 violent incidents during their careers. However, the perpetration of violence against sex workers is not predictable because "occupational studies of, and service for, prostitutes cannot be confined to the risks posed directly by exchanges with customers" (Day and Ward, 2001, p. 230). Other empirical findings demonstrate the high likelihood of street workers experiencing violence from other working women, predatory men involved in prostitution, the general public (Benson, 1998) and severe physical harm from boyfriend/pimps (May *et al.*, 2000, p. 18).

In addition, sex workers experience intimidation and harassment from the communities where they work and sometimes live. Several cities in Britain have experienced a clash between community groups and the sex work community (Leeds, Bradford and Bristol are described in Hubbard, 1998a; Liverpool and London are described by Campbell *et al.*, 1996). In Birmingham, street prostitution co-exists, albeit uncomfortably, alongside those who claim the same physical space as their community, territory or 'back-yard'. Over the past decade, Birmingham has experienced community campaigns against street prostitution that have been highly organised and intensively resourced.

Community action groups have several legitimate issues to raise about the implications of commercial sex advertised, negotiated and often delivered in the same spaces as those in which they live. Benson and Matthews (2000, p. 247) describe how residents complain about litter (particularly used condoms), noise and extra traffic from kerb-crawlers, while Hubbard (1998a, p. 272) explains how Birmingham residents felt that prostitution made their areas unsafe by attracting unscrupulous people, increasing drug-related crime and street robberies. The presence of men who cruise the streets looking to purchase sex has also caused problems for other women in the neighbourhood as they are inappropriately approached (see Brooks-Gordon and Gelsthorpe, 2003). These incidents invariably happen where children are present. However, while these concerns are real, the actions of protesters can also be understood as a public display of aggressive masculinity against a group of women that are understood as 'deviant others' and scapegoated for urban degeneration and wider political, economic and social change (Hubbard, 1999). Larsen (1992, p. 187) evaluates the anti-prostitution community groups in four Canadian cities and notes that there is a socioeconomic class bias regarding who is given powers to control street prostitution.

In addition to community policing, street sex workers are under constant pressure to avoid the police and the processes of criminalisation that pose severe risks to their livelihood. Although there is no law against two consenting adults swapping sex for cash or commodities, under the Street Offences Act, 1959, loitering and soliciting in a public place for the purpose of prostitution are illegal. As a response to complaints by local residents, the police are involved in a considerable amount of controlling and regulating street prostitution. Locally, there are several joint initiatives between the police and residents to monitor, track, record and report information about individual workers and their activities. Multiagency policing of prostitution has taken on a punitive agenda in response to changes in the law under the Crime and Disorder Act, 1998, that introduced Anti-social Behaviour Orders (hereafter, ASBOs) to be used against those who cause 'alarm, distress and harassment' to communities. Strengthened in their applicability under the Police Reform Act, 2002, ASBOs place geographical exclusion zones on women who are considered to be a persistent nuisance in an area. More seriously, if an ASBO is breached, women face up to five years' imprisonment. These more recent developments are rapidly changing the nature of risk in the street market, the practice of commercial sex and the organisation of street markets.

The Study

This paper draws on a 10-month ethnographic study of the sex industry in Birmingham, UK, during 2000–2001. Observations were made in both the street and indoor sex markets including saunas, working premises and escort agencies. Fifty interviews were conducted with sex workers and a further five interviews with women who owned or organised sex establishments. Fifteen interviewees had experience of working on the street and five currently remained in this market. The number of street workers formally interviewed is small due to the specific problems of accessing a group of women who live chaotic lives, on different time-frames, often with serious drug addictions and sometimes with a boyfriend-pimp (see Faugier and Sargeant, 1996; Maher, 2000; and Miller, 1995, for further details of the difficulties accessing this group). At least 20 other women agreed in principle to be formally interviewed, but in reality women either forgot the time and meeting-place, were too concerned with 'punting and scoring' or were agitated, depressed or had no time to talk to a researcher. The 15 interviews (which lasted between 45 minutes and 7 hours) were transcribed verbatim and analysed on the computer package Atlas.ti.

The interview data are strongly supported by over 400 hours of observations of the street market where informal conversations took place with many other street workers. Similar to other research (Porter and Bonilla, 2000; McKeganey and Barnard, 1996), this access was facilitated through a sexual health project that operates an outreach service specifically for women in prostitution. I was able to accompany the outreach workers on their nightly patrols of the street in their specialist equipped van, supplying condoms, needle syringes and hot chocolate to women on the street. This privileged position meant that I was privy to many incidents amongst the street sex work community that enabled me to piece together a picture of the types of risks that women faced. For instance, on several occasions I saw women arrested by police as they walked the beat, were hailing taxis to get a ride home or were approached when they were chatting with non-prostitute women in the neighbourhood. I saw groups of middle-class, middle-aged men and women protesting by congregating on street corners, communicating on two-way radios, taking down notes and even photographing women. Male vigilantes were also observed shouting verbal abuse to street workers, intimidating them with their physical presence and dogs. On one occasion, a mob of approximately 20 men chased a lone woman out of a particular neighbourhood because they assumed she was loitering for business. As Porter and Bonilla (2000, p. 106) note from their research methods, observing and interviewing sex workers in their working environment has specific advantages for the outsider as the contexts of the women's lives and activities inform and enrich the data collected.

Sex workers used the private 'women only' space in the van as their safe haven on the beat. Sometimes, the mobile unit really was a safe place to hide, seek protection and assistance. On six occasions, women came into the van after clients, pimps or dealers had attacked them. Equally, the van was a space where women could openly be themselves, free from the fear that they would be judged, recorded or condemned. At times, the van was a 'hot house' of gossip, laughter and information-sharing. Women would also share their latest encounters with 'dodgy punters', up-to-date reports on policing activity, or the whereabouts of the residents. This enabled me to witness first-hand the effective and speedy communication network that women use to pass on crucial information (such as description of men, cars and registration numbers) to keep themselves safe and stay out of danger.

Half way through the fieldwork, the sexual health project obtained a 'drop-in' centre in the neighbourhood where the women worked. This was open between 7pm and 11pm before the mobile outreach service began. Women would use the facility to discuss issues with the various drug specialists, community nurses, sexual health practitioners, housing officers and domestic violence workers who were present. More informally, the drop-in was a place where the women, like in any work environment, shared a communal space to have a chat over coffee, exchange working tips, talk about their day, their children and family. Also it was a place where women sought emotional refuge: women were often in dire conditions from heroin and crack withdrawal, rock-bottom from having children removed by state authorities, in the aftermath of abortions and miscarriages, homeless and excluded from family, clinging onto life, using prostitution to survive.

Over the period, I built up strong relationships with women, especially those who I saw several times a week. Ethnography is a powerful tool for accessing women's lives and, although "it is a messy business" (Maher, 2000, p. 232) it enables a process of representation of those who often have little voice. Questions may be asked about the representativeness of this case study due to the small numbers and the purposive nature of the sample selection. However, Maher (2000, p. 29) criticises the emphasis on representativeness because it "obscures what the anomalous or the marginal can reveal about the centre … and perhaps most of all, strategies of resistance". The street sex market I observed and the women I spoke with were a snapshot of other similar markets across Britain and perhaps other Western countries. This can be said with conviction because the sex workers, many of whom had worked in several cities, and experts representing national networks, confirmed that my findings were reflective of other street markets. The accounts and evidence in this paper are taken from this ethnographic process and it is a truthful account of my observations and the women's stories.

As Figure 1 identifies, the location of this research was the city of Birmingham, the UK's second-largest city. The location of street prostitution at the time of the fieldwork was a square mile of residential streets in Edgbaston, south of the city. It consists of a number of tree-lined streets, less than two miles from the city centre and adjacent to one of the main entertainment strips (hotels, conference centres, casinos, restaurants, bars, lap dancing clubs) running into the city. The location has a reservoir and a park where sex workers sometimes take their clients. Edgbaston is an affluent, leafy suburb, with Victorian period houses and a high proportion of home-ownership. The residents are mainly White, middle-class professionals or retired, older people. Due to house prices and the increasing number of students (the university is within one mile) this is changing as large houses become multioccupancy dwellings for students and those on welfare benefits and low wages. However, Edgbaston still remains a haven amidst areas of increased poverty, high-rise social housing and social exclusion.

This area can be contrasted to the previous location of street prostitution throughout the 1990s. Balsall Heath is only two miles south of Edgbaston and is markedly different in population. Balsall Heath has been home to South Asian immigrants for the past 50 years and has a strong community with mosques and many Asian businesses and shops. However, similar to Edgbaston, there is a large park and a central road that runs to the city centre. The original community action group 'Streetwatch' campaigned relentlessly to remove prostitution from the windows and streets of Balsall Heath. Traffic-calming measures, urban regeneration and a degree of vigilantism eventually moved prostitution to the neighbouring district of Edgbaston. Although the boundaries of the commercial area are in constant flux as women move their working territory in response to local politics and dynamics, Edgbaston is still the predominant place for trading sex.

Rationality, Strategy and Public Space

This section explores the direct responses sex workers create, implement and share to resist some of the occupational hazards of the urban market-place. Women use space strategically to avoid physical violence, arrest, criminalisation and harassment. The debates of agency and victimhood within prostitution have set out the complex parameters of whether a woman can consent to sell access to her body parts or whether all forms of prostitution are exploitative (for reviews, see Gulcur and Ilkkaracan, 2002; Kesler, 2002; O'Connell Davidson, 2002). However, my research and this paper argue that although women may not rationally decide to enter prostitution, their responses to the daily hazards they face are often calculated strategies of resistance. In this sense, the space on which women rely to advertise, negotiate and supply commercial sex is strategically used to their advantage in order to make cash and minimise chances of harm. This section builds on the arguments presented by Hubbard (1999, pp. 180–209) that flesh out how resistance is underpinned by intentionality. Powerless groups such as sex workers "rework and divert these spaces to create an alternative meaning of space—a space that has its own morality, rhythms and rituals which are often invisible to outsiders" (Hubbard, 1999, p. 183).

Managing Physical Harm

During the period 1989–2002, sex workers in

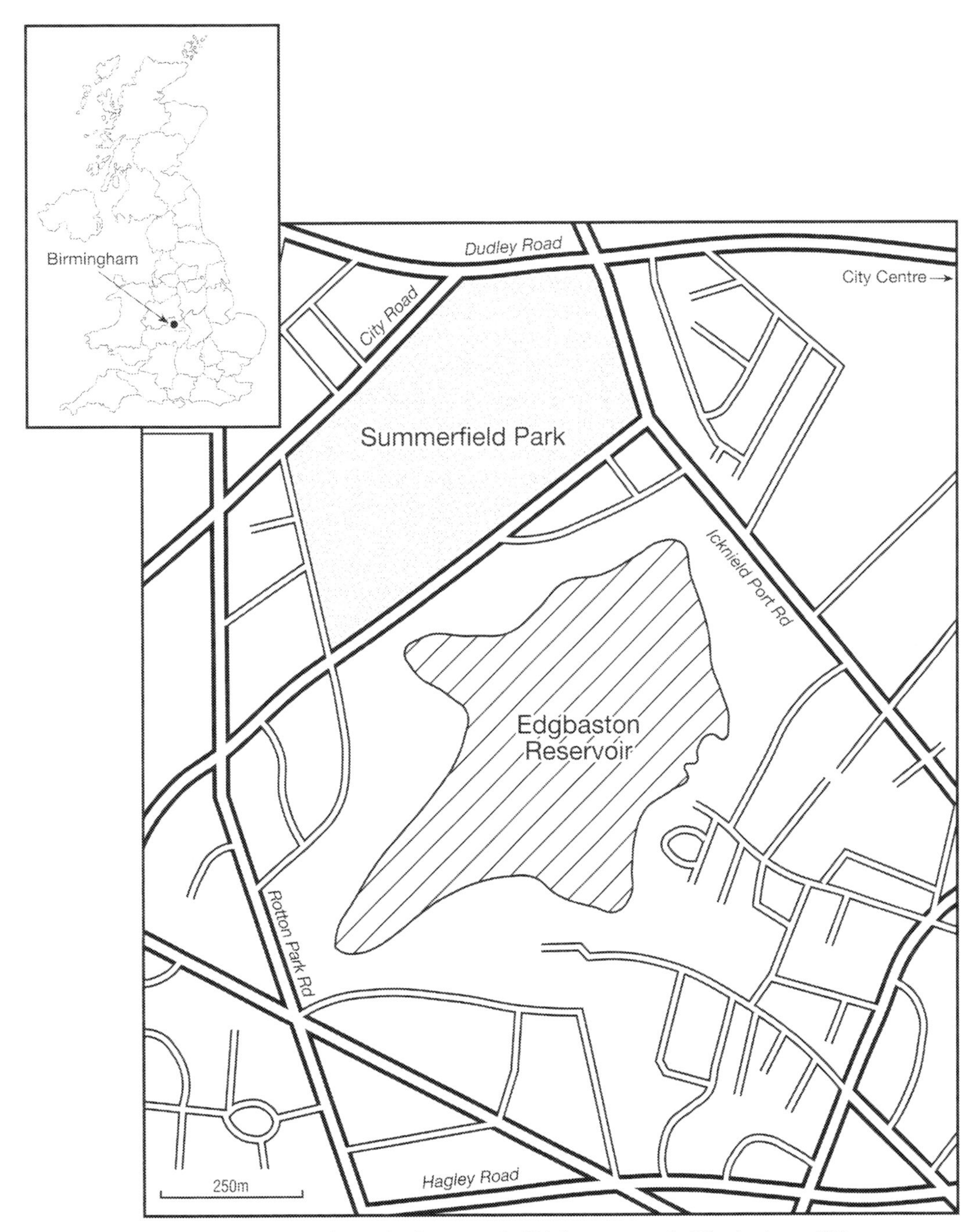

Figure 1. Location of prostitution area in Edgbaston, south Birmingham, UK.

Birmingham, the majority of whom worked on the street, reported over 400 separate incidents of physical or sexual attacks by clients to a local health organisation. Of the street workers I spoke with, those who had not been attacked were in the minority. In all, 15 women retold their experiences of multiple violence from clients: 7 had been raped; 14 had been physically beaten; 5 had been robbed at knife-point; 3 had been confined against their will or kidnapped; a further 3 had been robbed at gun-point; and 2 had been drugged by a client.

Violence, as an occupational hazard, has varying consequences for the women. Some decide that prostitution is too dangerous and leave all together, or leave the street for a safer working environment. Debbie explains how violence prompted her to work from a private rented flat

> I worked on the streets in 1993 and there was a girl killed by a punter. I knew her really well. I was only speaking to her a couple of hours before she was killed. I thought no, I better come off these streets (Debbie, 28, worked on the street for 4 years).[3]

Others learn quickly by their mistakes and change their working practices to reduce the risks of violence

> When I went out there to begin with I just used to get into any car, with two or three men and never used to check or anything. I got into quite a lot of trouble doing that ... I was raped and kidnapped and had to spend time in hospital (Nicky, 22, worked in different sectors for 5 years).

The strategies that women devise to protect themselves mean that in the majority of transactions women are not attacked, raped or robbed. Controlling the environment is a tactic sex workers use to prevent attacks and to ensure that sexual negotiations and transactions occur without incident. Sex workers said that the stages of the commercial transaction where they feel most vulnerable were during the negotiations with a client in his car and when engaged in the sexual activity. In response, women do not haphazardly go about 'doing business' without thinking about the safety issues. The geographical location of the commercial sex act is a crucial aspect of the precautions, deterrent and protection strategies that sex workers create and implement to keep themselves safe. Women have clear plans regarding the location of each stage of the transaction and stringently apply their working rules

> You never go where they want you to go, you always take them to somewhere that you know is safe. Never let them take you to another town as they will kidnap you ... Never go with two men ... always check that there is no-one hiding in the back ... don't get into vans (Melissa, 25, worked on the street for 7 years).

All of the respondents insisted the client drove to a predetermined location for the service: "If they want to take me to their spot then no way, I have been taken to their spots before and left stranded, and I have not had a clue where I am" (Nicky, 22, worked in different sectors for 5 years). If women are familiar with the physical lay-out of the area then they have more chance of escaping or calling for help. Women try to work away from dead-end roads or cul-de-sacs that may prevent their escape but instead work close enough to residential houses that may be called upon for assistance.

On the street, sex workers construct working practices that rely on technology as a deterrent to a client who may intend to harm or rob. Annie explains how she chooses the location for administering the sexual service with care and will only provide sexual services in public areas where CCTV is in operation

> I always take them to places where there are cameras. I don't normally point them out unless they are going to start to be funny then I will say to them that there are cameras out there and it is all being filmed and if they don't want business then to drop me off. I normally take them to the hospital, as it is all camera-ed there (Annie, 22, worked on the street for 4 years).

All of the street workers that I met carried some form of implement that could be used as a weapon in case they were attacked. Knives, blades, CS spray and lighters were said to be the best forms of weapons. One woman hid a metal bar in the bushes where she took her clients and had used it on several occasions to beat off attackers. As previously reported by other studies (Dunhill,

1989, p. 205; McKeganey and Barnard, 1996), to keep track of the client, street workers note car registration numbers: "One girl will take the registration number of the car … Where I worked we all stuck together on the beat which was like a little community" (Katrina, 32, reflecting on street experiences). Sometimes this is done so that the client can see their vehicle is being recorded. Often women would inform friends and boyfriends where they take clients in case they did not return when expected.

As a reaction to the increasing levels of violence, drug-related crimes and policing, it was becoming increasingly popular to combine working on the street with working from indoor locations: "Before I was out there every day but I have cut it down and work from the phones" (Nicky, 22, worked in different sectors for 5 years). Two routes of mobility were noted amongst street workers: either women continued to use the streets to attract clients but performed the service at an indoor location (hired flat, home, hotel, client's home); or they found permanent work indoors (often a rented establishment with others), only venturing onto the street to attract new clients or when business was quiet. This can be understood as a positive displacement of prostitution from the street to indoor locations which are generally safer and tolerated by law enforcers.

Avoiding Protesters

Hubbard (1997) describes the policing of prostitution as a spatial process that perpetuates the marginal status of those involved. At a local level, the community politics of direct action against prostitution in Birmingham over the past decade have determined official responses. During the mid 1990s, the South Asian community of Balsall Heath initiated direct action tactics (nightly patrols, pickets, media campaigns and car registration recording scheme) to remove prostitution from its streets. Many of the sex workers contacted in this fieldwork remembered the pressure from the vigilantes

> If ever I went out and the vigilantes were out or the media I would just go back home … half of the vigilantes were hypocrites because they were punters themselves. Well a few times they would try and push you about. A mate of mine, they would batter her with sticks and things and they were always giving verbal abuse. Or you maybe crossing the road and they would put their foot down on the car and you would have to run (Sally, 34 years old, worked on the street for 10 years).

As a direct result of displacement due to intense community action and state policing, sex workers moved their site of work to another area, one mile from the old district. This prompted the revival of the original Streetwatch action group under new management. The membership of mainly South Asian males was replaced by White, middle-class, older professional males who rapidly won the support of politicians, local counsellors, law enforcement agencies and the city council. Intense patrolling of the main streets was reinstated on a nightly basis. The police I interviewed described close partnerships with the Streetwatch community group who acted as informers, monitors and direct surveillance. One Inspector commented: "We work as a team with members of Streetwatch, who have their own patrols monitoring the area and logging car registration numbers, which are then passed on to us" (*Birmingham Evening Mail*, 8 May 1998, p. 43). Indeed the police created a special 'hot line' telephone service so that protesters could inform the police of individuals' activities. The police and local authority had jointly funded a specially equipped vehicle for use by the Streetwatch group to patrol the area. This van was fitted with holding cells.

During my observations and discussion with street workers, the intimidation and direct harassment of the women were noted. Protesters followed individual women to their homes, took photographs, repeatedly quizzed women about their intentions on the street (What were they doing? Where were they going? Who were they going to visit?).

Annie felt particularly harassed by the Streetwatch members as she lived and worked in the same area

> I will be walking up the road with my daughter and they will stop me. I say to them at night when I am on my own and I am dressed in mini skirts then they have got a right to stop me but not at 3 o'clock in the afternoon, with my baby in the push chair with about 6 carrier bags. I mean am I really doing business in that state? They still stop me (Annie, 22, worked the street for 4 years).

The direct action tactics employed by the protesters impinged on the women's personal space as much as it did on their commercial activities. Sex workers found it difficult to maintain any sense of privacy as the public nature of their work led to intrusion and interference. Nevertheless, methods of monitoring, observing and sharing information used by the protesters informed the sex workers' strategies for resisting state and community control

> I work in the day time to avoid the vigilantes. They are normally in the Streetwatch van, and when I see them coming round I jump into the bushes until they have gone. They usually drive round a set route and you know if they drive one way that you have got 15 minutes until they come back. You either walk the opposite way or you get a client and move off. The best thing to do is find out which way they are going and walk the opposite way and then get a customer as quickly as possible. Then I am away with the customer for 20 minutes and then they have not seen me as I have been in a car (Annie, 22, worked on the street for four years).

Women perceptively recreated tactics that were used to survey and control their behaviours to inform when and how they conducted business. A calculated cost and benefit analysis was performed to determine whether the likelihood of harassment was greater than attracting clients. Often, if the risk was too high, they would abandon their plans. Instead, they would wait until later, move to another town, or simply get the often desperately needed cash through another acquisitive crime. Benson and Matthews (2000, p. 249) suggest that geographical mobility is common amongst street workers as a reaction to policing methods, leading to what one police officer in their study described as 'national displacement'. As discussed below, often the displacement is more than geographical but to other forms of crime.

Managing Policing

Avoiding the police was a daily hazard for most street workers. Of the 15 participants who had worked on the street, 8 women had between 1 and 5 prosecutions for prostitution-related offences. Several other women contacted in this study had received between 20 and 30 arrests within a year, incurring fines impossible to pay unless they increased prostitution activities or other acquisitive crimes. In the fieldwork site, 2 police officers were designated specifically for street prostitution. Birmingham City Council has recently set a legal precedent by issuing 19 women whom they consider to be persistent offenders with ASBOs, prohibiting them from entering certain geographical areas.[4] Other street workers in London have reported up to 217 arrests over a 3-year period (personal communication). Inconsistent policing practices and arbitrary 'crackdowns' mean that women have no clear sense of what is acceptable and work under the fear of arrest

> Last Friday night, the police were everywhere. I got pulled because I was on warrant and they called it up and I had given them his name [boyfriend's surname] and they checked and let me go … So in the end I thought it was not worth working so I walked off up the road to get a taxi planning to go home. Got in the taxi and got nicked by vice. They would not even accept that I was going home, they had me for loitering. They said the taxi wasn't

> prebooked. Different coppers from earlier, but they said they watched me walk up the road, but if they had they would have seen me get out of the first police car. Then the copper started to push me and I went mad at him. They took me to the police station and I refused to give details because I was on warrant ... So I was locked up in the cell and I was there for hours and hours ... Couple of hours later they let me go and I was out back at work (Steph, 20 years old, worked in different sectors for 3 years).

Apart from arrest and criminalisation, sex workers were concerned about the effect of police presence on their ability to make money. The presence of police patrol cars in the area where women advertise is an obvious deterrent for those who are looking to buy sex. The reduction in potential custom has repercussions on working practices and the safety of women. Respondents report that, before the continual police presence, at least half of their clientele each night were regular customers. During the fieldwork, this drastically changed as Nicky recounts: "You know once they [police and residents] are there you are not going to be getting into any car as the clients won't stop for you as they don't want their reg taken". With increased policing, women are taking even less time to assess the client. Often, when a car slows down, the woman will jump in within seconds to move off the street and away from police attention (Barnard, 1993). O'Kane (2002) also found that new powers for arresting kerb-crawlers put sex workers' lives in danger.

Despite vehement complaints about police activities, the same women saw no reason to be hostile to officers who arrested them: "If you run from the vice ... they get you straight away. So the rules are you don't run from the vice you just have to come in" (Katrina, 32, reflecting on street experiences). Some workers discussed the importance of showing compliance and co-operation to the police, sometimes developing effective working relationships with individual officers: "They are sweet the police, if you are sweet with them then they will be sweet back" (Lucy, 19, worked on the street for 1 year). These relationships were often a trade-off for lenient treatment in exchange for information about other, more serious crimes.

The Consequences of Controlling Space

From these empirical findings several points can be raised. First, the way in which police and protesters control the sites of street prostitution has significant implications for the way in which the street market is organised and for the safety of individual women. Secondly, solutions sought through community safety policing result in the geographical displacement of prostitution at both local and national levels. Thirdly, local policies that focus on the 'disorder' in relation to the presence of prostitution inevitably criminalise individuals rather than tackling the issues associated with making the sex market safer for those who sell and buy commercial sex. These contributions will be discussed in turn.

The implications of intense policing activities and persistent protesters mean that women have to adapt their working practices to avoid harassment and arrest. Increasingly, because of the pressures to move off the beat and not be seen with a client, sex workers are not making preliminary checks when a man approaches. This is dangerous, as it does not allow time for women to think and check any dangerous signs or what type of client the man is. In the car, when there is time to assess the client, it can often be too late. During the fieldwork, there was a notable change in the working patterns of sex workers. In direct response to the presence of protesters, women arrive on the streets much later (around midnight) and work into the dawn. Many of the women who have costly heroin and crack cocaine addictions avoid the streets by staying in crack houses in between attracting customers. Women who continue to use the streets to advertise are working in increasingly dangerous places, away from the safety of the public streets. This change in working hours and practices avoids the police and protesters but at the same time increases their exposure to danger from clients, pimps and dealers.

Another effect of intense policing is that there is less stability within the organisation of sex markets: "Everyday arrests, imprisonment, fines and police raids led women to move within the industry to minimise their risks" (Day and Ward, 2001, p. 230). Mobility has taken two forms. Street workers are moving between different geographical areas in order to avoid becoming known as 'a prostitute' by law enforcement agencies. This means that women work several beats in one night in order to reach their cash targets and avoid arrest. Secondly, the movement between street and indoor markets is becoming increasingly fluid as women rent premises or only rely on the streets for advertising. Mobility away from the street market could potentially reduce the risk of violence as the indoor environments tend to be safer. However, there is a strong likelihood that pushing sex markets further into an illegal and illicit economy only ostracises some of the most vulnerable women in society. Encouraging women to hide their involvement in prostitution creates difficulties for health care and support agencies to make relationships with individuals, preventing any chance of accessing drug rehabilitation.

The policing practices that have been adopted as solutions to street prostitution are essentially community safety strategies. This style of policing is written into the Crime and Disorder Act, 1998, as a form of crime prevention in urban areas. This has been criticised because "intensive policing may also produce harmful forms of social displacement" (Maher and Dixon, 1999, p. 503). Displacement of prostitution happens in terms of relocating the activity to other geographical areas which essentially only "shuffles crime from one area to the next but never reduces it" (Pease, 2003, p. 956). In terms of prostitution, 'crime shuffling' can at best dissuade women from earning cash through commercial sex but through acquisitive crimes such as shoplifting, forgery and selling drugs. The community safety policies' focus on removing the 'problem' in neighbourhoods, rather than finding wider solutions to tackle the underlying causes of the crimes. The most effect displacement policy can achieve is the reduction in criminal inclination, because high-profile policing may persuade new or novice sex workers to work indoors or use other localities. Arresting and prosecuting women or using ASBOs to prohibit women from certain areas only targets individuals and does nothing to find effective solutions to the place of prostitution in the city. Serving deterrents that are costly and difficult to police on women who are usually heavily entrenched in prostitution and street-related activities is neither offering a way out for women, nor addressing issues raised by the community. As individuals are removed, others will take their place because the factors that determine women's involvement in the street sex economy—namely, drug addiction and coercive male relationships—are not the centre of any cohesive policy approach.

Instead, the focus of urban prostitution policies is on the 'disorder' and 'nuisance' that are associated with this visible illicit sex economy. The nuclei of prostitution policy are on the dangers created to others in the community and the impact of punitive policies on the vulnerability levels of sex workers is not taken into consideration. Hubbard (1998b, p. 56) suggests that the "ordering (and representation) of urban space plays a crucial role in producing and reproducing gender, sex and bodily identities". In urban spaces, territory is marked out for those who can legitimately and safely use space, while groups who are outside the mainstream are confined, or at least attempts are made to confine them, to hidden shadows, away from a legitimate place in public and their rights to full citizenship. Pratt and Hanson (1994, p. 25) echo a similar point: "social boundaries are constructed and maintained through geographical ones that mark off distinctive ways of life". Indeed, society is concerned with regulating the type of sex and the types of sexual identity that are tolerated in public. The ways in which sex workers experience the street can be understood through wider issues of how space is used to reinforce the social conditioning of gender norms (Duncan, 1996). Although this paper has argued that sex workers create resistance strategies to manage occupational risks, doing business is increasingly dangerous because of community safety policies and policing practices. If the sites of prostitution are being increasingly reduced while at the same time made the target of surveillance and scrutiny then sex workers are given little option but to take risks. Remembering Douglas' emphasis on the need to understand the dilemma of risk-taking and risk-aversion from the individual's interaction with the community, there seems little reason in dividing those who have vested interests in the same streets. Work by O'Neill and Campbell (2001) demonstrates how sex workers and community residents can come together through community arts projects and research initiatives to establish ways to introduce effective urban policy that is neither punitive, moralistic nor biased. Until the implications of these so-called solutions are fully realised, the prostitute and the punter will continue furtively to use the city streets to match need with desire, while the police and protester chase them from one place to the other.

Notes

1. Douglas demonstrates this by an example of the homosexual man who is advised by his doctor to stem certain practices because of the dangers. The man's refusal to act on the advice because he enjoys the risky lifestyle is not a weakness of understanding or irrational behaviour, but a preference (Douglas, 1992, p. 103). Rhodes (1997, p. 215) explains this further with the example of unprotected sex with partners who are known to be HIV-positive can be understood as 'situated rationality' or 'informed-choice making'.
2. The prevalence of violence in prostitution is reported in the US (James, 1974; Maher, 2000, pp. 155–159; Miller, 1997; Silbert and Pines, 1985), Canada (Lowman, 2000), Europe (Hoigard and Finstad, 1992; Mansson and Hedin, 1999), South America (Downe, 1999; Nencel, 2001) and South Africa (Wojcicki and Malala, 2001).
3. All names have been changed to protect anonymity but the rest of the details are factually correct.
4. These civil actions were taken at the beginning of 2003 and a reliable source indicates that at least half of the women who were served these civil actions remain working the streets.

References

ASHWORTH, G., WHITE, P. and WINCHESTER, H. (1988) The red light district in the west European city: a neglected aspect of the urban landscape, *Geoforum*, 19, pp. 201–212.

BARNARD, M. (1993) Violence and vulnerability: conditions of work for street using prostitutes, *Sociology of Health and Illness*, 15, pp. 5–14.

BENSON, C. (1998) *Violence against female prostitutes*. Department of Social Sciences, Loughborough University.

BENSON, C. and MATTHEWS, R. (1995) *National Vice Squad survey*. School of Sociology and Social Policy, Middlesex University.

BENSON, C. and MATTHEWS, R. (2000) Police and prostitution: vice squads in Britain, in: R. WEITZER (Ed.) *Sex for Sale*, pp. 245–264. London: Routledge.

BOYNTON, P. (1996) *Beauty, envy, disease and danger: stereotypes of women in pornography*. Paper presented at '*Issues in Pornography*' Conference, University of West England.

BROOKS-GORDON, B. and GELSTHORPE, L. (2003) Prostitutes' clients, Ken Livingstone and a new trojan horse, *The Howard Journal*, 42, pp. 437–451.

CAMPBELL, R., COLEMAN, S. and TORKINGTON, P. (1996) *Street prostitution in inner city Liverpool*. Liverpool Hope University College.

CHURCH, S., HENDERSON, M., BARNARD, M. and HART, G. (2001) Violence by clients towards female prostitutes in different work settings: questionnaire survey, *British Medical Journal*, 322, pp. 524–525.

DAY, S. and WARD, H. (2001) Violence towards female prostitutes, *British Medical Journal*, 323, p. 230.

DOUGLAS, M. (1992) *Risk and Danger: Essays in Cultural Theory*. London: Routledge.

DOWNE, P. (1999) Laughing when it hurts: humour and violence in the lives of Costa Rican prostitutes, *Women's Studies International Forum*, 22, pp. 63–78.

DUNCAN, N. (1996) Renegotiating gender and sexuality in public and private spaces, in: N. DUNCAN (Ed.) *Bodyspace*, pp. 127–145. London: Routledge.

DUNHILL, C. (1989) Working relations, in: C. DUNHILL (Ed.) *The Boys In Blue*, pp. 205–208. London: Virago.

FAUGIER, J. and SARGEANT, M. (1996) Boyfriends, pimps and clients, in: G. SCAMBLER and A. SCAMBLER (Eds) *Rethinking Prostitution*, 121–136. London: Routledge.

GULCUR, L. and ILKKARACAN, P. (2002) The 'Natasha' experience: migrant sex workers from the former Soviet Union and eastern Europe in Turkey, *Women's Studies International Forum,* 25, pp. 411–421.

GYSELS, M., POOL, R. and NSALUSIBA, B. (2002) Women who sell sex in a Ugandan trading town: life histories, survival strategies and risk, *Social Science and Medicine,* 54, pp. 179–192.

HOIGARD, C. and FINSTAD, L. (1992) *Backstreets: Prostitution, Money and Love*. Cambridge: Polity.

HUBBARD, P. (1997) Red-light districts and toleration zones: geographies of female street prostitution in England and Wales, *Area,* 29, pp. 129–140.

HUBBARD, P. (1998a) Community action and the displacement of street prostitution: evidence from British cities, *Geoforum,* 29, pp. 269–286.

HUBBARD, P. (1998b) Sexuality, immorality and the city: red light districts and the marginalisation of female street prostitutes, *Gender Place and Culture,* 5, pp. 55–72.

HUBBARD, H. (1999) *Sex and the City: Geographies of Prostitution in the Urban West*. Aldershot: Ashgate.

HUBBARD, P. and SANDERS, T. (2003) Making space for sex work: female street prostitution and the production of urban space, *International Journal of Urban and Regional Research,* 27, pp. 73–87.

JAMES, J. (1974) Motivation for entrance into prostitution, in: L. CRITHES (Ed.) *The Female Offender*, pp. 177–205. London: Heath.

KESLER, K. (2002) Is a feminist stance in support of prostitution possible? An exploration of current trends, *Sexualities,* 5, pp. 219–235.

KINNELL, H. (1992) *Wolverhampton sex workers survey*. Birmingham: Safe Project, Birmingham Health Authority.

LARSEN, E. (1992) The politics of prostitution control: interest group politics in four Canadian cities, *International Journal of Urban and Regional Research,* 16, pp. 169–189.

LOWMAN, J. (2000) Violence and the outlaw status of (street) prostitution in Canada, *Violence Against Women,* 6, pp. 987–1011.

MAHER, L. (2000) *Sexed Work: Gender, Race and Resistance in a Brooklyn Drug Market*. Oxford: Oxford University Press.

MAHER, L. and DIXON, D. (1999) Policing and public health, *British Journal of Criminology,* 39, pp. 488–512.

MANSSON, S. A. and HEDIN, U. (1999) Breaking the Matthew effect: on women Leaving prostitution, *International Journal of Social Welfare,* 8, pp. 67–77.

MAY, T., EDMUNDS, M. and HOUGH, M. (1999) *Street business: the links between sex and drug markets*. Police Research Series Paper 118. London: Home Office.

MAY, T., HAROCOPOS, A. and HOUGH, M. (2000) *For love or money: pimps and the management of sex work*. Police Research Series Paper 134. London: Home Office.

MCKEGANEY, N. and BARNARD, M. (1996) *Sex Work on the Streets*. Buckingham: Open University Press.

MILLER, J. (1995) Gender and power on the streets, *Journal of Contemporary Ethnography,* 24(4), pp. 427–451.

MILLER, J. (1997) Researching violence against street prostitutes, in: M. SCHWARTZ (Ed.) *Researching Sexual Violence Against Women*, pp. 144–156. London: Sage.

NENCEL, L. (2001) *Ethnography and Prostitution in Peru*. London: Pluto Press.

O'CONNELL DAVIDSON, J. 2002. The rights and wrongs of prostitution, *Hypatia*, 17, pp. 84–98.

O'KANE, M. (2002) Prostitution: the Channel 4 survey (http:www.channel4.com/news/microsites/D/Dispatches/prostitution/survey.html).

O'NEILL, M. (2001) *Prostitution and Feminism*. London: Polity Press.

O'NEILL, M. and R. CAMPBELL. (2001) *Working together to create change*. Walsall Consultation Research, Staffordshire University & Liverpool Hope University.

PEASE, K. (2003) Crime reduction, in: M. MAGUIRE, R. MORGAN and R. REINER (Eds) *Oxford Handbook of Criminology*, pp. 948–979. Oxford: Oxford University Press.

PHOENIX, J. (1999) *Making Sense of Prostitution*. London: Macmillan.

PILE, S. (1996) *The Body and the City*. London: Routledge.

PORTER, J. and BONILLA, L. (2000) Drug use, HIV and the ecology of street prostitution, in: R. WEITZER (Ed.) *Sex for Sale*, pp. 103–121. London: Routledge.

PRATT, G. and HANSON, S. (1994) Geography and the construction of difference, *Gender, Place and Culture,* 1, pp. 5–29.

RAPHAEL, J. and SHAPIRO, D. (2004) Violence in indoor and outdoor prostitution venues, *Violence Against Women,* 10, pp. 126–139.

RHODES, T. (1997) Risk theory in epidemic times: sex, drugs and the social organisation of 'risk behaviour', *Sociology of Health and Illness,* 19, pp. 208–227.

SANDERS, T. (2001) Female street sex workers,

sexual violence and protection strategies, *Journal of Sexual Aggression*, 7, pp. 5–18.

SANDERS, T. (forthcoming) *Sex Work: A Risky Business*. Cullompton: Willan.

SHARPE, K. (1998) *Red Light, Blue Light: Prostitutes, Punters and the Police*. Aldershot: Ashgate.

SIBLEY, D. (1995) *Geographies of Exclusion: Society and Difference in the West*. London: Routledge.

SILBERT, A. and PINES, M. (1985) Sexual abuse as an antecedent of prostitution, *Child Abuse and Neglect*, 5, pp. 407–411.

SKEGGS, B. (1999) Matter out of place: visibility and sexualisation in leisure spaces, *Journal of Leisure Studies Association*, 18, pp. 213–232.

STERK, C. (2000) *Tricking and Tripping: Prostitution in the Era of AIDS*. New York: Social Change Press.

WILLIAMSON, C. and CLUSE-TOLAR, T. (2002) Pimp-controlled prostitution: still an integral part of street life, *Violence Against Women*, 8, pp. 1074–1092.

WOJCICKI, J. and MALALA, J. (2001) Condom use, power and HIV/AIDS risk: sex-workers bargain for survival in Hillbrow/Joubert Park/Berea, Johannesburg, *Social Science and Medicine*, 53, pp. 99–121.

Homosexuality and the City: An Historical Overview

Robert Aldrich

Since the time of the Biblical Sodom and Gomorrah and classical Athens, homosexuality has been associated with the city.[1] This article surveys historical studies of homosexuality in urban milieux published over the past decade, suggesting the multiple links between homosexuality and the city evident since Antiquity. It argues that urban centres have been conducive to homosexual expression, whether integrated into or transgressive against social norms. Recent books not only record the history of homosexual groups in cities, but show the ways in which homosexual life has intersected with other aspects of urban life. Particular themes that reoccur are the emergence of homosexual cultures and types of sociability—from clandestine encounters to brazen parades—which cities have hosted, the impact of gay men and lesbians on urban environments and the spread of Western-style gay urban life around the world. My intention is not to articulate a general theory of urban homosexuality, but to illustrate sources and research strategies employed by contemporary historians and to argue the value to social scientists of an historical perspective in studies of homosexuality in the city.

The case studies discussed will demonstrate how those who experienced sexual desire for others of the same gender have been both marginal and central to the city in history. In some societies, homosexual acts were condoned, as in ancient Athens; in others, as in Renaissance Florence, they were condemned but nevertheless omnipresent. In later periods, homosexuals—in a similar way to ethnic and religious minorities—were consigned to the fringes. Yet, paradoxically, they always commanded attention: subverting normative standards of behaviour, carving out social niches, fertilising cultural life, demanding political changes.

The approach will be largely chronologi-

Robert Aldrich is in the Department of History, University of Sydney, Sydney, NSW 2006, Australia, Fax: +61 2 9351 3918. E-mail: Robert.Aldrich@arts.usyd.edu.au.

cal, looking first at classical, early modern and late modern Europe. I will then examine extra-European regions, including settler societies—the US, Australia and Brazil in particular—followed by Asia, before drawing general conclusions about urban gay history and historiography. However, I shall begin with sections that look at the city in the gay imaginary and at general social science theories of homosexuality and the city.

The City in the Gay Imaginary

The persistence of a vernacular linkage between homosexuality and the city is illustrated in titles of novels, which chart the changing position of homosexuality in Western society. In 1948, Gore Vidal's *The City and the Pillar* depicted the homosexual underworld in Hollywood and New York, but challenged stereotypes. *City of Night*, in 1963, presented John Rechy's gritty street-prowling hustlers and their clients. Armistead Maupin's multivolume *Tales of the City*, which appeared from 1978 to 1989, chronicled gay life in San Francisco, from the coming-out of small-town boys come to enjoy the Bay City in an ebullient period of gay liberation to the physical and emotional trauma of AIDS.

Certain cities have been particularly linked with gay life. From the 1600s onwards, homosexually inclined tourists journeyed to Florence, Venice and Rome to tour archaeological sites, Renaissance palaces and Baroque churches and, not coincidentally, to enjoy the company of Italian men (Aldrich, 1993). Others travelled further afield to the colonies. Sydney in the early 19th century—home to transported convicts, footloose sailors and rambunctious colonists—was the 'Sodom of the South Seas'. A raffish character in one 1905 French novel set in Saigon remarked cheerfully, 'We're in Sodom here', and the hero of an Italian novel which takes place in 1930s Ceylon branded its capital an 'equatorial Athens' (Aldrich, 2003). Sometimes, a particular individual, city and homosexual milieu have been conjoined. The poems of Constantine Cavafy linked romantic and cosmopolitan Alexandria to historical Byzantium and later encounters with shop-assistants, clerks or workers (Keeley, 1976; Liddell, 1974). E. M. Forster and Laurence Durrell added to the mythification of a city where seemingly any liaison was possible—as evidenced in Forster's affair with an Egyptian tram-conductor during the First World War. In collective gay memories of the inter-war period, no link is stronger than the one between Christopher Isherwood and Weimar Berlin, the city of strapping Germans, English expatriates, naughty cabarets and Magnus Hirschfeld's sexological institute—and of Nazis waiting to destroy such decadence (Isherwood, 1935a, 1935b, 1978).[2] Following the Second World War, attention switched to the Tangiers of Paul Bowles: another multicultural Mediterranean city where young men were as much on offer as strong coffee, hashish and Orientalist eroticism (M. Green, 1991). In more recent times, after flocking to New York's Christopher Street and San Francisco's Castro district during the era of gay liberation, travellers and migrants have swarmed to London's Soho, the Marais *quartier* of Paris and Oxford Street in Sydney, while those with more exotic tastes fly to Bangkok.

Travellers' accounts, as well as novels, evoke gay urban life. Neil Miller's *Out in the World* recounts peregrinations from Johannesburg to Cairo, Tokyo to Buenos Aires, Sydney to Prague, while Lucy Jane Bledsoe's anthology excerpts gay travels to Beijing and Tangiers, Berlin and Mexico City, Paris and even Sacramento (Miller, 1992; Bledsoe, 1998). Some have tried to capture the mystique of favourite cities: Edmund White in Paris and Robert Tewdwr Moss in Damascus (White, 2001; Moss, 1997). Proliferating gay and lesbian guides provide essays and practical details of where to go and what to do.[3]

The city has been a magnet for homosexuals, as for others made 'deviant' by the law, the church, medicine and social opprobrium. In 19th- and 20th-century Europe, the city was a lodestone for dissidents and rebels, particularly 'sites' of pleasure, such as Montmartre in Paris, that beckoned to migrants, foreigners, *avant-garde* artists, entertainers, prostitutes and criminals (Chevalier, 1980). Proust's gay Paris and, despite the prudish streak that finally did him in, Wilde's libertine London, beamed the bright lights drawing people to the big city.

The reasons are not difficult to discern. Cities offered a larger selection of partners than smaller towns and villages. Crowds provided anonymity and, where homoseuxal acts remained illegal, a measure of safety. Migrants could break out of the strictures imposed elsewhere, locating new 'sub-cultures' to satisfy reprobate desires. Soldiers, workers, tourists, students swelled city populations, joined by women domestic servants, factory workers, clerks and prostitutes. Libido, hope for friendship and romance, and a need for money, drove them to search out casual, situational or long-term partners or patrons. Cities have provided venues where men who have sex with men (and women who have sex with women) can meet: pubs and clubs, cafés and cabarets. In times of clandestine homosexuality, public baths and toilets, parks and back streets were especially hospitable to trysts.

Social Science and the Gay City

If the presence of homosexuals in cities can be taken for granted, the relationship between homosexuality and the city is more complex, a dynamic that has inspired considerable research. Some have looked at gay and lesbian migration, in reality and the imaginary (Weston, 1995). Urban geographers and sociologists have investigated 'gay space', from anecdotal older accounts of the 'tea-room trade' (men seeking sex in public toilets) (Humphreys, 1970) to sophisticated theoretical formulations of demographic patterns, social networks and commercialisation of sex and sexually linked entertainment and leisure. In the 1970s, the concept of 'ghetto' was extended to homosexuals in cities. Martin Levine commented on aggregations of homosexual residents and venues in particular districts, considering how such 'ghettos' might be defined and measured (Levine, 1979). Manuel Castells and K. Murphy pioneered a theory of sociocultural identity and urban structure (Castells and Murphy, 1982). Social analysts, and many lesbians and homosexual men, contended that conglomerations, and the institutions they fostered, formed an urban 'sub-culture' or, in even broader (though debated sense), a 'community' bound by shared interests and hoped-for solidarity.

In addition to looking at gays and lesbians *in* the city, researchers have analysed their impact *on* the city. A collection edited by Stephen Whittle inventories the ways in which the homosexual and the city intersect: the marking out of homosexual spaces in Manchester and Newcastle, gentrification in Toronto, political activism in Britain and North America, language (urban gay slang) and sub-groups (for example, sadomasochists) within gay culture (Whittle, 1994). Lawrence Knopp has examined the "homosexualisation of gentrified areas in cities by both dominant interests and gays seeking economic and political power as well as sexual freedom". Case studies include gentrification in Louisiana and homophobia in Scotland, with cross-cultural comparisons of the US, the UK and Australia (Knopp, 1995, 1997, 1998). The impact of gay and lesbian issues on local government and municipal politics has been studied by Robert W. Bailey. Arguing that cities are privileged arenas for identity development and expression, Bailey focuses on four American cases: political organisation and sexual identity in Birmingham, municipal districting in New York, police practices in Philadelphia, and school politics and domestic partnerships in San Francisco. Regime theory, the notion of neighbourhood activism, theories of identity formation and economic development perspectives are marshalled to evaluate the imprint of lesbians and gay males on urban America (Bailey, 1999). Such studies connect with broader theorisation of sexuality in space by geographers (Bell and Valentine, 1995; Betsky, 1997; Ingram *et al.*, 1997; Fincher and Jacobs, 1998). They link, too, with studies of commercialism and consumerism, and attempts by the market economy to capitalise on the 'pink dollar' (Hennessy, 2000; Badgett, 2001).

Combining psychology and sociology, Henning Bech has gone further in suggesting organic links between homosexuals and the city. "The city is the social world proper of the homosexual, his life space", theorised Bech, the place where the homosexual comes out and where "the homosexual can *be*". The natural environment for the homosexual 'species' (Bech's word) is the city. Homosexual cruising Bech described as a "combination of gazes and movements, which at gay bars takes place in an enclosure and finds its proper territory out in the city". The 'gaze' itself, the strategy for establishing contacts, 'belongs to the city'. Furthermore, "the city is not merely a stage on which a pre-existing, preconstructed sexuality is displayed and acted out; it is also a space where sexuality is generated". What is it about the city that stimulates? "Surely that altogether special blend of closeness and distance, crowd and flickering surface and gaze, freedom and danger". Homosexuality as "a form of existence" is "essentially social", "essentially urban". Without denying homosexual *milieux* outside urban areas, Bech adds that "homosexual existence is a phenomenon *of* the city and not just something occuring *in* the city" (Bech, 1997).

Bech's insistence on clandestine sites of encounter, such as toilets and rail-stations, seems time-bound given the proliferation of open venues—backrooms have displaced backstreets for *rendez-vous*. However, his work retains interest for pointing to the symbiotic partnership of urban space and homosexual life in the culture of modernity. Bech's study is also incisive in drawing on historical materials on urban–homosexual links, for instance, Hirschfeld's 1914 enumeration of venues in Berlin. As George Chauncey has shown for New York, late 19th-century observers in America likewise associated homosexuality with the city. "Only in a great city", testified a migrant in 1882, could a homosexual

> give his overwhelming yearnings free rein *incognito* and thus keep the respect of his every-day circle. ... In New York one can live as Nature demands without setting every one's tongue wagging (Chauncey, 1994, p. 42).

Chauncey's *Gay New York* (discussed below) investigates 'urban culture and the making of the gay male world' in a period when ideas and contours of the city and gender were both changing rapidly.

Historians and the Gay City

Historians have provided the most complex case studies of homosexuality and the city. Some accept the idea that homosexuality developed as a particular sexual identity only in the late 19th century. Others have pushed the 'invention' of homosexual cultures further into the past—Rictor Norton's study of London molly-houses argues the existence of a clearly perceived homosexual culture in 18th-century England (Norton, 1992, 1997). Luiz Mott has isolated a 'gay' sub-culture in 17th-century Rio de Janeiro (Mott, 2003) and John Boswell famously if controversially used 'gay' to describe same-sex unions in the Middle Ages (Boswell, 1980, 1994). Arguments between 'essentialists' and 'constructionists' have run their course over the past 20 years, with a safe consensus that the shapes of same-sex attractions and relationships, present in almost all societies and epochs, have metamorphosed over the *longue durée* and continue to do so. The exact nature of situations and transformations calls for historical nuance rather than reductionist generalisation, as the following survey demonstrates.

Homosexuality in European Cities: From Ancient to Modern

In the Greek world, as shown by Kenneth Dover, David Halperin and Bernard Sergent, sexual intercourse between males formed an accepted part of life, although surrounded with restrictions relating to age, social position, sexual repertoire, marriage and fatherhood (Dover, 1978; Halperin, 1990; Sergent, 1984). In Athens, Sparta and other Greek city-states, or in the urban centres of the Hellenistic or Roman empires, same-sex activity was integrated into public and private life. Vase-painters pictured sodomy; dramatists praised (and satirised) the love of boys; male friends such as Patroclus and Achilles consorted in mythology; and Hadrian and Antinous flaunted their compansionship throughout the Roman empire. Exercise-grounds and gymnasia counted among sites of seduction, and symposia (such as Plato's philosophical banquet) stimulated discussions of love and longing.

Such integrative same-sex behaviour and attitudes in Antiquity are too well known to require further discussion, but it should be remembered how much a part of city life man-to-man liaisons were. Sexual meeting spots were public gathering places; most positions of homosexual fornication were not legally punishable; homosexual 'ghettos' did not exist. Present-day homosexuals would receive a rude shock if transported to a classical city, but what would now be seen as 'homosexual' sex was embedded in a nexus of condoned conduct and representations inseparable from the city. Some scholars, notably Boswell, have argued that this sexual culture persisted well past the fall of the Roman empire, though Christianity became increasingly violent in condemnation of homosexual behaviour, forcing men who wanted sex with other men into subterfuge (Boswell, 1980, 1994).

The homoerotic nature of urban Renaissance culture—art works, neo-Platonic evocations of comradely love—has long been familiar, but contemporary historians have more clearly delineated the extent. Guido Ruggiero's chapter on Venice provided a pioneering account (Ruggiero, 1985), followed by Michael Rocke's study of Florence. Rocke's specific subject is a tribunal established in 1432 to investigate sodomitical offences and its extraordinary archive of prosecutions down until 1502. The thrust is how many people were involved

> In this small city of around only 40,000 inhabitants, every year during roughly the last four decades of the fifteenth century an average of some 400 people were implicated and 55 to 60 condemned for homosexual relations (Rocke, 1996).

The Office of the Night examined 17 000 men on charges of sodomy and convicted 3000. Rocke extrapolates:

> In the later fifteenth century, the majority of local males at least once during their lifetimes were officially incriminated for engaging in homosexual relations (Rocke, 1996).

Illegality did not deter homosexual coupling, and those so inclined found ample opportunities to indulge their passions. Labour relations (master craftsmen mentoring apprentices), living arrangements (apprentices or servants boarding with employers), demographic movements (travelling students and artists), prevailing cultural norms, the status of women: all encouraged homosexual dalliances. The geography of the city collaborated, with the Florentine street of the furriers known as a place to pick up 'trade'. Rocke notices that some families facilitated relatives' homosexual activities and professional corporations served as networks of homosexual recruiting and partner-swapping.

With the repressive turn of secular rulers and Protestant and Catholic clerics in early modern Europe, homosexual behaviour and attitudes were increasingly transgressive. From at least the 16th century onwards, homosexuality in the city was driven into semi-clandestinity. This situation endured until late in the 20th century in many regions, although Napoleonic Europe saw decriminalisation of homosexual acts in countries (such as Italy and the Netherlands) with law codes rewritten under the French *imperium* and, in Britain and Germany, penalties were lightened. However, homosexual life remained part of the public domain, although in less evident fashion. Throughout Europe, and in settler societies overseas, cities played host to sites of homosexual meeting. In Britain, the courts of Elizabeth I and James I were rife with sexual innuendos and practices (Manzione, 1996; Shephard, 1996; Smith, 1991; Bray, 1982). French courts, such as that of the Henri III—large courts functioned as miniature cities—were hotbeds of sexual misbehaviour (Merrick and Ragan, 2001). Caravaggio and Marlowe, James I and Frederick the Great were deviant celebrities in an urban and courtly world of sexual ambivalence.

Modern Gay Centres in Urban Europe

Continuities in homosexual history from early modern Europe to the present are illustrated in *Queer Sites*, edited by David Higgs, a valuable (if uneven) collection on 'gay urban histories since 1600' (Higgs, 1999). Five chapters concern European cities. Similar patterns emerge—the existence of well-known homosexual venues from the 1600s, networks of sodomitical sociability, the evolution of same-sex cultures even in the face of disapproval—although each city displays particular traits. Randolph Trumbach argues that, by the first generation of the 1700s, three genders were identifiable in England, masculine-acting men who usually slept with women but sometimes with men, women, and a third-sex of male but effeminate inverts. Sodomites met each other and 'hetero-sexual' men in 'molly-houses' and the third sex also mixed and mingled with prostitutes. The nature of 'homosexual' relations remained constant for two and a half centuries, changing only after the Second World War with the cleavage of the worlds of gay men and prostitutes, decrease in persecution of homosexuals, a new gay sub-culture and the 'domestication of gay sex' into quasi-conjugal arrangements.

For Gert Hekma, the relatively open presence of homosexuals in early modern Amsterdam is explained by religious toleration in the 1600s and libertinage in the subsequent century. There existed a network of men who had sex with men—and the latrines, inns and parks where they gathered—although they did not self-consciously embrace a homosexual identity until the 1950s. Decriminalisation of sodomy in 1811 and later failure to impose comprehensive anti-homosexual laws allowed homosexuality to flourish with limited harassment. Medico-legal debates about deviance did take place, prosecutions for indecency occurred and Amsterdam authorities in the 1880s designed public urinals with separated stalls to preclude sexual activities. By the 1950s, Amsterdam boomed as a homosexual capital: bars with rent-boys, the first leather bar in 1952 and the first publicly gay sauna in 1962; meanwhile, one of Europe's first post-war gay organisations, COC, was established in 1946.

In other cities, public gay life has not been so continuous as in London and Amsterdam. Higgs' chapter on Lisbon shows the concern of the Inquisition with sodomitical acts; from 1536 to 1821, 5000 men were denounced, many ritually humiliated and flogged, or put to death. By the early 1600s, networks of sodomites nevertheless operated—lackeys and priests enjoyed reputations as procurers. For many males, sodomy represented 'a temporary and occasional transgression'. Higgs discerns tension between the 'feudal model' of age- and class-based sexual relations and an emergent subcultural identity—a paradigm seen elsewhere. In the late 1800s, a renascent homosexual life extended from a 'toilet culture' through the first homosexually themed novels, and homosexual cafés opened in the Bairro Alto district, still the heart of Lisbon's gay scene.

Dan Healey's essay on Moscow provides a comparative perspective of a city not generally associated with gay history. Evidence of sodomitical contacts survives from the 1600s, often linked, at least in plebian society, with excessive consumption of alcohol. A sub-culture emerged in the second half of the 1800s (several centuries after its development in western Europe): coach-drivers, waiters and apprentices made themselves available for 'gentlemen's mischief', street-cruising was obvious and bath-houses, a key institution in Russian life, were important sexual sites. In the early 20th century, drag balls were held in Moscow. With Bolshevism, homosexual acts were legalised from 1922 to 1934, and the public sub-culture temporarily blossomed. Stalinism cracked down on dissidence and deviance and, for the rest of the Communist period, homosexual meetings remained furtive.

Michael Sibalis, arguing that "urbanisation is a precondition to emergence of a significant gay sub-culture", traces their evolution in Paris, a city for centuries associated with licence. Changing topography mirrored a metamorphosis in sexual identities. From the late years of the *ancien régime*, a sodomitical map of Paris can be drawn showing men cruising in such areas as the Tuileries Gardens. Through the 1800s, new sites were added, including the Champs-Elysées, the Palais-Royal and the Seine embankments, while the hundreds of urinals that punctuated the streetscape hosted casual encounters. The sexual map grew denser in the early 20th century with the artistic cafés of Montmartre and the louche bars of the Rue de Lappe. In the post-Second World War period, homosexuals along with bohemians and existentialist philosophers haunted Saint-Germain-des-Prés. Homosexual life shifted to the more commercialised and exclusively gay venues of the Rue Sainte-Anne in the 1970s and, finally, to the Marais in the 1980s.

Paris since the 1700s has thus been a gay capital (Merrick and Ragan, 1996; Merrick and Sibalis, 2001). Michel Rey used police archives to investigate Enlightenment sexual attitudes and sub-cultures (Rey, 1985, 1987, 1989) and Jeffrey Merrick has also explored old regime Paris, homosexual scandals and the policing of homosexuals (Merrick, 1996, 1997, 1999). William Peniston has reconstructed a network of homosexuals around a waiter in the 1870s, and examined homosexuality at the trial of a socialite (Peniston, 1996, 1999). Leslie Choquette has studied creative works about Paris lesbians from the 1830s to 1890s (Choquette, 2001). Francesca Canadé Sautman has revealed the existence of working-class lesbian culture in the early 1900s (Sautman, 2001), and there have been studies of lesbian and reactionary politics (Benstock, 1989), as well as élite lesbian circles in interwar Paris (Jay, 1988).

A picture of interwar homosexual sub-culture of Paris, including one of the first homosexual magazines and the Magic City Ball, has been painted by Gilles Bardedette and Michel Carassou (Barbedette and Carassou, 1981). Gay Saint-Germain-des-Prés, the "*folles*, swells, effeminates and homophiles in the 1950s", is the subject of Georges Sidéris (Sidéris, 2001). Sibalis has published articles on homosexual space in Paris in the 1800s, and images of homosexuality throughout the 19th-century city (Sibalis, 1999, 2001a, 2001b). Paris illustrates the sources and themes available to historians of gay cities. It also shows how the 'marginal' world of homosexual culture played a vital cultural role in the circles of Montesquiou, Cocteau and other luminaries. Homosexual groups were a transgressive part of the urban population, but nevertheless integrated into the cultural, recreational and commercial environment: indeed, they formed a key part of it. Long before rebels founded a 'Homosexual Front for Revolutionary Action' in 1971 or trendy cafés hung gay flags in the 1980s, homosexuals played a wide-ranging and fertile role in urban life. Work on other cities confirms these patterns, as in the case of homosexuality in London's Bloomsbury (Stansky, 1996).

Florence Tamagne's *Histoire de l'homosexualité en Europe* from 1919 to 1939 stands out as a comparative approach (Tamagne, 2000). The sub-title, which names Berlin, London and Paris, emphasises the centrality of the capitals in the evolution of homosexuality: political movements, social life, literature, medico-legal debates, scandals and the place of individual lesbians and gay men. Tamagne surveys these areas, summarises interwar publications, explores police archives, pays close attention to lesbianism and proposes provocative interpretations about the relationship between homosexuality and national cultures. A more overt homosexuality emerged during a period liberated from pre-war social mores and avid for pleasure. The efflorescence of a cult of the beautiful body, androgynous and adolescent youth, and unbridled enjoyment, represented reaction against the death and destruction of wartime. These decades offered 'liberation', but hardly '*insouciance*': law, public opinion and police harassment weighed heavily, despite the myth of the *années folles* that outlived the interwar period.

Tamagne adumbrates differences between national models of homosexuality. Berlin witnessed a 'communitarian' approach, the collective nightlife for which the city became famous, and rising political consciousness in the face of anti-homosexual laws. In Paris, decriminalisation meant that homosexual life was less politicised and more individualistic. London displayed a veritable 'cult of the homosexual' with 'homosexualisation' of the élite induced by public schools and sex-segregated sociability, combined with an identity crisis of post-war British youth. Paris and London were more of a magnet for lesbians than Berlin; homosexuality in the German capital was especially linked to working-class culture. Tamagne characterises political differences as 'contestatory militancy' in Germany, 'subversive integration' in Britain' and 'pleasure-seeking individualism' in France. While the situation changed in the 1930s, with horrific consequences in Berlin, these models provided prototypes for post-1945 gay cultures.

Homosexuality in American Urban History

Westernised cities outside Europe also engendered gay cultures. Homosexuality in the city provided the subject for an early work of American gay history, John Gerassi's 1966 book on a 1955 scandal in Boise, Idaho (Gerassi, 1996). One of the first of a new breed of gay histories of America—embodying a lesbian- and feminist-affirmative and communitarian perspective—was Elizabeth Lapovsky Kennedy and Madeline D. Davis' study of lesbians in Buffalo, New York, from the 1930s until the 1960s (Kennedy and Davis, 1993). *Boots of Leather, Slippers of Gold* traces "the roots of gay and lesbian liberation to the resistance culture of working-class lesbians", when "butch-fem roles coalesced an entire culture into the prepolitical, but none the less active, struggle against gay and lesbian oppression" (Kennedy and Davis, 1993).

Another classic work, which acknowledged a debt to Kennedy and Davis, is George Chauncey's study of New York from the *fin de siècle* to the Second World War (Chauncey, 1994). Abolishing myths that New York's homosexuals were isolated one from another, that they were invisible and that they internalised social opprobrium, Chauncey uncovered an immensely lively and varied gay life in Manhattan, especially in Greenwich Village, Harlem and Times Square. 'Homosexual' and 'heterosexual' held little meaning from 1890 to 1940, as sexual categories were far more complex: 'fairies', men who inverted genders by acting effeminately (sometimes dressing as men), 'wolves' who preferred sex with men though occasionally with women, young male 'punks' who provided sexual favours for reward, and 'queers' who identified as homosexual. ('Hetero-homosexual binarism' is a relatively recent invention.) Well into the 20th century, homosexual practices were widespread, firmly rooted among the working-class men on whom Chauncey focuses. 'Real men', including Irish and Italian immigrants, and sailors, labourers and hoboes, saw no wrong in 'masculine' same-sex acts and suffered little loss of peer esteem for engaging in them. Chauncey also charts interracial and middle-class homosexual cultures, with homosexuals so conspicuous by 1930 that a 'pansy craze' swept entertainment and an annual gay ball in largely Black Harlem drew thousands. *Gay New York* maps a 'sexual topography', types of 'everyday resistance' to homophobia, a non-binary 'sexual regime' and 'sub-cultural codes' used to reinforce social networks. Chauncey's book serves as a theoretical template, just as he argues that New York offered a prototype for American gay cultures.

Charles Kaiser's *The Gay Metropolis* aspires to be a sequel to Chauncey's book (Kaiser, 1997). The introduction promises

> the story of ... how America's most despised minority overcame religious prejudice, medical malpractice, political persecution and one of the worst scourges of the 20th century to stake its rightful claim to the American dream (Kaiser, 1997).

The triumphalist view continues in chapters devoted to each decade since the 1940s, but the book does not engage with theoretical or historiographical issues. Kaiser looks at the rich and famous, and on key events that marked gay history. However, the crucial 1969 Stonewall episode is more extensively covered by Martin Duberman (Duberman, 1993). Kaiser shows that political activism actually pre-dated the Greenwich Village uprising; even in the conservative 1950s, "gay life acted like a bracing undertow, exerting a powerful opposite pull beneath waves of conformity", oppression and persecution. Readers await Chauncey's own second volume for a more interpretive account of New York's recent gay past.

San Francisco is sometimes considered the 'gay capital of the world'. Les Wright's chapter provides an overview (Wright in Higgs, 1999), *Gay by the Bay* is an illustrated survey with a lucid text (Stryker and van Kuskirk, 1996); and *Out in the Castro* includes chapters on celebrations, AIDS, policing, gay institutions and symbols (Leyland, 2002). Two model studies provide full-scale academic treatment. Nan Alamilla Boyd traces the origins of 'queer' San Francisco to the 19th century, arguing that the Bay City was a "queer town because sex and lawlessness were sewn into the city's social fabric", a product of the Pacific frontier and the 1849 Gold Rush (and wealth and migration that followed) (Boyd, 2003). In the 20th century, the repeal of Prohibition in 1933 fuelled tourism and settlement. Boyd's chapters, each opening with an oral history, examine cabarets and male impersonation, the demarcation of lesbian space, activities of the police and alcohol surveillance authorities, 'homophile' movements after the Second World War and activism in the 1960s. She finishes with protests against a police raid on a costume ball in 1965, a landmark rebellion pre-dating the New York riot by four years. Boyd thus examines two strands of gay history. One is sociability, particularly the bar culture, which she argues was not so much apolitical as a kind of "politics of everyday life", a "marketplace activism which ... served the interests of queer community development". The other is the formal political movement, including the Mattachine Society, Daughters of Bilitis and Society for Individual Rights. The two first stood at loggerheads, but came together effectively in the 1960s, obtaining reforms even before the advent of Gay Liberation.

Elizabeth A. Armstrong takes up where Boyd concludes (Armstrong, 2002). She concentrates on politics, but does not exclude gay sociability, in a study of organisations active since 1945, their strategies and ideologies, culminating in election of the openly gay Harvey Milk to the city council in 1977 and the gay and lesbian march on Washington the following year (the year, too, of the assassination of Milk). The political movement then disaggregated, replaced by different sorts of AIDS-related activism. Armstrong treats each stage of the evolution and devotes a chapter to groups distanced from the activist mainstream, including lesbian separatists, and men and women of colour. Using 'field formation' theory, she demonstrates the "transformation of an underground sub-culture into a culturally vibrant, politically ambitious, organisationally complex gay identity movement" (Armstrong, 2000) later contested, though with limited success, by queer politics.

Marc Stein's book on Philadelphia offers a study of the 'everyday geography', public cultures and political movements of what he claims is probably the biggest American city without a gay reputation (Stein, 2000). One specific trait is that homophile activism did not see the separation of gay men and lesbians (with male domination) that occurred elsewhere, a feature explained by the 'sex egalitarianism' and religious tolerance of Quaker-founded Philadelphia. Stein provides a comprehensive survey of gay and lesbian life from 1945 until 1972, the date of the first Gay Pride parade. In addition to bars, bathhouses and other forums, he discusses how homosexuality appeared in periodicals, especially the pioneering *Drum* magazine. One section presents a mid 1950s controversy about the naming of the Walt Whitman Bridge over the Delaware River—the proposal proved anathema to those who labelled the poet a 'pervert'. Stein's study, with treatment of racial and gender issues, and the role of the police and moral puritans, analyses many features crucial in the American gay and lesbian past.

Although a work in urban development rather than history, Moira Rachel Kenney's *Mapping Gay L.A.* displays fine historical sense in reading the city 'both as space and text' (Kenney, 2001; see also Quimby and Williams, 2000; Retter, 2000). She sustains two assumptions: "First, place and the city have importance for culture and politics; and second, place is created and altered through our interactions in and with it" (Kenney, 2001). She argues that Los Angeles is "the greatest hidden chapter in American gay and lesbian history", where key homophile organisations and the first national lesbian publication were founded. Specificities include a multicentred geography, lack of street life and the gay sociability it produced, relatively underdeveloped radical activism and the pronounced impact of Hollywood. She introduces the useful ideas of 'mental maps' that combine real and symbolic sites, and 'place claiming' that contributes to group identity. Citizens' efforts to secure municipal independence for West Hollywood, a largely gay area, allow investigation of local politics. Kenney also treats provision of gay-related social services, gentrification and the 'inclusive separatism' of lesbian groups in Los Angeles.

A collection edited by Brett Beemyn has inaugurated gay 'community histories' of American cities, including localities such as Detroit and Flint in Michigan (Beemyn, 1997). The editor provides a salutary reminder that "gay history in the United States is not limited to New York and San Francisco", and underlines the benefits of studies of 'middle America'. Some of the same themes of gay life in the 'heartland' and in coastal megalopolises reoccur: resistance against homophobic law codes, use of commercial and non-commercial venues for socialising, issues of identity and community. These community histories reveal research sources, ranging from police records to oral histories, available to uncover a neglected past, which has been exploited in several other grassroots studies (The History Project, 1998; Johnson, 1994–95; Howard, 1995; Thorpe, 1996; Freeman, 2000; Ward, 2003). Each shows the particular combination of demography, social and religious attitudes and political tolerance (or lack of it) that contours local gay life.

Variations on Urban Gay Themes: From Australia to the Americas

A plea for more community histories can extend beyond America, as such works underline the different constructions of sexuality outside that country. They also explore cultures that continue to exist, despite what Dennis Altman has called 'the Americanisation of the homosexual' (and the concomitant 'homosexualisation of America') (Altman, 1982).

Australia has been well served by historians, with a pioneering work by Garry Wotherspoon, *City of the Plain*, on the birth and development of an 'urban sub-culture' in Sydney (Wotherspoon, 1991). Wotherspoon delves into the colonial history of Australia's major city to identify strands of male-bonding. 'Situational homosexuality' flourished among convicts transported to Botany Bay, as did liaisons in a frontier society with an enormous imbalance in the sex ratio. The arrival of mostly male migrants from Asia and the South Pacific added to social and sexual cosmopolitanism. The promiscuity of colonial boarding houses, the lubrication of morals by alcohol and the model of (theoretically platonic) 'mateship' encouraged renegade encounters. These strands became woven together into a new homosexual culture during and after the Second World War, as automobiles and varied housing stock promoted social mobility and contestatory groups influenced by the New Left and American Gay Liberation challenged persistant puritanism and antipodean McCarthyism. So successful was the onslaught that, by the 1980s, Sydney was promoted as a gay mecca, typified by the annual Mardi Gras.

Other studies on Sydney include a history of Mardi Gras, a collection of essays on gay politics and general volumes on Australia that give much consideration to urban homosexuality (Carbery, 1995; Johnston and van Reyk, 2001; Phillips and Willett, 2000; Willett, 2000; Wafer *et al.*, 2000; Reynolds, 2002; Moore and Saunders, 1998; Coad, 2002). Particularly noteworthy is a study of the gay precinct of Sydney, *Street Seen: A History of Oxford Street*, by Clive Faro (Faro, 2000). First an Aboriginal pathway, Oxford Street became the main conduit from the colonial city centre to the Pacific Ocean. A thoroughfare by the mid 19th century, it was also the location of a jail and a barracks, and a developing reputation for dissident groups. The opening of department stores attracted shoppers, while busy commercial activities and inexpensive housing brought in office-workers, labourers and bohemians, as well as Mediterraneans, Chinese and other 'ethnic' residents. Already at the end of the 1800s, numerous pubs catered for male drinkers, a Turkish *hammam* hosted homoerotic dalliances and newspapers caricatured 'sissy' shop assistants. In the 1970s, with sex-on-premises venues and gay emporia, Oxford Street remained Sydney's 'golden gay mile'. In model fashion, Faro carefully connects these phenomena to the evolution of the housing market, transport and planning, demographic changes and political decision-making. Rather than isolating gay history as a specific niche of urban studies, therefore, Faro uses a particular precinct to underline the interrelated urban history of Australia's premier metropolis.

In a second Australian case, regional rather than exclusively urban, Clive Moore examines Brisbane and Queensland (Moore, 2001). Many Sydney traits reappear, though modified by the tropical environment of Australia's north-east. Here, too, casual homosexual venues can be traced back into the 19th century and the presence of a multicultural population (including Aborigines) made for diverse contacts—Moore identifies several 'beats' popular for a century.

Another example of homosexuality and the city far away from Europe concerns Brazilian cities, particularly Rio de Janeiro, covered succinctly by David Higgs in *Queer Sites* (1999). More extensively, James Green uses medical records, newspapers and interviews to trace an identifiable gay culture in Rio back to the late 19th century (Green, 1999). A same-sex erotic culture existed well before the invention of the 'homosexual'. From the 1870s, the Praça Tridentes offered a cruising-ground so familiar that newspapers openly satirised the park and its homosexual denizens, including a famous journalist, the Wildean João do Rio. An interracial homosexual novel, *Bom-Crioulo,* appeared in 1895, and illustrated homosexual porn was sold by 1914. Medico-legal interest in homosexuality grew in the 1920s and 1930s with doctors sending students to measure homosexuals to confirm theories about somatotypes. Psychiatrists submitted homosexuals to electroshock therapy, while the Catholic church tried to hide priests' pederastic peccadillos. A 'homosexual topography' was known in Rio, 'urban territories' of homosexuality that expanded rapidly over the next decades. Green also shows how homosexuals invested the city and appropriated the Carnival, increasingly viewed as a gay festival.

In addition to gay biographies of cities, historians and other social scientists have investigated specific aspects of gay urban life. Space lacks for detailed discussion, but examples show the varieties of research. On the basis of court and police records and newspaper accounts, Steven Maynard has discussed homosexual cruising and police entrapment in public lavatories in Toronto in the late 19th and early 20th centuries (Maynard, 1994). Arne Nilsson, using oral histories from 1930 to 1960, describes male gay encounters around the busy port of Gothenburg, Sweden (Nilsson, 1998). Ross Higgins has demonstrated how public sex, and supposedly anonymous encounters in saunas and parks, helped to define a gay 'community' in Montreal in the 1950s and 1960s (Higgins, 1999). Also on Montreal, Line Chamberland reflects on types of socialising, appropriation of space, class dimensions of drinking establishments and lesbian cultures from 1955 to 1975 (Chamberland, 1993). Ira Tattelbaum uses advertisements, personal recollections, floorplans and computer-created graphics to write the history of St Mark's Baths, an iconic sauna in New York City (Tattelbaum, 2000). Other researchers have examined minority lesbian and gay groups—for instance, Asians in Sydney and London, and similar minorities in America (Jackson and Sullivan, 1996; Leong, 1996).

'Gay cities' are not confined to the Western world. Same-sex activities are common around the world and anthropologists have discovered the permutations of sexual behaviours which seldom corresponded to modern Western profiles of homosexuality. In recent years, cities outside Europe and North America have seen more open, self-identified and consumerist gay cultures. Rio de Janeiro provides one example. The gay life of other South American centres, such as Mexico City and Havana, draw particular qualities from cultural and ethnic *métissage* of European, American and African traditions (Lumsden, 1996; Prieur, 1998). Different cultural traditions have also combined in South Africa, notably in Cape Town (Lewis and Loots, 1994). Studies of the particular situations provide a warning against assumptions of simple replications of European and American gay life.

Homosexuality and the City in Asia

Further away from European traditions, similarities and differences stand in higher relief. Bangkok is a magnet for tourists who crowd 'boy bars'; gay cabarets and saunas are more numerous, and gay life is less restrained, than in many Western cities. Homosexual behaviour is not foreign to Thailand, as Peter A. Jackson and others have shown; the *kathoey*, a man of intermediate gender who engaged in sexual relations with other men or boys, held an established place in Thai life (Jackson, 1995, 1997a, 1997b; Jackson and Sullivan, 1999). Buddhist religion preaches tolerance of diverse sexual desires and Thai law does not criminalise same-sex acts. Homosexual, just as heterosexual, relations intertwined with other social and cultural interactions in a hierarchical society. A variation has come with the opening of Western-style pubs and bath-houses and networks that involve male prostitution. Thai tradition and Western commercialism have combined to create a new gay life, although Jackson warns against perceptions of Thailand as a 'gay paradise' (Jackson, 1995, 1997a, 1997b).

Homosexuality is remarkably public in broad-minded Bangkok, much less accepted in Singapore (Hiang Khng, 1998). In the 1950s, tranvestite prostitutes frequented infamous Bugis Street, but with its transformation into the powerhouse of south-east Asia, Singapore exchanged a reputation as a city of sin for a button-down image. The first gay bar in Singapore opened only in the late 1970s, although not until 1983 did a disco allow same-sex dancing. Affluence and a liberalisation in authoritarian policies heralded greater tolerance towards sexual dissidence until in 1989 police raided the disco. In 1994, a gay activist organisation formed, which Russell Heng Hiang Khng interprets as a move from a simple gay 'scene' to a gay 'community', although the government refused to register the group and continued harassment. Homosexual acts remain criminal, and a 'don't ask, don't tell' rule prevails, but Singaporeans have developed a quietly active gay culture.

Gay life is even more closeted in Hanoi, although homosexual acts were criminal under neither the French colonial admistration nor the post-independence Communist government (Aronson, 1999). As far back as 1900, visitors wrote about men meeting other men for sex at Hanoi's Hoan Kiem lake. One French writer considered 'pederasty' an urban phenomenon in Vietnam and others attributed its prevalence to deracinated villagers or to Chinese and European migrants. A century later, Hoan Kiem lake remains a site for homosexual meetings, as young men lounge on benches, striking up sexually inflected conversations. Hanoi has at least one Western-style bar patronised by homosexuals, although according to Jacob Aronson, most encounters proceed with ambiguity.

Yet another example, here from South Asia, is provided by Jeremy Seabrook's book on India, based on conversations in 1997 with 75 men in Delhi (Seabrook, 1999). Seabrook met his interlocutors in a public park that the Indians visited for homosexual encounters. Many are migrants from other parts of the country, the largest number neither very rich nor very poor, but "somewhere on the borderline of traditional Indian society, on the threshold of the modernizing, liberalizing shift which has accelerated in the past few years". Few identify as gay and often consider sexual position—active versus passive—more significant than a partner's gender:

> The great majority of men who have sex with men do not, on the whole, identify with Western cultural normals [*sic*] of being gay or bisexual. To impose such categories—except upon a small minority who have been much influenced by Western gay experience—is to bring alien concepts to the people involved (Seabrook, 1999).

The existence of a more Westernised gay culture in Mumbai, with gay bars, a newspaper and a political organisation, foreshadows a new sexual culture.

Bangkok, Singapore, Hanoi and Delhi are culturally and politically disparate places, distant from the gay capitals of America, Europe or Australia. Each nevertheless demonstrates the city as a catalyst for homosexual activity because of the size of the population, long-standing same-sex traditions and, in varying fashion, the commercialisation of gay sociability. These examples point to the recent export of Western identities, commodification of sex and the spread of AIDS, as pertinently analysed by Dennis Altman (Altman, 2001). Hanoi, and particularly Delhi, however, show the diversity of sexual cultures and mark the limits—at least for the present—of sexual globalisation with typecast identities, behaviours and institutions.

Urban Gay Strategies: Historical and Historiographical

The appearance of new gay cultures alongside the persistence of traditional ones of 'men who have sex with men' prolongs a phenomenon of urban homosexuality stretching back to Antiquity. They show the city as a site of homosexual life *par excellence*, a privileged place of constantly changing same-sex cultures. The chronology of their appearance, their contexts, social and spatial segregation, legal situations, political entitlements, the degree of public visibility: all vary. The exact contours of gay cultures are formed by the particular urban (and national) cultures. The city has shaped the homosexual from molly-houses in early modern London to the culture of 'fairies' and 'wolves' in working-class New York in the early 20th century, from the carnivalesque tradition in Rio to the 'multicentred geography' of Los Angeles and the cohabitation of traditions in Thailand and Vietnam.

Homosexuality is manifest in the city in identifiable ways. One is sexual *topography*. This includes generic venues of sociability, but also gay centres, such as Greenwich Village in New York, the Castro in San Francisco, West Hollywood in Los Angeles, Oxford Street in Sydney and the Marais in Paris. Also important are particular urban *occasions*: the balls of Magic City in Paris, the Harlem drag ball, carnival in Rio, Mardi Gras in Sydney and 'Pride' parades around the world. *Organisations* with an urban base have particular significance, from Hirshfeld's gay emanicipation committee in Wilhelmine Berlin, through homophile associations in post-Second World War Amsterdam and Paris, to Gay Liberation movements in 1970s America. Case studies of homosexuality, furthermore, suggest the *geographical extension*, and permutation, of gay and lesbian urban communities, from Europe and America to the cities of Asia, the advent of urban sexual cultures in smaller provincial towns as well as larger cities. They underline the *cultural and social role* of homosexuals in the city, notably in challenging received axioms about gender, disorienting divisions between public and private space, and inventing new forms of sociability and eroticism.

As well as themes about urban homosexuality, the books discussed here inventory materials available for study: criminal records, newspaper reports, archives of associations, business, political and demographical data, oral histories, creative literature. They suggest ways in which historians can write gay urban 'biographies', or concentrate on specific events, personalities or themes, undertaking micro-histories or broad-brush surveys. Perhaps most importantly, they emphasise how homosexuality intersects with the city, how the urban environment moulds and often promotes homosexual expression. Homosexuals also influence the city through gentrification, pressure on municipal authorities, a not inconsiderable economic and electoral power, new networks and venues of sociability. Recent gay urban histories are important precisely because they do not excise the 'gay community' from the urban body, but treat the way in which it forms an integral part of the city, from Renaissance Florence to 21st-century Sydney.

Segregation into 'gay ghettos' may have marked a passing phase in homosexual urban history, crucial in establishing an identitarian homosexual culture. Increasingly, however, homosexual life blends with other aspects of the city, as it did—in a different way—in Antiquity and the Renaissance. Castro is only one of the social and ethnic districts of San Francisco; gay venues fly flags next to coffeeshops, Turkish restaurants and *patisseries* in Amsterdam. Paris and Berlin have openly gay mayors. Los Angeles and San Francisco have gay archives; Sydney and Amsterdam have gay monuments. Sexuality couples with urban life in diverse relationships and with mutually experienced effects.[4]

Historians have moved from simply 'discovering' homosexuals in cities, then writing in a 'gay is proud' tradition about the formation of affirmative sub-cultures to a more complex understanding of the links between sex and the city. Future research may study more fully how homosexual groups have made their presence felt. Further studies of municipal politics and of the formation of largely gay precincts will provide insight into the political, economic and social role of special-interest-groups. There is a need for more historical study of specific gay organisations and venues, from political groups to such quotidian meeting-places as baths and bars. Historians might profitably pay greater attention to specific groups within urban gay society, from ethnic groups—who have received relatively little attention—to those with particular sexual interests (such as S and M *afficianados*). Urban intergenerational sex ought not to remain the taboo subject that many consider it. The homosexual activities of priests and soldiers should be more fully investigated as historians disaggregate urban communities. Historical studies of male prostitution have yet to be done. Links between crime and homosexuality could be investigated.

The manifold nature of sexual desire in urban space might be a particularly fertile area for work. As Chauncey has shown for late 19th-century New York (and scholars have underlined for earlier societies), sexual categories were generally not fixed, with no great divide between heterosexuals and homosexuals. Even in the 20th century, sexual categories have often been breached and the polymorphous sexual life of youths and workers, clear from Chauncey's work, might be more fully explored. Greater study of cities where 'traditional' sexual cultures persisted (for example, in the Mediterranean) would reveal constructions of both urban and gay life different from ones familiar in western Europe.

Studies of smaller cities and those without a gay and lesbian reputation will illustrate nuances in gay urban history, the particular traits that mark each city and the varying trajectories of development. The biographies of certain gay metropolises have yet to be written; the absence of comprehensive works on London and Berlin is remarkable, as is the relatively paucity of gay historical works on Canada, southern and eastern Europe and Third World countries. Studies of these areas would continue (to paraphrase Bech) to reveal the construction of homosexual groups and the appropriation of homosexual space *in* the city, the creation and identification of homosexuals as creatures *of* the city, and the economic, political and cultural impact of homosexuals *on* the city.

Notes

1. This essay concentrates on book-length works in English and on male homosexuality.
2. Isherwood's autobiographical *Mr Norris Changes Trains* and *Good-bye to Berlin*, published in 1935, achieved even greater renown when made into the play *I Am a Camera* in 1951, the musical *Cabaret* in 1966 and the film of the same name in 1972.
3. Not all of homosexual myth-making has been urban. The search for pastoral Arcady has signalled a contervailing direction for the gay imaginary since Theocritus, most famously revisited by Whitman. A severely transmuted version is the gay beachside resort, from the genteel circles of Capri and Taormina at the time of Baron von Gloeden to the drug- and dance-driven rages of Mykonos and Ibiza, or the North American versions of Provincetown and Russian River (Phillips *et al.*, 2000). Deserts and sheiks, ranches and cowboys, boarding schools and rosy-cheeked lads, ships and seamen, and industrial suburbs and working-class blokes have all figured among gay fantasies.
4. A recent example: the concept of 'metrosexual'. A 'metrosexual' is a heterosexual man who is nevertheless 'in touch' with his feminine side, or with activities and practices (such as cosmetic care) once considered feminine. Such 'metrosexuals' flirt with sexual ambivalence, court gay attention and become gay icons (see *Guardian Weekly*, 2003).

References

ALDRICH, R. (1993) *The Seduction of the Mediterranean: Writing, Art and Homosexual Fantasy.* London: Routledge.

ALDRICH, R. (2003) *Colonialism and Homosexuality.* London: Routledge.

ALTMAN, D. (1982) *The Homosexualization of America, the Americanization of the Homosexual.* New York: St Martin's Press.

ALTMAN, D. (2001) *Global Sex.* Chicago, IL: University of Chicago Press.

ARMSTRONG, E. A. (2002) *Forging Gay Identities: Organizing Sexualities in San Francisco, 1950–1994.* Chicago, IL: University of Chicago Press.

ARONSON, J. (1999) Homosex in Hanoi: sex, the public sphere, and public sex, in: W. LEAP (Ed.) *Public Space/Gay Sex.* New York: Columbia University Press.

BADGETT, M. (2001) *Money, Myths, and Change: The Economic Lives of Lesbians and Gay Men.* Chicago, IL: University of Chicago Press.

BAILEY, R. W. (1999) *Gay Politics, Urban Politics: Identity and Economics in the Urban Setting.* New York: Columbia University Press.

BARBEDETTE, G. and CARASSOU, M. (1981) *Paris Gay 1925.* Paris: Presses de la Renaissance.

BECH, H. (1997) *When Men Meet: Homosexuality and Modernity.* Chicago, IL: University of Chicago Press.

BEEMYN, B. (Ed.) (1997) *Creating a Place for Ourselves: Lesbian, Gay, and Bisexual Community Histories.* New York: Routledge.

BELL, D. and VALENTINE, G. (1995) *Mapping Desire: Geographies of Sexualities.* New York: Routledge.

BENSTOCK, S. (1989) Paris lesbianism and the politics of reaction, 1900–1940, in: M. DUBERMAN *ET AL.* (Eds) *Hidden from History*, pp. 332–346. New York: Penguin.

BETSKY, A. (1997) *Queer Space.* New York: William Morrow and Co.

BLEDSOE, L. J. (Ed.) (1998) *Gay Travels: A Literary Companion.* San Francisco, CA: Whereabouts Press.

BOONE, J. *ET AL.* (Eds) (2000) *Queer Frontiers: Millennial Geographies, Genders, and Generations.* Madison, WI: University of Wisconsin Press.

BOSWELL, J. (1980) *Christianity, Social Tolerance, and Homosexuality: Gay People in West-*

ern Europe from the Beginning of the Christian Era to the Fourteenth Century. Chicago, IL: University of Chicago Press.

BOSWELL, J. (1994) *Same-sex Unions in Premodern Europe*. Chicago, IL: University of Chicago Press.

BOYD, N. A. (2003) *Wide Open Town: A History of Queer San Francisco to 1965*. Berkeley, CA: University of California Press.

BRAY, A. (1982) *Homosexuality in Renaissance England*. London: Gay Men's Press.

CARBERY, G. (1995) *A History of the Sydney Gay and Lesbian Mardi Gras*. Melbourne: Australian Gay and Lesbian Archives.

CASTELLS, M. and MURPHY, K. (1982) Cultural identity and urban structure: the spatial organisation of San Francisco's gay community, in: N. FEINSTEIN and S. FEINSTEIN (Eds) *Urban Policy under Capitalism*. Beverly Hills, CA: Sage.

CHAMBERLAND, L. (1993) Remembering lesbian bars: Montreal, 1955–1975, in: R. MENDÈS-LEITE and P.-O. DE BUSSCHER (Eds) *Gay Studies from the French Culture*, pp. 231–270. Binghamton, NY: Harrington Park Press.

CHAUNCEY, G. (1994) *Gay New York: Gender, Urban Culture, and the Making of the Gay Male World, 1890–1994*. New York: Basic Books.

CHEVALIER, L. (1980) *Montmartre du plaisir et du crime*. Paris: Robert Laffont.

CHOQUETTE, L. (2001) Homosexuals in the city: representations of lesbian and gay space in nineteenth-century Paris, in: J. MERRICK and M. SIBALIS (Eds) *Homosexuality in French History and Culture*, pp. 149–168. New York: Harrington Park Press.

COAD, D. (2002) *Gender Troubles Down Under: Australian Masculinities*. Valenciennes: Presses Universitaires de Valenciennes.

DOVER, K. J. (1978) *Greek Homosexuality*. London: Duckworth.

DUBERMAN, M. (1993) *Stonewall*. New York: Dutton.

El Dorado: Homosexuelle Frauen und Männer in Berlin 1850–1950: Geschichte, Alltag und Kultur (1984) Berlin: Frölich und Kauffmann.

FARO, C., with WOTHERSPOON, G. (2000) *Street Seen: A History of Oxford Street*. Melbourne: Melbourne University Press.

FINCHER, R. and JACOBS, J. M. (Eds) (1998) *Cities of Difference*. New York: Guilford Press.

FREEMAN, S. K. (2000) From the lesbian nation to the Cincinnati lesbian community: moving toward a politics of location, *Journal of the History of Sexuality*, 9(1/2), pp 137–174.

GERASSI, J. G. (1996) *The Boys of Boise: Furor, Vice and Folly in an American City*. Seattle, WA: University of Washington Press.

GLUCKMAN, A. and REED, B. (Eds) (1997) *Homo Economics: Capitalism, Community and Lesbian and Gay Life*. New York: Routledge.

Goodbye to Berlin? 100 Jahre Schwulenbewegung (1997) Berlin: Verlag Rosa Winkel.

GREEN, J. M. (1999) *Male Homosexuality in Twentieth-century Brazil*. Chicago, IL: University of Chicago Press.

GREEN, M. (1991) *The Dream at the End of the World: Paul Bowles and the Literary Renegades in Tangier*. New York: Harper Collins.

Guardian Weekly (2003) Man of the moment is a "metrosexual", 7–13 August

HALPERIN, D. M. (1990) *One Hundred Years of Homosexuality and Other Essays on Greek Love*. New York: Routledge.

HENNESSY, R. (2000) *Profit and Pleasure: Sexual Identities in Late Capitalism*. New York: Routledge.

HIANG KHNG, R. H. (1998) Tiptoe out of the closet: the before and after of the increasingly visible gay community in Singapore, in: G. SULLIVAN and P. A. JACKSON (Eds) *Gay and Lesbian Asia: Culture, Identity, Community*, pp. 81–97. New York: Harrington Park Press.

HIGGINS, R. (1999) Baths, bushes, and belonging: public sex and gay community in pre-stonewall Montreal, in: W. L. LEAP (Ed.) *Public Sex/Gay Space*, pp. 187–201. New York: Columbia University Press.

HIGGS, D. (Ed.) (1999) *Queer Sites: Gay Urban Histories since 1600*. London: Routledge.

THE HISTORY PROJECT (1998) *Improper Bostonians: Lesbian and Gay History from the Puritans to Playland*. Boston, MA: Beacon Press.

HOWARD, J. (1995) The library, the park, and the pervert: public space and homosexual encounter in post-World War II Atlanta, *Radical History Review*, 62, pp. 166–187.

HUMPHREYS, L. (1970) *Tearoom Trade*. London: Duckworth.

INGRAM, G. B., BOUTHILLETTE, A.-M. and RETTER, Y. (1997) *Queers in Space: Communities/Public Places/Sites of Resistance*. Seattle, WA: Bay Press.

ISHERWOOD, C. (1935a) *Good-bye to Berlin*. London: Hogarth Press.

ISHERWOOD, C. (1935b) *Mr Norris Changes Trains*. London: Hogarth Press.

ISHERWOOD, C. (1978) *Christopher and His Kind*. London: Hogarth Press.

JACKSON, P. A. (1995) *Dear Uncle Go: Male Homosexuality in Thailand*. Bangkok: Bua Long Books.

JACKSON, P. A. (1997a) *Kathoey*—gay—man: the historical emergence of gay male identity in Thailand, in: M. JOLLY and L. MANDERSON (Eds) *Sites of Desire/Economies of Pleasure: Sexualities in Asia and the Pacific*, pp. 166–190. Chicago, IL: University of Chicago Press.

JACKSON, P. A. (1997b) Tolerant but unaccepting: correcting misperceptions of a Thai 'gay paradise', in: P. A. JACKSON and N. COOK (Eds) *Genders and Sexualities in Modern Thailand*, pp. 226–242. Chiang Mai: Silkworm Books.

JACKSON, P. A. and SULLIVAN, G. (Eds) (1996) *Multicultural Queer: Australian Narratives*. New York: Haworth Press.

JACKSON, P. A. and SULLIVAN, G. (Eds) (1999) *Lady Boys, Tom Boys, Rent Boys: Male and Female Homoseuxalities in Contemporary Thailand*. New York: Harrington Park Press.

JAY, K. (1988) *The Amazon and the Page: Natalie Clifford Barney and Renée Vivien*. Indianapolis, IN: Indiana University Press.

JOHNSON, D. K. (1994–95) 'Homosexual citizens': Washington's gay community confronts the civil service, *Washington History*, 6(2), pp. 44–63, 93–96.

JOHNSTON, C. and REYK, P. VAN (Eds) (2001) *Queer City: Gay and Lesbian Politics in Sydney*. Annandale, NSW: Pluto Press.

KAISER, C. (1997) *The Gay Metropolis, 1940–1996*. Boston, MA: Houghton Mifflin.

KEELEY, E. (1976) *Cavafy's Alexandria: Study of a Myth in Progress*. Cambridge, MA: Harvard University Press.

KENNEDY, E. L. and DAVIS, M. D. (1993) *Boots of Leather, Slippers of Gold: The History of a Lesbian Community*. New York: Penguin.

KENNEY, M. R. (2001) *Mapping Gay L.A.: The Intersection of Place and Politics*. Philadelphia, PA: Temple University Press.

KNOPP, L. (1995) Sexuality and urban space: a framework for analysis, in: D. BELL and G. VALENTINE (Eds) *Mapping Desire: Geographies of Sexualities*, pp. 149–161. New York: Routledge.

KNOPP, L. (1997) Gentrification and gay neighborhood formation in New Orleans: a case study, in: A. GLUCKMAN and B. REED (Eds) *Homo Economics: Capitalism, Community and Lesbian and Gay Life*, pp. 45–63. New York: Routledge.

KNOPP, L. (1998) Sexuality and urban space: gay male identity politics in the United States, the United Kingdom, and Australia, in: R. FINCHER and J. M. JACOBS (Eds) *Cities of Difference*. New York: Guilford Press.

LEONG, R. (Ed.) (1996) *Asian American Sexualities: Dimension of the Gay and Lesbian Experience*. New York: Routledge.

LEVINE, M. P. (1979) *Gay Men: The Sociology of Male Homosexuality*. New York: Harper and Row.

LEWIS, J. and LOOTS, F. (1994) Moffies and manvroue: gay and lesbian life histories in contemporary Cape Town, in: M. GEVISSER and E. CAMERON (Eds) *Defiant Desire: Gay and Lesbian Lives in South Africa*, pp. 140–157. New York: Routledge.

LEYLAND, W. (2002) *Out in the Castro: Desire, Promise, Activism*. San Francisco, CA: Leyland Publications.

LIDDELL, R. (1974) *Cavafy: A Critical Biography*. London: Duckworth.

LUMSDEN, I. (1996) *Machos, Maricones and Gays: Cuba and Homosexuality*. Philadelphia, PA: Temple University Press.

MANZIONE, C. K. (1996) Sex in Tudor London: abusing their bodies with each other, in: J. MURRAY and K. EISENBLICHLER (Eds) *Desire and Discipline: Sex and Sexuality in the Premodern West*, pp. 87–100. Toronto: University of Toronto Press.

MAYARD, S. (1994) Through a hole in the lavatory wall: homosexual subcultures, police surveillance, and the dialectics of discovery, Toronto, 1890–1930, *Journal of the History of Sexuality*, 5(2), pp. 207–242.

MERRICK, J. (1996) Commisioner Foucault, Inspector Noël and the 'pederasts' of Paris, 1780–1783, *Journal of Social History*, 32, pp. 287–307.

MERRICK, J. (1997) Sodomitical inclinations in early eighteenth-century Paris, *Eighteenth-century Studies*, 30, pp. 289–295.

MERRICK, J. (1999) Sodomitical scandals and subcultures in the 1720s, *Men and Masculinities*, 1, pp. 365–384.

MERRICK, J. and RAGAN, B. T. JR (Eds) (1996) *Homosexuality in Modern France*. Oxford: Oxford University Press.

MERRICK, J. and RAGAN, B. T. JR (Eds) (2001) *Homosexuality in Early Modern France: A Documentary Collection*. Oxford: Oxford University Press.

MERRICK, J. and SIBALIS, M. (Eds) (2001) *Homosexuality in French History and Culture*. New York: Harrington Park Press.

MILLER, N. (1992) *Out in the World: Gay and Lesbian Life from Buenos Aires to Bangkok*. London: Vintage Books.

MOORE, C. (2001) *Sunshine and Rainbows: The Development of Gay and Lesbian Culture in Queensland*. St Lucia, Qld: University of Queensland Press.

MOORE, C. and SAUNDERS, K. (Eds) (1998) *Australian Masculinities: Men and Their Histories*. Special issue of the *Journal of Australian Studies*, 56.

MOSS, R. T. (1997) *Cleopatra's Wedding Present: Travels through Syria*. London: Duckworth.

MOTT, L. (2003) Crypto-sodomites in colonial Brazil, in P. SIGAL (Ed.) *Infamous Desire: Male Homosexuality in Colonial Latin America*, pp. 168–196. Chicago, IL: University of Chicago Press.

MURRAY, J. and EISENBLICHLER, K. (Eds) (1996) *Desire and Discipline: Sex and Sexuality in the Premodern West*. Toronto: University of Toronto Press.

NEWTON, E. (1993) *Cherry Grove, Fire Island: Sixty Years in America's First Gay and Lesbian Town*. Boston, MA: Beacon Press.

NILSSON, A. (1998) Creating their own private and public; the male homosexual life space in a Nordic city during high modernity, in: J. LÖFSTRÖM (Ed.) *Scandinavian Homosexualities: Essays in Gay and Lesbian Studies*, pp. 81–116. New York: Haworth.

NORTON, R. (1992) *Mother Clap's Molly House: The Gay Subculture in England, 1700–1830*. London: Gay Men's Press.

NORTON, R. (1997) *The Myth of the Modern Homosexual: Queer History and the Search for Cultural Unity*. London: Cassell.

PENISTON, W. A. (1996) Love and death in gay Paris: homosexuality and criminality in the 1870s, in: J. MERRICK and B. T. RAGAN JR (Eds) *Homosexuality in Modern France*, pp. 128–145. Oxford: Oxford University Press.

PENISTON, W. A. (1999) A public offense against decency: the trial of the Count of Germiny and the 'moral order' of the Third Republic', in: G. ROBB and N. ERBER (Eds) *Disorder in the Court: Trials and Sexual Conflict at the Turn of the Century*, pp. 12–32. New York: New York University Press.

PHILLIPS, D. L. and WILLETT, G. (2000) *Australia's Homosexual Histories*. Sydney: Australian Centre for Lesbian and Gay Research.

PHILLIPS, R., WATT, D. and SHUTTLETON, D. (Eds) (2000) *Decentring Sexualities: Politics and Representations beyond the Metropole*. New York: Routledge.

PRIEUR, A. (1998) *Mama's House, Mexico City: On Transvestites, Queens and Machos*. Chicago, IL: University of Chicago Press.

QUIMBY, K. and WILLIAMS, W. L. (2000) Unmasking the homophile in 1950s Los Angeles: an archival record, in: J. BOONE *ET AL.* (Eds) *Queer Frontiers: Millennial Geographies, Genders, and Generations*, pp. 166–195. Madison, WI: University of Wisconsin Press.

RETTER, Y. (2000) Lesbian activism in Los Angeles, 1970–1979, in: J. BOONE *ET AL.* (Eds) *Queer Frontiers: Millennial Geographies, Genders, and Generations*. Madison, pp. 196–221. WI: University of Wisconsin Press.

REY, M. (1985) Parisian homosexuals create a lifestyle, 1700–1750: the police archives, in: R. MACCUBIN (Ed.) *'Tis Nature's Fault: Unauthorized Sexuality during the Enlightenment*. Cambridge: Cambridge University Press.

REY, M. (1987) Justice, police, et sodomie à Paris au XVIIIe siècle, in: J. POUMARÈDE and J.-P. ROYER (Eds) *Droit, histoire et sexualité*, pp. 175–184. Lille: Université de Lille II.

REY, M. (1989) Police and sodomy in eighteenth-century Paris: from sin to disorder, in: K. GERARD and G. HEKMA (Eds) *The Pursuit of Sodomy: Male Homosexuality in Renaissance and Enlightenment Europe*, pp. 129–146. New York: Harrington Park Press.

REYNOLDS, R. (2002) *From Camp to Queer: Remaking the Australian Homosexual*. Melbourne: Melbourne University Press.

ROCKE, M. (1996) *Forbidden Friendships: Homosexuality and Male Culture in Renaissance Florence*. Oxford: Oxford University Press.

RUGGIERO, G. (1985) *The Boundaries of Eros: Sex Crime and Sexuality in Renaissance Venice*. Oxford: Oxford University Press.

SAUTMAN, F. C. (2001) Invisible women: lesbian working-class culture in France, 1880–1930, in: J. MERRICK and B. T. RAGAN JR (Eds) *Homosexuality in Early Modern France: A Documentary Collection*, pp. 177–201. Oxford: Oxford University Press.

SEABROOK, J. (1999) *Love in a Different Climate: Men Who Have Sex with Men in India*. London: Verso.

SERGENT, B. (1984) *L'Homosexualité dans la mythologie grecque*. Paris: Payot.

SHEPHARD, R. (1996) Sexual rumours in English politics: the cases of Elizabeth I and James I, in: J. MURRAY and K. EISENBLICHLER (Eds) *Desire and Discipline: Sex and Sexuality in the Premodern West*, pp. 101–122. Toronto: University of Toronto Press.

SIBALIS, M. (1999) Paris-Babylone/Paris-Sodome: images of homosexuality in the nineteenth-century city, in: J. WEST-SOOBY (Ed.) *Images of the City in Nineteenth-century France*, pp. 13–22. Mooroka, Qld: Boombana Publications.

SIBALIS, M. (2001a) Les Espaces des homosexuels dans le Paris d'avant Hausmann, in: K. BOWIE (Ed.) *La Modernité avant Haussmann: formes de l'espace urbain à Paris, 1801–1853*, pp. 231–241. Paris: Editions Recherches.

SIBALIS, M. (2001b) The Palais-Royal and the homosexual subculture of nineteenth-century Paris, in: J. MERRICK and M. SIBALIS (Eds) (2001) *Homosexuality in French History and Culture*, pp. 117–130. New York: Harrington Park Press.

SIDÉRIS, G. (2001) *Folles*, effeminates and homophiles in Saint-German-des-Prés of the 1950s: a new 'precious' society, in: J. MERRICK and M. SIBALIS (Eds) (2001) *Homosexuality in French History and Culture*, pp. 219–232. New York: Harrington Park Press.

SMITH, B. P. (1991) *Homosexual Desire in Shakespeare's England: A Cultural Poetics*. Chicago, IL: University of Chicago Press.

STANSKY, P. (1996) *On or about December 1910: Early Bloomsbury and its Intimate World*. Cambridge, MA: Harvard University Press.

STEIN, M. (2000) *City of Sisterly and Brotherly Loves: Lesbian and Gay Philadelphia, 1945–1972*. Chicago, IL: University of Chicago Press.

STRYKER, S. and KUSKIRK, J. VAN (1996) *Gay by the Bay: A History of Queer Culture in the San Francisco Bay Area*. San Francisco, CA: Chronicle Books.

TAMAGNE, F. (2000) *Histoire de l'homosexualité en Europe: Berlin, Londres, Paris, 1919–1939*. Paris: Editions du Seuil.

TATTELBAUM, I. (2000) Presenting a queer (bath)house, in: J. BOONE *ET AL.* (Eds) *Queer Frontiers: Millennial Geographies, Genders, and Generations*, pp. 220–260. Madison, WI: University of Wisconsin Press.

THORPE, R. (1996) 'A house where queers go': African-American lesbian nightlife in Detroit, 1940–1965, in: E. LEVINE (Ed.) *Inventing Lesbian Cultures*, pp. 40–61. Boston, MA: Beacon Press.

WAFER, J., SOUTHGATE, E. and COAN, L. (2000) *Out in the Valley: Hunter Gay and Lesbian Histories*. Newcastle, NSW: Newcastle Region Library.

WARD, J. (2003) Producing 'pride' in west Hollywood: a queer cultural capital for queens with cultural capital, *Sexualities*, 6(1), pp. 65–94.

WESTON, K. (1995) Get thee to a big city: sexual imaginary and the great gay migration, *GLQ*, 3, pp. 253–277.

WHITE, E. (2001) *The Flâneur: A Stroll through the Paradoxes of Paris*. London: Bloomsbury.

WHITTLE, S. (Ed.) (1994) *The Margins of the City: Gay Men's Urban Lives*. Aldershot: Arena.

WILLETT, G. (2000) *Living Out Loud: A History of Gay and Lesbian Activism in Australia*. Sydney: Allen and Unwin.

WOTHERSPOON, G. (1991) *City of the Plain: History of Gay Sub-culture*. Sydney: Hale and Iremonger.

Urban Space and Homosexuality: The Example of the Marais, Paris' 'Gay Ghetto'

Michael Sibalis

Introduction

During recent decades, 'gay ghettos' have emerged in many large cities in North America and western Europe. The word 'ghetto' originated in 16th-century Venice and initially referred to an area of a city where local authorities forced Jews to reside. American sociologists of the Chicago School appropriated the word in the 1920s to designate urban districts inhabited predominantly by racial, ethnic or social minorities, whether by compulsion or by choice. By the 1970s, sociologists were applying the term 'gay ghetto' to neighbourhoods characterised by the presence of

> gay institutions [like bars, bookstores, restaurants and clothing stores] in number, a conspicuous and locally dominant gay sub-culture that is socially isolated from the larger community, and a residential population that is substantially gay (Levine, 1979, p. 364),

such as West Hollywood in Los Angeles and the West Village (part of Greenwich Village) in New York.

The classic (and most studied) gay ghetto is, of course, San Francisco's. A homosexual community first emerged in that city in the 1920s and 1930s, most notably within the bohemian atmosphere of North Beach. In the early 1960s gay men began moving into Eureka Valley, a working-class Irish-Catholic neighbourhood whose inhabitants were deserting the city's centre for the suburbs. The gay men renovated the dilapidated (and

Michael Sibalis is in the Department of History, Wilfrid Laurier University, Waterloo, Ontario, Canada, N2L 3C5. Fax: 1 519 746 3655. E-mail: msibalis@wlu.ca. *The author gratefully acknowledges the financial support for this research received from a grant partly funded by Wilfrid Laurier University operating funds and partly by an Institutional Grant awarded to the university by the Social Sciences and Humanities Research Council of Canada. Pam Schaus of the Department of Geography and Environmental Studies drew the maps.*

therefore still affordable) Victorian houses, opened new businesses, including gay bars, and created a visibly gay neighbourhood (soon dubbed 'the Castro' after Castro Street) that urban sociologist Manuel Castells has described as "not only a residential space but also a space for social interaction, for business activities of all kinds, for leisure and pleasure, for feasts and politics" (Castells and Murphy, 1982, p. 246; Castells, 1983, pp. 138–170; Duggins, 2002; Stryker, 2002). The pattern apparent in San Francisco has been replicated (with variations) elsewhere, leading to the formation of gay ghettos in many other large North American cities, such as Bay Village in Boston (Pattison, 1983), the Marigny neighbourhood in New Orleans (Knopp, 1990) and Cabbagetown in Toronto (Bouthillette, 1994). By constructing their own urban enclaves, gay men have come to "figure prominently" in the "urban renaissance"—which is to say in the redevelopment and gentrification of the inner city (Lauria and Knopp, 1985). An attractive and centrally located but rundown neighbourhood ripe for gentrification draws in gays who are not only responding to economic incentives (low rents and real-estate prices), but also seeking to create a territory which they can inhabit and control and where they can feel at home within a self-contained community set apart from a world perceived as indifferent or even hostile. Their presence encourages the opening of bars and other businesses that cater to a gay clientele. A gay ghetto provides them with a territorial base for the development of a gay movement, which can then become a force in municipal politics.

Great Britain, too, has its gay ghettos, although residential enclaves have been slower to appear there than in North America. In London, the 'gay village' of Soho and especially Old Compton Street ("the gayest 100 yards in Britain") is a commercial and not a residential neighbourhood (Binnie, 1995, pp. 194–198). Newcastle's gay scene is "predominantly non-residential inner-city apart from a large block of housing association flats on Waterloo Street, many residents of which are gay" (Lewis, 1994, p. 90). Manchester's 'gay village', centred on Bloom Street and Canal Street and reputedly the largest in Britain outside London, consists of bars, clubs, businesses and a community centre that serve the city's gay population, but once again it is primarily a social scene rather than a residential district, although single men have been moving into the city centre since the 1990s (Hindle, 1994, pp. 17–22; Quilley, 1997).

France has only one gay ghetto, in the historical Marais quarter of central Paris. This paper is a case study of the Marais. It seeks not only to throw light on the similar processes by which gay ghettos everywhere tend to emerge and grow, but also to examine those features that are unique to the French experience. The gay Marais shares certain characteristics of both the British gay village, which is primarily commercial, and the North American gay ghetto, which is commercial and residential. Like its British and American counterparts, the Marais came into being largely as a product of impersonal economic forces (the real-estate market) and contemporary social change (the emergence of a significant urban gay population with its own distinctive sub-culture). There are also notable differences, however. Paris' gay ghetto resulted to a large extent from politically motivated decisions made by a few businessmen who intentionally set out in the late 1970s to promote a more open gay lifestyle in France. Another key difference is the hostile reaction provoked by the ghetto in France. While the development of urban gay enclaves has everywhere brought some degree of social and political tensions in their wake, only in France, where the dominant political discourse rejects multiculturalism and minority rights in favour of 'universal' values presumably shared by all citizens, has the existence of the gay ghetto been perceived as a threat to the very foundations of national solidarity and become an issue of broad ideological significance.

This paper begins with a description of the distinctive character of the historical Marais quarter and how it has been shaped by urban development and economic change over many centuries. Gentrification and the prolif-

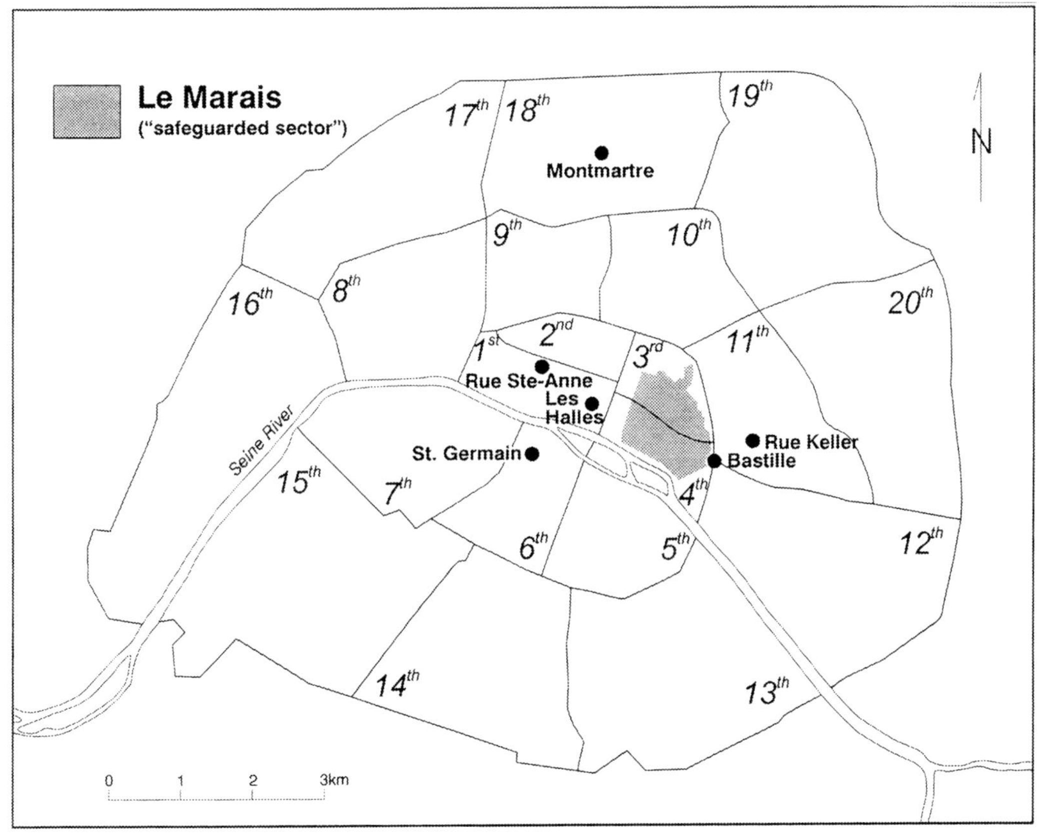

Figure 1. The location of the Marais within Paris.

eration of gay venues constitute only the most recent phase in this neighbourhood's very long history. After examining the relationship between homosexual men and Parisian space during the 20th century, the paper looks at the social and economic factors that fostered the development of the Marais as the site of a gay ghetto in the 1980s and 1990s, and most notably the motives and role of certain gay businessmen who financed the transformation. The paper then turns to the relationship in France between the gay ghetto (territoriality) and the emergent gay community's new sense of identity. It ends with a detailed account of the disputes that have raged around the very existence of the Marais and the 'ghettoisation' of homosexual life that it purportedly represents, issues that have made the Marais the target of virulent criticism from both outside and inside the gay community.

The Marais

The Marais, situated in central Paris on the Left Bank of the Seine (see Figure 1), is the oldest quarter of the city to have survived the centuries relatively intact (Chatelain, 1967). 'Marais' means 'marshland' and most of the Marais was indeed swamp until drained in the 8th century, but in medieval times the word also referred to land used for growing vegetables; the one-time prevalence of market gardening in the area most likely explains its name. The Marais straddles 2 of Paris' 20 *arrondissements* (administrative districts), encompassing most of the 3rd

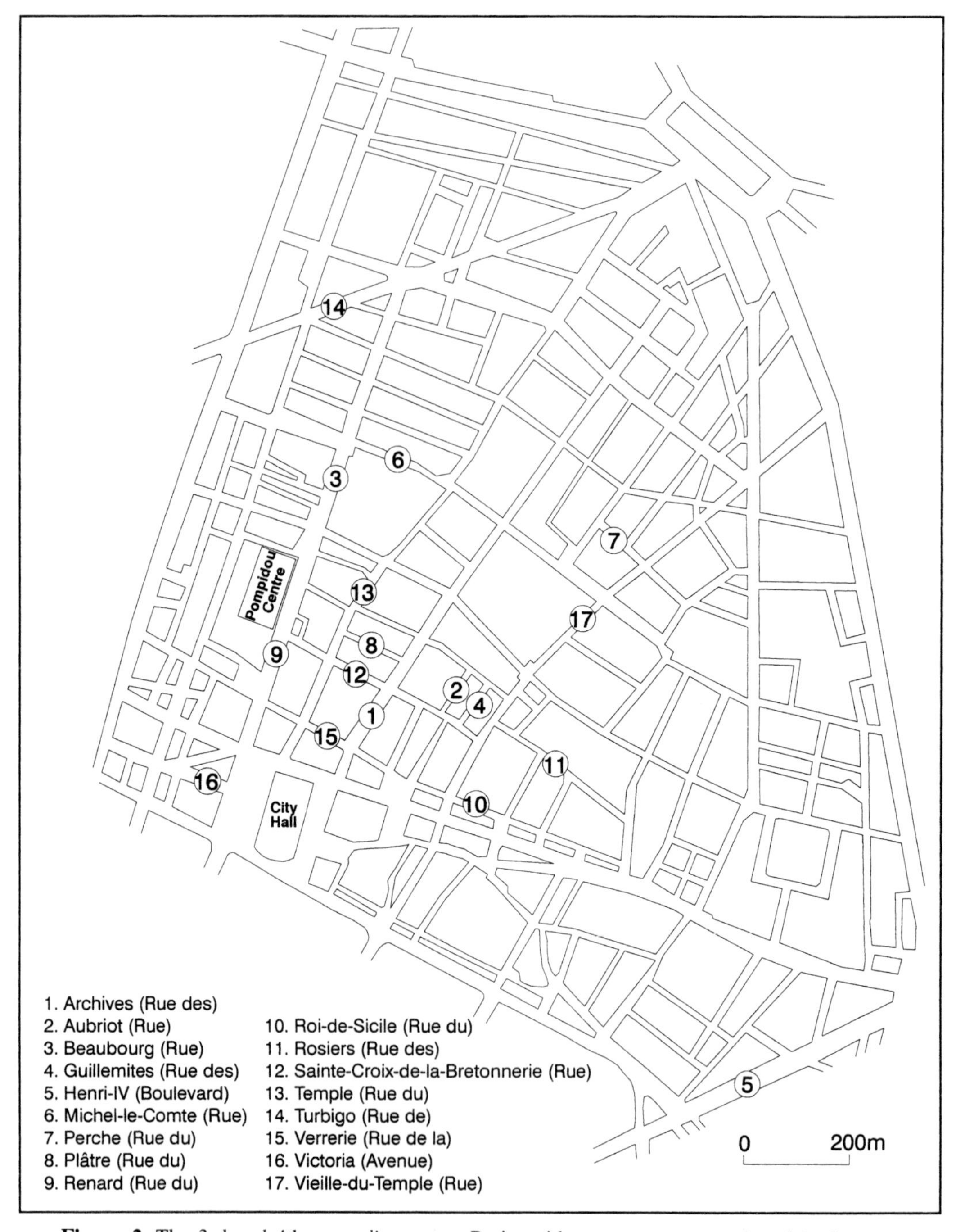

Figure 2. The 3rd and 4th *arrondissements*, Paris, with street names mentioned in the text.

arrondissement (everything but those parts west of the Rue Beaubourg or north of the Rue de Turbigo) and about half of the 4th *arrondissement* (excluding what lies west of the Rue du Renard, south of the Seine's Right Bank, or east of the Boulevard Henri-IV) (see Figure 2). King Philippe-Augustus' fortified wall (built 1190–1215) took in only the southern part of this area and religious orders built convents and monasteries in the fields beyond. Further urban development followed upon the construction of a new wall by Charles V in the mid 14th century, which put the entire Marais within city limits. In the first decade of the 17th century, Henry IV decided to reshape the Marais as a luxurious residential quarter. At its apogee in the mid 17th century, the Marais boasted numerous palaces and town houses inhabited by wealthy aristocrats, high state officials and financiers (Babelon, 1997; Faure, 1997, pp. 7–51; Gady, 2002, pp. 9–21). The result was a

> relatively homogeneous townscape at least as far as age and style are concerned. ... The *hôtels particuliers* [mansions] of the aristocracy were set among the lesser buildings of their socially-inferior dependents—the whole ensemble an appropriately splendid setting for seventeenth-century life (Kain, 1981, p. 209).

The social élite began abandoning the Marais after Louis XIV moved the royal court to Versailles in the 1680s. The process continued apace in the 18th and 19th centuries, when aristocrats preferred to live in the new western quarters of Paris: the Faubourg Saint-Germain on the Left Bank and the Faubourg Saint-Honoré on the Right (Le Moël, 1997). Their departure "left room for a new social occupation of the space" by shopkeepers, craftsmen and wage-earners (Prigent, 1980, p. 19). By the late 19th century, the installation of small industry and commerce in the quarter and the sub-division of its mansions into apartments had turned most of the formerly aristocratic Marais into an overcrowded and rundown slum. In 1965, the Marais was still home to 7000 businesses (especially manufacturers and wholesalers in jewellery, optics, leather goods and ready-made clothes) employing 40 000 people (Kain, 1981, p. 213).[1] In 1975, only 17.3 per cent of all Parisian housing dated from before 1871, but the figure was 65.1 per cent in the Marais; 1 in 5 Parisian apartments had been constructed since 1948, but only 1 in 20 in the Marais (Prigent, 1980, p. 32).

> It was perhaps inevitable that the Marais, with its architectural beauty, its calm ambience, and its relatively central location, would one day revert to its original status as a quarter of fashion and wealth (Evenson, 1979, p. 320).

This resulted from the Malraux Law of 4 August 1962 (Stungo, 1972). This was "one of the most important and influential pieces of European conservation legislation", which "laid down a 'grand design' for a renaissance of the historical quarters of French towns" (Kain, 1981, p. 200). The goal was no longer to preserve only individual buildings and monuments, but rather an entire urban site, to maintain a given neighbourhood's traditional character while modernising living conditions within it. In 1964/65, the City of Paris, with the support of the national government, designated 126 hectares of the Marais a "safeguarded sector" for preservation and renovation (Kain, 1981, p. 201).

Gentrification thus began in the 1960s and took off rapidly in the late 1970s and early 1980s. In fact, the Marais had the highest gentrification rate of any neighbourhood in the capital in the period 1975–82 (Winchester and White, 1988, p. 47). The population decline (already evident in the 1950s) accelerated and, as the working class left, the middle class and white-collar workers moved in. The Marais lost about 40 per cent of its inhabitants between the 1960s and the end of the century, as indicated by the population figures for all of the 3rd and 4th *arrondissements*: 1968: 110 281; 1975: 82 172; 1990: 68 903; 1999: 65 979 (Le Clère, 1985, p. 649; INSEE, 2000, p. 75/3). The safeguarded sector accounts for about half this population.[2]

The national and municipal governments promoted the transformation of the Marais by renovating the many public buildings in the sector and by providing owners with grants to improve their properties. Investment by real-estate developers, commercial companies and individual citizens also played an important role (Kain, 1981, p. 214; Carpenter and Lees, 1995). What was once an "uncelebrated area of extreme overcrowding and urban poverty" thus became "a gentrified landscape of consumption" in which, moreover, "consumerism ... is associated only with the 'best' or most fashionable" (Noin and White, 1997, pp. 212–213), a change that dismayed some people nostalgic for the colourful past of "Algerian workers in the small hotels of the Rue du Roi-de-Sicile, or Yiddish- or Polish-speaking Jews in and around the Rue des Rosiers" and "the small workshops that cluttered the courtyards of seventeenth-century town houses and palaces" (Cobb, 1985, p. 193). What no one anticipated was that the "aesthetic oasis reserved for the bourgeoisie" created by urban renewal (Prigent, 1980, p. 96) would also draw a flood of gay men and even some lesbians into the quarter. But in fact Parisian gays, like gays elsewhere in the world, had their part in the gentrification process, as one businessman recently recalled with some exaggeration:

> I've seen how in twenty years real estate prices have been multiplied by ten. ... I knew the Marais when everything was neglected and there were not even mailboxes in the buildings. If the quarter has changed, it's undoubtedly because there has been a municipal effort, but also and above all the investment of gays. The Parisian example resembles other capital cities: gays have always taken over the most decayed, the oldest and at the same time the prettiest quarters (Garcia, 2002, p. 14).

Gay Men and Urban Space in Paris

Gay men have a special relationship to urban space. Only in cities are there enough homosexually inclined men to permit the emergence of a self-aware community with its own commercial venues, social and political organisations and distinctive sub-culture (Harry and DeVall, 1978, pp. 134–154). In the words of the Danish sociologist Henning Bech, "being homosexual ... is ... a way of *being*, a *form of existence*". Homosexuals belong to one of a number of social worlds (Bech does not identify the others) that are all

> essentially urban: they are largely worlds of strangers and not just of personal acquaintances; they depend in part upon the non-personal, urban free flow of signs and information, as well as upon the pool of strangers, for recruitment and reproduction; they occupy time–space slices of the city and need urban stages to be enacted on (Bech, 1997, pp. 153–156).

We know a great deal about the urban spaces used by Parisian homosexuals (generally called 'sodomites' or 'pederasts' before 1900) since the early 1700s, both outdoor ones (parks, gardens, riverbanks, quays and streets) and indoor ones (taverns, bars, clubs and restaurants). In the 18th and 19th centuries, these were spread across the city, but were usually situated on its margins, either literally (on its physical periphery) or figuratively (in poorer and seamier districts) (Sibalis, 2001). Beginning in the 1880s, however, commercial venues catering to homosexuals clustered in the Montmartre quarter of northern Paris, known for bohemianism and illicit sexuality, including female prostitution. In the 1920s and 1930s, other districts, like the Rue de Lappe near the Bastille or Montparnasse in the south, also became important to Paris' homosexual sub-culture. After the Second World War, homosexuals frequented the bars, clubs and cafés of the Left Bank district of Saint-Germain-des-Prés, the centre of post-war intellectual life and non-conformity. In the 1970s, homosexual nightlife migrated across the Seine to the streets between the Palais-Royal and the Opera House and, most famously, to the Rue Sainte-Anne. In marked contrast to Montmartre and Saint-Germain, this was a quiet residential and business neighbourhood, almost deserted after the workday ended; the possibility of going out in relative secrecy is probably what attracted gay customers to its venues (Sibalis, 1999, pp. 26–31).

The popularity of the Rue Sainte-Anne

lasted hardly more than a decade. In June 1983, a gay journalist observed that

> the homosexual geography of the capital has changed dramatically. Saint-Germain and the Rue Sainte-Anne are out. Les Halles and especially the Marais are in (Jallier, 1983, p. 35).

Several factors explain the shift. First of all, there was the accessibility of the Marais, which is centrally located and easily reached by public transport. A few hundred metres to the west lies Les Halles, former site of Paris' wholesale food market, which was transferred to the suburbs in 1969. In the 1970s, Les Halles underwent major commercial redevelopment, which included construction of an underground station (opened in December 1977) to link the subway system and the RER (Réseau Express Régional), a network of suburban trains that served 60 per cent of the population of the Paris region (Michel, 1988). The nearby Avenue Victoria, running between City Hall and Châtelet, is also the main terminus for the city's night buses, which operate from 1.30a.m. to 5.30a.m.

Secondly, the renovated Marais had an undoubted aesthetic appeal. In the overblown rhetoric and rather stilted English of a recent bilingual guidebook:

> No other area of Paris has such a strong personality in spite of its [architectural] diversity. The same beauty of its dwellings can be seen in every street, the same refinement of the stones, the same warmth of the thoroughfares and everywhere the same poetic poetry [*sic*]. The Marais ... has a spirit, a soul, an immaterial existence beyond the mirror of life (Auffray, 2001, p. 8).

The attractiveness increased in the 1970s and 1980s, when the Marais was turned into an important cultural and artistic quarter. The Pompidou Centre (a new national museum of contemporary art) opened on its western edge in 1977 and the opening or refurbishing of other museums and the proliferation of commercial art galleries soon followed.

But there is a third factor that explains how and why the Marais became the centre of Parisian gay life. Gay businessmen recognised that the Marais, with its low rents and real-estate prices, was ripe for investment. In this respect, the gay Marais, like gay villages and ghettos in Britain and North America, developed spontaneously in response to favourable market conditions. But gay investors in Paris were concerned with more than the balance sheet. They consciously set out to create a new gay quarter as much because of their personal convictions as from their desire to benefit financially from an evident commercial opportunity.

Businessmen and the 'Gay Marais'

Joël Leroux launched the first gay bar in the Marais in December 1978. An accountant bored with his job, Leroux decided "to change [his] skin" (as he put it) and bought "for a song" a small café on the Rue du Plâtre, which he renamed Le Village after New York City's Greenwich Village and reopened as a gay bar. Le Village was something quite new to Paris. Whereas most gay venues did business only in the late evening and at night, its hours were noon to 2a.m. Le Village also opened directly onto the street, just like any other café in the city, and it charged regular prices for coffee and beer. Gay bars and clubs more usually protected themselves with locked doors guarded by doormen; customers rang for admittance, then paid a cover charge and exorbitant prices for the privilege of entering and consuming. "Starting from the principle that we [gays] had nothing to hide", Leroux has explained, "I wanted people inside to be able to see what was happening outside and vice versa" (*Le Parisien*, 2001). His bar was an immediate success and doubled its turnover within a year: "There was the clientele of the clubs of Saint-Germain mixed in with another clientele that went out less often and with heterosexual clients who stayed or returned" (*Le Parisien*, 2001). In 1980, Leroux sold out (Le Village still survives under another name) and opened a larger gay bar, Le Duplex on the nearby Rue Michel-le-Comte, which he still owns today (Jallier, 1983, p. 36).

Maurice McGrath, a former sailor in the Royal Navy and owner of a Parisian travel agency, soon noticed that "the bar 'Le Village' was starting to do very well and the Marais was promising to become a French 'Greenwich village' " (Le Douce, 1983, p. 40). Eager to embark on a new business venture, he has explained, "I discovered in the Marais, a great many establishments that had been for sale for a long time. These cafés were no longer frequented, because poorly situated, and the quarter's population was changing" (Roland-Henry, 1983). McGrath and eight associates opened a bar on the Rue du Perche in November 1979, but in September 1980 he branched out on his own with the Bar Central, at the intersection of the Rue Vieille-du-Temple and the Rue Sainte-Croix-de-la-Bretonnerie (Le Douce, 1983). Like Leroux, McGrath believed that "It was necessary to change the gay scene in France. ... The idea of a daytime bar had been launched with Le Village and I took the plunge. ... One of the goals I set myself in opening Le Central was to make homosexual life part of everyday life" (Roland-Henry, 1983). "My ambition then was to make homosexuality commonplace, to make it visible in broad day" (Chayet, 1996).

For men like Leroux and McGrath, opening a gay venue in the Marais was evidently both a business decision and a political statement. Their bars embodied a new kind of gay culture patterned on the contemporary American scene: militant and self-assertive; the days of clandestinity and internalised shame were definitively over. But militancy did not preclude shrewd business sense and an eye for financial opportunity. As one journalist has put it:

> In creating establishments run by and for themselves, gays ... have grouped together in the same sector interdependent activities, for practical reasons, not without self-interested motives on the part of the businesses: to bring together in the same place offer and demand (Madesclaire, 1995, p. 48).

Bernard Bousset is today the most successful of this breed of gay entrepreneurs who built up the gay Marais venue by venue. He began his business career in St Tropez in the 1960s and eventually acquired a gay bathhouse in Paris, the IDM, in the 9th *arrondissement*. In April 1987, he opened Le Quetzal, a gay bar on the Rue de la Verrerie in the Marais, and he soon acquired other businesses in the quarter. In 1990, he founded the SNEG (Syndicat National des Entreprises Gaies, or National Syndicate of Gay Enterprises), a lobby group for gay businessmen that he would lead during the first decade of its existence (Neuville, 1995).

Gradually more and more bars, cafés and restaurants catering to a predominantly gay clientele appeared in the Marais (Martel, 1999, pp. 171–173), while other gay-owned or gay-friendly businesses sprang up to sell books (Les Mots à la Bouche, the city's gay bookstore since 1980, moved to the Marais in 1983), clothing, furniture, art, antiques, home decorations and so on. There has even been a gay pharmacy on the Rue du Temple since the mid 1990s. Its owner has explained, in words that any gay businessman could echo, that

> la Pharmacie du Village uses its geographical position, at the heart of the Marais, to target a gay clientele by winning their confidence and establishing a reassuring complicity with its clients (Laforgerie, 1997, p. 26).

The presence of such establishments inevitably had repercussions for other venues in the quarter as "homosexual visibility indeed spreads into public space but also into nearby businesses" (Bordet, 2001, p. 136). A good example of this spreading out occurred on the Rue des Archives. In 1995, the owners of a traditional café at 17 Rue des Archives, on the corner of the Rue Sainte-Croix-de-la-Bretonnerie, worried by a decline in their regular business, set out to attract the new gay clientele present in the neighbourhood by changing its decor and renaming it the Open Bar, "to show that we are open to everyone: homos, lesbians, heteros, without distinction" (Ulrich, 1996). Bernard Bousset soon bought them out, renovated the place and renamed it the Open Café: "In summer it overflows and on some evenings the Rue des Archives seems to have become the terrace of the Open Café" (Garcia, 2002, p. 11). The year 1995 also saw the opening of the (gay) Café Cox next door at 15 Rue des Archives. Largely because of the proximity of these two gay cafés, four non-gay venues across the street (a Chinese restaurant at No. 16, a pizzeria at No. 12 and two ordinary cafés at Nos 8 and 18) soon found themselves welcoming throngs of gay customers all day and into the early hours of the morning.

The development of the gay Marais coincided with the burgeoning of the 'pink economy' in France—the gay market that business is reportedly eager to tap (Wharton, 1997). By the turn of the century, French gay men had become "a much courted clientele" (Revel, 2001), averaging 25 to 40 years in age, with a purchasing power estimated to be 30 per cent higher than that of the heterosexual consumer (Cornevin, 1996). The Marais has become "a considerable magnet" whose 184 gay or gay-friendly bars, restaurants and shops attract an average 20 000 clients a day. This reportedly generates 1000 jobs directly and another 1500 indirectly, "which makes the gay businesses as a whole the principal employer of the [fourth] *arrondissement*" (Garcia, 2002, p. 10). If one newspaper reporter is to be believed:

> From the grocer to the restauranteur, all rub their hands and try to win over this clientele known for its high purchasing power. "They buy without looking at the price" marvels Maryse, saleswoman at the furniture store Maison de Ville (*Le Parisien*, 2001).

As the above quotations imply, the vast majority of the Marais' gay clientele is male. There are at present no more than three or four lesbian bars in the Marais and, while women can enter most (but not all) of the men's bars, they are rarely made to feel welcome there. Although cafés, restaurants and shops do welcome women, female customers are nonetheless clearly in a minority. Generally speaking, lesbian communities are less territorially based than gay male communities and lesbians socialise far less in bars and clubs than do male homosexuals (Lockard, 1985; Retter, 1997). A study published 15 years ago, and therefore rather outdated by now, suggests that while lesbians are most probably overrepresented among the residents of inner Paris, 'lesbian facilities' (bars, restaurants and nightclubs, but also social centres, cinemas and bookshops) are less geographically concentrated than those serving gay men (Winchester and White, 1988).

Even taking into account only homosexual men, however, the expression 'gay Marais' is somewhat misleading. First of all, gays have not taken over the entire Marais. Gay businesses cluster along relatively few streets, principally in the south-western corner of the quarter, like the first 200 metres of the much longer Rue des Archives or the relatively short (300 metres long) Rue Sainte-Croix-de-la-Bretonnerie, which one newspaper has called "gay Paris' display window" (Baverel, 1996). Secondly,

> If homosexuals come here to consume and to seduce each other, only an infinitesimal minority have moved into the Marais. The ghetto is primarily commercial (Chayet, 1996).

Actually, observation and anecdotal evidence suggest that many gay men do in fact live in the Marais, but rising rents and real-estate prices in central Paris make this difficult for all but the relatively well-to-do. As a result, probably more gay men live in adjacent (and somewhat cheaper) districts, like the 11th *arrondissement*, than in the Marais itself.[3] Moreover, many gays prefer to put distance between where they live and where they go out to socialise and homosexuals can certainly be found in every neighbourhood of the city (Bordet, 2001, p. 116). Thirdly, the Marais is not the only gay scene in Paris.

> The Marais had the pretension to be Castro Street [in San Francisco] or Christopher Street [in New York City]. It has never entirely succeeded. ... Gay life is spread out and several decades have scattered meeting places to the four corners of the capital (Vanier, 1991, p. 56).

To take a few of the more obvious examples: Le Palace, which opened in 1978 and became the most fashionable Parisian gay club of the 1980s, was situated on the Rue du Faubourg-Montmartre, well outside the Marais, while Le Queen, opened in 1994 and the most chic gay club of the 1990s, is even further away, on the Champs-Élysées. The Rue Keller, in the 11th *arrondissement* and some 1500 metres east of the gay bars of the Marais, has developed quite independently since the late 1970s into a small but distinct centre of gay bars and clubs; the city's Lesbian and Gay Centre even moved there in the early 1990s.

Emmanuel Redoutey, in studying the geographical distribution of gay and gay-friendly spaces across Paris, has used the image of a cone.

> Like the tip of an iceberg, the concentration of establishments in the Marais quarter, where self-identified homosexuals exercise a kind of supremacy over businesses and over the animation of several streets that are also treasured by tourists, plays a central role [in gay life] (Redoutey, 2002, p. 60).

(By one recent estimate, 40 per cent of Paris' gay or gay-friendly venues are located in the 3rd and 4th *arrondissements*.)[4] About 40 establishments "occupy a broader zone in the heart of Paris" that takes in *arrondissements* adjacent to the Marais. These venues, usually less obvious to passers-by than those located in the Marais, are mainly bathhouses and 'sex bars' with 'back-rooms' or 'dark-rooms' where clients can engage in sexual relations. (There are also nightclubs and discotheques in this zone, which Redoutey neglects to mention.)[5] Finally, the wide base of the cone comprises outdoor spaces (public toilets, streets, quays along the Seine and canals, and public parks) that homosexual men use for pick-ups and anonymous semi-public sexual relations. This broad base covers the entire city, as well as two vast wooded parks on its outskirts: the Bois de Boulogne to the west and the Bois de Vincennes to the east. For Redoutey, the Marais showcases a socially "acceptable" homosexuality in contrast to "an underground and disparaged homosexuality [in bathhouses, sex clubs and outdoor cruising-grounds] whose diffuse expression occupies the dark corners of the city" (Redoutey, 2002, p. 63). But it is the Marais, precisely because it is more visible and more acceptable, that draws public attention, represents gay life to the straight world and has served as the territorial base for the construction of a gay community.

The Marais and the Emergence of a Community

The Marais has thus become a clearly delineated gay space in the heart of Paris, where gay men and lesbians can stroll hand-in-hand or kiss in the street without embarrassment or risk of harassment. In the convoluted jargon of a geographer, such public displays of affection constitute an

> appropriation and territorialisation [of a quarter] through the street behaviour of the clientele of gay establishments [who] challenge the hetero-centric character of public spaces and thus give the Marais a conspicuous territoriality (Bordet, 2001, p. 119).

The average homosexual would put it more simply. According to one gay man,

> One feels more among family here [in the Marais] than anywhere else in Paris. Perhaps that's what we mean by the [gay] community (Darne, 1995).

And for another, who recently moved from Lille to Paris, the Marais represents his community's financial clout:

> I was glad to see that *les pédés* ['fags' or

'poofters'] had money and could open stylish establishments. I was glad to belong to something organized, which represented a certain economic power (Laforgerie, 1998, p. 20).

Their enthusiastic appreciation of the ghetto is a relatively recent attitude and even today is not shared by all gays and lesbians. As long ago as 1964, the monthly magazine *Arcadie*, organ of France's politically conservative 'homophile' association, the Club Littéraire et Scientifique des Pays Latin (Literary and Scientific Club of the Latin Countries), warned French homosexuals against copying what was occurring in the US by creating

> a little artificial world, enclosed and suffocating, where everything would be homosexual: not only the bars, restaurants and movie theatres, but also the houses, the streets (in New York several streets are already almost entirely inhabited by homosexuals), the neighbourhoods .... A world where one could live one's entire life without seeing anything other than homosexuals, without knowing anything other than homosexuality. In Europe that is called *ghettos*. ... We hate this false, harmful and grotesque conception of homosexuality (Daniel, 1964, p. 387).

Radical gay militants of the 1970s had little in common with their homophile elders, but they too denounced gay ghettos—both the 'commercial ghetto', meaning the bars at Saint-Germain-des-Prés or on the Rue Sainte-Anne, and the 'wild ghetto' constituted by the parks, gardens and public urinals where homosexual men hunted for sexual adventure (Martel, 1999, p. 77).

Radicals believed that ghettos encouraged a separatist homosexual identity (J. Girard, 1981, pp. 132–133), whereas they wanted homosexuals to participate in the revolutionary transformation of society as a whole: "Instead of shutting everybody up in their own space, we need to change the world so that we find ourselves all mixed together" (Boyer, 1979/80, p. 74).

Some gay radicals, however, eventually changed their minds and came to recognise the political potential of the gay ghetto. Guy Hocquenghem (1946–88), the emblematic radical militant of the 1970s, told an American interviewer in 1980:

> We don't have a gay community in France. That is, we have a gay movement—with several organisations actively working for political rights, as in all the Western countries—but people do not feel part of a *community*, nor do they live together in certain parts of the city, as they do here in New York City or in San Francisco—for example. And this is the most important difference and the most significant aspect of gay life in the U.S.: not only having a 'movement', but having a sense of community—even if it takes the form of 'ghettos'—because it is the basis for anything else (Blasius, 1980, p. 36).

The relationship evoked here by Hocquenghem—linking territory, collective identity and political activism—is a complex one. Veteran militant Jean Le Bitoux, for instance, has argued that the gay community appeared first and then produced the gay Marais:

> The homosexual community that was successfully emerging ... most likely wanted to complete this social emergence in the 1980s with a space 'for expressing an identity' [*un espace 'identitaire'*]. An emergent community needed a new geographical anchorage (Le Bitoux, 1997, p. 49).

Other analyses invert the equation, however, insisting that the Marais created a gay community and not the other way around. For example, Yves Roussel has noted that, whatever their political camp, homosexual activists of the 1950s–1970s rejected the formation of a distinct gay community (conservatives advocating assimilation into society, radicals wanting to overthrow it), but that by the 1990s a new generation had come to embrace 'identity politics'.

> Many are the men and women who see themselves as belonging to a minority group, which is the victim of a process of exclusion; this sentiment of exclusion has combined with the intense desire to constitute and to structure a homosexual community (Roussel, 1995, p. 85).

He has attributed this shift to several factors, including the need to mobilise against the AIDS epidemic, but one particularly significant determinant has been "the emergence of a vast ensemble of gay commercial enterprises [that] have allowed for the constitution of a community of homosexual consumers with characteristic lifestyles" (Roussel, 1995, p. 107). Jan-Willem Duyvendak has similarly concluded that "in the middle of the 1980s, the concentration of gay clubs and bars, such as in the Marais in Paris, provided a certain 'infrastructure' for a community", although he minimises this community's political activism: "the militants took the occasion to go dancing rather than to demonstrate" (Duyvendak, 1993, p. 79).

Gay businessmen share this view that their venues have contributed to the growing sense of community among French gays. In the mid 1980s, the gay entrepreneur David Girard (1959–90) responded to those activists who criticised him for his brazen capitalist spirit by declaring that

> The bar owner who, in the summer, opens an outdoor terrace where dozens of guys ... meet openly, is at least as militant as they are. ... I think that I have done more for gays than they ever have (D. Girard, 1986, p. 164).

He even told his customers: "This gay life that is ever more present and diversified in Paris, ... it is first of all you who create it by consuming" (D. Girard, 1983). This was precisely the message put out in an advertising campaign by the SNEG in 1996: "To consume gay is to affirm one's identity" ("*Consommer gay, c'est s'affirmer*"). The campaign's avowed purpose was to promote its members, but "it is equally a communitarian campaign, a way to bring home to people the visibility of gay establishments" (Primo, 1998).

Arguments like these are certainly self-interested on the part of the businessmen who advance them, but that does not mean that they are without merit. As Scott Gunther has recently pointed out,

> The transformed Marais of the 80s provided a space for the development of a gay identity that had not existed before in France. As the community grew, gays themselves gained a reputation as respectable, resourceful, and affluent. ... Throughout the 80s, the emerging gay identity and geographical space of the Marais became increasingly inseparable and by the early 90s it seemed impossible to imagine the existence of one without the other. The resulting community, which may initially have been defined by a sexual orientation, became increasingly united by shared tastes, cultural preferences in music and food, and even by a distinct 'Marais look' among the gay male inhabitants (Gunther, 1999, p. 34).

Not surprisingly, the proliferation and increasing visibility of gay establishments in the Marais and the concomitant development of a self-conscious gay community have resulted in conflict with some long-time residents who resent the on-going influx of gays and the dramatic changes that they have brought about in the quarter. There is also discord among homosexuals and lesbians themselves, many of whom disapprove of the Marais or feel excluded by its dominant cultural values.

Disputed Territory

Sudden change in a neighbourhood often alarms its residents, a problem that has by no means been unique to the Marais. In the Butte-aux-Cailles district in Paris' 13th *arrondissement*, the artists, writers and middle class who took over this once working-class neighbourhood in the 1980s now complain that it

> has become of the new meeting places for Parisian youth. … All year long, music coming from the bars and noisy laughter invade the streets … into the early hours of the morning. This nightlife has become a nightmare for certain inhabitants (Chenay, 2003).

In the Popincourt district of the 11th *arrondissement*, it is the 'Chinese invasion' by Chinese-born wholesalers in the clothing trades that upsets residents, who find the immigrants "discreet, kind, likeable", but insist that "they have killed the quarter" by taking over every shop that comes onto the market, with the result that there are fewer stores, bakeries and restaurants (Goudet, 2003). But nowhere in Paris has the changing character of a neighbourhood aroused more sustained animosity and acrimony than in the Marais.

Homosexuals have had relatively little trouble with the Marais' other significant minority, the orthodox Jews who live on and around the Rue des Rosiers on the very edge of the gay neighbourhood. When the only Jewish tobacconist's on the Rue Vieille-du-Temple became a gay bar in 1983, "[the Jews] were furious that we had taken over this sacred place on their territory", Maurice McGrath has recalled, but "very quickly, we fraternized". Gay businesses even lent their support to protect the quarter in 1986 during a wave of anti-Semitic terrorism (Chayet, 1996). In contrast, there has been a long-festering dispute with some of the Marais' middle-class residents who, unhappy with the gay influx, have sometimes expressed themselves in words that carry an explicit or implicit homophobic message.

The Association Aubriot-Guillemites has been at the forefront of the struggle against gay businesses. Residents founded the association, named for two small streets in the Marais, in 1978

> to protect the area … from damage to the architectural quality of buildings and to the environment, noise, especially at night, nuisances, various inconveniences, and generally anything that might trouble the peace and comfort of the inhabitants of the aforementioned area (Association circular, *c.* 1996, reproduced in Méreaux, 2001, appendix).

By the mid 1990s, the association was denouncing "the major alteration in the atmosphere of the quarter brought about by the proliferation of homosexual businesses" and calling on its members to report "the multiple incidents that bear witness to the accelerated degradation of daily life in our quarter: noise pollution, solicitation of young boys, prostitution, sexual relations on the public street". (There is in fact little or no evidence for most of these charges.) The association warned that "a small group dreams of making this quarter the equivalent of the homosexual quarters of certain large American cities, which the inhabitants do not want at any price" (Razemon and Galceran, 1996a and 1996b; Rémès, 1996). In 1997, it went so far as to declare that

> no normally constituted citizen, whether homosexual or heterosexual, can approve of the multiplication of these specialised bars … the inevitable result of which is segregationist and discriminatory and the sole purpose of which is nothing other than the economic exploitation of homosexuality" (e-m@le magazine, 1998).

Under pressure from the Association Aubriot-Guillemites, the mayor of the 4th *arrondissement*, Pierre-Charles Krieg, told his constituents in 1996 that "a structured homosexual community has recently received coverage in the media that is disproportionate and dangerous to the harmony of local life". While deploring the "simplistic and racist ideas of thoroughgoing homophobes", he also criticised the "proselytism, ostentation or virulence" allegedly manifested by the neighbourhood's gay men (Krieg, 1996). Krieg was a conservative gaullist, but even a socialist municipal councillor declared herself "generally hostile to communitarianism, *a fortiori* if there is risk of [creating] a ghetto" (Pitte, 1997, pp. 51–52). The dispute crystallised in the 'Affair of the Flags'. In July 1995, Bernard Bousset, as president of SNEG, suggested that gay and gay-friendly businesses in the Marais display the 'rainbow flag' (the internationally recognised symbol of gays) on their façades and some 15 businesses did so. The Association Aubriot-Guillemites objected and in April 1996 the police invoked an ordinance issued by the prefect of police in 1884 and ordered the removal of the flags (which Mayor Krieg contemptuously dismissed as "multicoloured rags") on the grounds that

> the grouped and quasi-systematic display of overly large emblems risks arousing hostile reactions. And in these circumstances it is not necessary to wait for trouble to occur before imposing a ban (Baverel, 1996; Berthemet, 1996; Razemon and Galceran, 1996a).[6]

The flags soon reappeared or were replaced with more discreet rainbow decals stuck on windows and doors, but the dispute was symbolic of deeper and more persistent concerns. In late 1990, a new police commissioner, determined to put an end to "ten years of laxness", began a crack-down on noise in the quarter which, bar owners claimed, amounted to "police harassment" of their establishments (Rouy, 1990; *Illico*, 1991). In 1995/96, gay bars in the Marais deplored another round of harassment by police, including reports for "disturbing the peace at night" actually drawn up in the late morning or early evening (Rémès, 1996). A series of meetings held in early 1997 by police, municipal officials and representatives of SNEG produced an agreement for concerted efforts "to diminish the nuisances for residents from 'nighttime' establishments" (*Paris Centre*, 1997). Another series of discussions followed in the spring of 1999, after renewed complaints about noisy bars and customers who gathered outside the doors of certain establishments, blocking the pavement and sometimes impeding the circulation of automobiles in the street. The Association of Co-owners in the Marais Quarter (Association des Copropriétaires du Quartier du Marais) even advocated closing bars at 11p.m. (instead of 2a.m.). One gay publication observed that "these problems raise the issue of a veritable redevelopment of the quarter by residents, businesspeople, police and administrators", but its proposed solution ("the creation of districts designated for partying, leisure and places for socializing") was probably unrealistic (Abal, 1999). In 2003, the municipality had to put metal rivets in the pavement to delimit the outdoor terraces of restaurants and cafés that were encroaching too far onto the pavement along the Rue des Archives (Laforgerie, 2003, pp. 31–32)

Although in the past decade or so the French press has been generally favourable to gay demands for equal rights, coverage of gay issues has rarely been free of unconscious prejudice, especially where the Marais is concerned. As David Caron has pointed out, the heading above one newspaper article about the Marais—"The gay flag hangs over the Rue Sainte-Croix-de-la-Bretonnerie" (Baverel, 1996)—typically used "a metaphor for foreign invasion" to describe the gay presence in the Marais (Caron, 2001, p. 151). Newspapers tend to ignore lesbians, portray gay men as hedonistic and sex-obsessed revellers, equate homosexual "visibility" with "provocation" (of heterosexuals) and depict the Marais as a geographical and metaphorical "ghetto" and the headquarters of gay "corporatism", "communitarianism" or "militant *apartheid*" (Huyez, 2002). As these charges indicate, the conflict within the neighbourhood over the use of urban space has much broader ideological implications arising from the particular way that the French conceptualise their society.

The Issue of 'Ghettoisation'

In other words, recent developments in the Marais raise the spectre of 'ghettoisation', a term that has evident application to many different minority groups but is most systematically applied to the concentration of gays and gay businesses in a single quarter of Paris (Sibalis, 2003; Caron, 2003). The geographer Jean-Robert Pitte, professor at the

Sorbonne, has been one of the most forceful critics of the 'ghettoisation' of the Marais:

> Born in San Francisco, Amsterdam and London, [homosexual] ghettoisation has reached Paris. … The development of ghettos is obviously dangerous, so much does it undermine sociability and urbanity. … Just like a nation, a city can endure only by allowing varied populations to live together and by assimilating newcomers or minorities. … The very notion of a city is negated when authorities permit and even encourage ethnic or, more generally, cultural groupings. … Nobody dares any longer to venture outside his own minuscule territory (Pitte, 1997, p. 52).

But what one journalist has called the "temptation of the [homosexual] ghetto" (Guichard, 1996) is much more than a question of territory. It is also the purported desire of homosexuals to cut themselves off from the heterosexual world. Frédéric Martel (himself openly gay) has denounced "this hypermodern folly that consists of creating a cultural [gay] ghetto", "a project … for putting up barriers and not for opening up to the world" (Martel, 1997). Martel sees the gay-ghetto-dweller as someone who "lives, in short, on a gay planet that is inaccessible to heteros" (Guichard, 1996, p. 93).

The debate over the gay ghetto in France is part-and-parcel of a far wider controversy over 'universalism' (a word almost inevitably followed by the tag '*à la française*',—i.e. French-style) versus 'communitarianism' (usually dubbed '*à l'américaine*', or American-style) (see Caron, 2001, pp. 149–161). The French consider race, ethnicity, religion and sexual orientation to be strictly private matters that have no legitimate role in the public sphere. Individuals are expected to live as free, equal and autonomous citizens under the authority of the nation-state and to embrace the nation's dominant cultural values. However, instead of merging into the national collectivity, French minorities are now demanding official recognition of their communities and seeking special rights (or so the argument runs). According to one polemicist, "We are witnessing the victory … of community over society. Community: that is the new idol before which we must bow". And who are these aggressive "cultural minorities"? The feminists first, then gays, who are "second in the rank of influence", followed by French-born Arabs, Blacks, Corsicans, the disabled and so on. By giving in to their demands, France is progressively abandoning "the values of the Republic", "the power of citizenship" and "equality" (Minc, 2003, pp. 17–18).[7] The Marais, as a gay ghetto in the historical centre of the national capital, has become the physical embodiment of forces that allegedly threaten to undermine the very foundations of the French Republic. These dangerously subversive forces are closely identified with the US, which is (as many people in France accept as undisputed fact) a dysfunctional and fragmented multicultural society whose racial, ethnic, religious and sexual communities have come to reject the assimilation of the once-vaunted 'melting pot' and now lobby in the political arena for special rights and privileges.

Gay men and lesbians are themselves sharply divided over the ghetto. While many celebrate the Marais as a sheltered space for fun and socialising, others reject it and the superficial and separatist lifestyle that they believe that it represents, striving instead to live 'outside the ghetto' and within the wider world. For instance, Patrick Schindler, an anarchist and veteran militant, has stated that

> The Marais is a place of profit. It's a drinking establishment. It's nothing else …. I aspire … above all not to ghettoise my friendships and my encounters …. It's not where my life happens (Bordet, 2001, p. 110).

And radio journalist Gérard Lefort has said, "There are people who are alone and whom it suits to go see people who are just like them. But … it's a bit sad". He describes the Marais as "a game preserve" and "a little like Jurassic Park: one keeps the monsters together" (Gac, 2000, p. 25). Even some gays in the French provinces, where homosexuals once dreamed of moving to Paris, have joined the chorus of dissent. In September 1998, a gay magazine published in western France headed one article (in English): "Paris is Dead!": "The Marais is a market opportunity like any other. ... Gay businesses are, for the most part, in the hands of racketeers who are as shrewd as they are greedy". The article decried the conformity that allegedly reigns in the Marais: "If, by misfortune for himself, a gay in the Marais happens to deviate ... from the 'clonically recognised' look, he is immediately condemned to general reprobation and studied indifference". And it boasted that gays had more fun in the provinces because "our regional bars are convivial, familial and friendly" and "the provincial nightclubs do not sink into the spirit of the ghetto, they welcome cool heteros with open arms" (Fauconnier, 1998).

The most acerbic comments have come from militant lesbians and from minorities within the gay community. In the course of a recent debate, the lesbian sociologist Marie-Hélène Bourcier maintained that middle-class gays seek to mimic the heterosexual organisation of society (for example, by demanding the right to adopt children) and that this middle class is "rather well symbolised by the Marais". For her, "visibility in the Marais ... serves a single type of identity, masculine, bourgeois and white". Fouad Zéraoui, a French Arab, added: "the Marais symbolises a culture that is young, white, muscled, virile to the point of fascist. ... For me, the Marais, ... prevents our community from raising the real questions concerning the subversive and essential role of homosexuality in society" (Garcia, 2002, p. 14). A handicapped gay man, who uses a wheelchair, has even blamed the Marais for the exclusion he experiences: "I would like to have a social life, a love life, which is denied me because I do not enter into the stereotypical mould of the Marais" (Berger, 2003).

Exasperated by such criticism of the Marais, Christophe Girard, who is currently in charge of culture at City Hall, remarked that

> We cannot accept being restricted to clichés. ... Stop giving so much importance to the Marais, which is only a quarter of Paris where people go to have a drink, dine, walk about and have fun. Nothing more (Royer, 1998).

But if clichés are by definition simplistic, they often contain some element of truth. Do the gays who reside and socialise in the Marais really want to live cut off from the wider (heterosexual) world? Improbable. Are they all male, young, White, muscle-bound and monied consumers? The Marais' gay venues certainly do promote a male-oriented lifestyle that has little place for women, the poor (who cannot spend) or the old, the unattractive, the flabby and the effeminate, who violate current canons of homosexual desire, but even the most casual observer cannot fail to notice that lesbians, older men and visible minorities are increasingly present in the quarter's bars or sitting on its café terraces.

Conclusion

As this paper has demonstrated, the Marais is both a real neighbourhood, with all the conveniences and inconveniences associated with modern urban life, and an imagined cityscape that people interpret in the light of their own prejudices, wishes and desires. A gay ghetto has formed in the Marais over the past 25 years because of the confluence of multiple factors: a physically attractive historical site, a successful programme of urban renewal by national and municipal governments, a strategic location in the centre of Paris, rents and real-estate prices initially low enough to draw gay investors, a growing gay market available to be tapped (the 'pink economy'), the determination of certain businessmen to promote a more open gay lifestyle and the eagerness of a new generation of homosexuals to embrace it. The fate of the Marais over the coming years will similarly be determined by the interplay of many complex economic and social forces and is therefore difficult to predict.

Financial pressures may very well destroy (or at least disperse) the gay Marais by pushing more and more gay businesses into other neighbourhoods. Indeed, a recent magazine article has suggested that the Marais is today "in crisis" and that its gay venues are "victims of the quarter's success". Because of their notoriously fickle clients, who are always on the look-out for something new, gay businesses need to recuperate their investment quickly (over five years). But today's high cost of buying an existing gay business or establishing a new one in the Marais (estimated at a minimum of 450 000 euros, including renovation and redecoration) tends to discourage new investors, especially since they must face intense competition from existing businesses. For his own part, however, the president of the SNEG insists confidently that the future of the Marais as Paris' gay ghetto is secure:

> It is not possible to reproduce in another quarter of Paris what happened in the Marais. What quarter today is central, accessible by the RER and the subway, decrepit without too much insecurity and very cheap? These are the particular conditions that permitted the birth of the [gay] Marais (Laforgerie, 2003, p. 32).

In contrast, there are many people who believe (or even hope) that the ghetto will gradually become irrelevant and dissolve of itself as gays become progressively accepted by and integrated into society at large (for instance, Paris has had an openly gay mayor since 2001). In addition, the unwillingness of many homosexuals to frequent the ghetto—indeed, the virulent hostility to the Marais manifested by a significant number of gays and lesbians—could eventually undermine its economic viability. On the other hand, the ghetto may well continue to prosper and even expand by diversifying its venues in order to appeal to those who now feel excluded.

The point is that, in the final analysis, the continued existence of a Parisian gay ghetto—whether in the Marais or perhaps one day in some other neighbourhood—will ultimately depend on factors more intangible than transport links and real-estate prices. The emergence of the Marais as a gay ghetto in the heart of Paris cannot be understood by taking into account only material factors, which have produced more or less similar gay ghettos in most Western cities. The motives of business investors, the attitudes of their customers and the subjective perceptions of people both inside and outside the gay community have been equally (and sometimes even more) important in giving shape to Paris' gay ghetto and in determining the public response to it. These things all derive from French national culture. This case study therefore suggests that each and every gay ghetto has to be studied not only in terms of broad social and economic trends, but also within the context of a particular cultural, social and political environment.

Notes

1. As compared with 22 351 establishments employing 72 374 in 1860 and 9721 establishments employing 44 637 in 1956 (Benedetti, 1960, pp. 9–27).
2. The available figures for the safeguarded sector alone are 1918: 100 000; 1954: 79 000; 1962: 75 000; 1968: 66 000; 1982: 40 000 (Audry and Starkman, 1987).
3. These remarks are based on personal observation and on interviews with officers of several gay associations that draw their membership from across the city. The official census does not categorise the French by religion, race or sexual orientation.
4. In October–December 2001, according to listings in e-m@le magazine, of 230 gay or gay-friendly establishments in the city, 77 (33.5 per cent) were in the 4th *arrondissement* and 18 (7.8 per cent) in the 3rd (Bordet, 2001, p. 64).
5. The SNEG claims that more than half of Paris' gay businesses (252 of 488) are situated in central Paris (the entire 1st, 2nd, 3rd and 4th *arrondissements*, which it misleadingly labels 'the Marais'), including 52 gay bars (of 88 in all Paris), 69 gay restaurants (of 99), 11 gay discotheques (of 17) and 5 bathhouses (of 16) (Laforgerie, 2003, p. 32).
6. The 1884 ordinance declared that "the individual act of manifesting religious, political or other opinions can be regu-

lated ... whenever circumstances transform the individual act into a collective act of the kind likely to trouble the public peace".

7. Another manifestation of this concern is the current French debate over whether Muslim girls should be permitted to wear headscarves to class in public schools.

References

ABAL, O. (1999) Trop de bruits dans le Marais?, *CQFG* 3 (22 April), pp. 12–13.

AUDRY, J.-M. and STARKMAN, N. (1987) L'évolution récente du Marais, in: *Le Marais, mythe et réalité*, pp. 264–269. Paris: Picard.

AUFFRAY, M.-F. (2001) *Le Marais: la légende des pierres, if stones could speak*, trans. by E. POWIS. Paris: Hervas.

BABELON, J.-P. (1997) L'urbanisation du Marais, *Cahiers du Centre de Recherches et d'Études sur Paris et l'Ile-de-France,* 9, pp. 17–34..

BAVEREL, P. (1996) Le Drapeau gay flotte rue Sainte-Croix-de-la-Bretonnerie, *Le Monde*, 22 June, p. 11.

BECH, H. (1997) *When Men Meet: Homosexuality and Modernity*, trans. by T. MESQUIT and T. DAVIES. Chicago, IL: University of Chicago Press.

BENEDETTI, J. (1960) *Préfecture de la Seine: Le Quartier du Marais*. Paris: Imprimerie Municipale.

BERGER, Y. (2003) Interview, in: *Marche des Fiertés LGBT*, free supplement to *Nova Magazine*, 102(June), p. 5.

BERTHEMET, T. (1996) Querelle de drapeau dans le Marais, *Le Figaro*, 25 April, p. 24.

BINNIE, J. (1995) Trading paces: consumption, sexuality and the production of queer space, in: D. BELL and G. VALENTINE (Eds) *Mapping Desire: Geography of Sexualities*, pp. 182–199. London: Routledge.

BLASIUS, M. (1980) Interview: Guy Hocquenghem, *Christopher Street,* 4(8), pp. 36–45.

BORDET, G. (2001) *Homosexualité, altérité et territoire: les commerces gais sur le bas des pentes de la Croix-Rousse et dans le Marais*. Unpublished Mémoire de maîtrise de géographie, Université Lumière-Lyon 2.

BOUTHILLETTE, A.-M. (1994) Gentrification by gay male communities: a case study of Toronto's Cabbagetown, in: S. WHITTLE (Ed.) *The Margins of the City: Gay Men's Urban Lives*, pp. 65–83. Aldershot: Arena.

BOYER, J. (1979/80) Quand les homosexuels se lancent à la conquête de l'espace, *Masques*, 3(Winter), pp. 73–74.

CARON, D. (2001) *Aids in French Culture: Social Ills, Literary Curses*. Madison, WI: University of Wisconsin Press.

CARON, D. (2003) Ghetto, in: D. ERIBON (Ed.) *Dictionnaire des cultures gays et lesbiennes*, pp. 218–219. Paris: Larousse.

CARPENTER, J. and LEES, L. (1995) Gentrification in New York, London and Paris: an international comparison, *International Journal of Urban and Regional Research,* 19, pp. 286–303.

CASTELLS, M. (1983) *The City and the Grassroots: A Cross-cultural Theory of Urban Movements*. Berkeley, CA: University of California Press.

CASTELLS, M. and MURPHY, K. (1982) Cultural identity and urban structure: the spatial organization of San Francisco's Gay Community, in: N. I. FAINSTEIN and S. S. FAINSTEIN (Eds.) *Urban Policy under Capitalism*, pp. 237–259. Beverly Hills, CA: Sage Publications.

CHATELAIN, P. (1967) Quartiers historiques et centre ville: l'exemple du quartier du Marais, in: *Urban Core and Inner City: Proceedings of the International Study Week, Amsterdam, 11–17 September 1966*, pp. 340–355. Leiden: E. J. Brill.

CHAYET, S. (1996) Marais, le triangle rose, *Le Point,* 1232(27 April), p. 96.

CHENAY, C. DE (2003) A Paris, la Butte-aux-Cailles, des communards aux 'bobos', *Le Monde*, 29 July, p. 9.

COBB, R. (1985) *People and Places*. Oxford: Oxford University Press.

CORNEVIN, C. (1996) 100,000 gays dans la capitale, *Le Figaro*, 25 April, p. 24.

DANIEL, M. (1964) Le plus grave danger, *Arcadie,* 11(129), pp. 385–389.

DARNE, R. (1995) Ghetto? Milieu? Communauté? Un débat ouvert, *Exit, le journal*, 21 July, pp. 8–9.

DUGGINS, J. (2002) Out in the Castro: creating a gay subculture, 1947–1969, in: W. LEYLAND (Ed.) *Out in the Castro: Desire, Promise, Activism*, pp. 17–28. San Francisco, CA: Leyland Publications.

DUYVENDAK, J.-W. (1993) Une 'communauté' homosexuelle en France et aux Pays-Bas? Blocs, tribus et liens, *Sociétés: Revue des Sciences Humaines et Sociales*, 39, pp. 75–81.

e-m@le magazine (1998) Pas de quartier pour les gays!, *e-m@le magazine,* 23(12 March), p. 15.

EVENSON, N. (1979) *Paris: A Century of Change, 1878–1978*. New Haven, CT: Yale University Press.

FAUCONNIER, T. (1998) Paris is dead!, *West & Boys,* 27(September), pp. 34–35.

FAURE, J. (1997) *Le Marais: organisation du cadre bâti*. Paris: L'Harmattan.

GAC, J. (2000) Gerard Lefort: pour ou contre le poil à gratter?, *Le Chevalier de la Rose,* 28(May–July), pp. 24–25.

GADY, A. (20002) *Le Marais: guide historique et architectural*. Paris: Le Passage.

GARCIA, D. (Ed.) (2002) Dossier: le gay Marais, ghetto ou village? *Le Nouvel Observateur: Paris Île-de-France,* 1947(February–March), pp. 8–16.

GIRARD, D. (1983) Édito, *5 sur 5,* 1 (September), p. 1.

GIRARD, D. (1986) *Cher David: les nuits de citizen gay*. Paris: Ramsey.

GIRARD, J. (1981) *Le Mouvement homosexuel en France 1945–1980*. Paris: Syros.

GOUDET, A. (2003) Pas de racisme, mais trop de Chinois dans le sape, *Marianne,* 329(11–17 August), p. 59.

GUICHARD, M.-T. (1996) Homosexuels, la tentation du ghetto, *Le Point,* 1232(27 April), pp. 93–95.

GUNTHER, S. (1999) The indifferent ghetto, *Harvard Gay and Lesbian Review,* 6(1), pp. 34–36.

HARRY, J. and DEVALL, W. B. (1978) *The Social Organization of Gay Males*. New York: Praeger.

HINDLE, P. (1994) Gay communities and gay space in the city, in: S. WHITTLE (Ed.) *The Margins of the City: Gay Men's Urban Lives*, pp. 7–25. Aldershot: Arena.

HUYEZ, G. (2002) Dix ans de 'ghetto': le quartier gay dans les hebdomadaires français, *Pro-Choix,* 22(Autumn), pp. 59–81.

Illico (1991) Le Marais en alerte, 3 January, p. 7.

INSEE (INSTITUT NATIONAL DE LA STATISTIQUE ET DES ÉTUDES ÉCONOMIQUES) (2000) *Populations légales: recensement de la population de 1999: France*. Paris: INSEE.

JALLIER, G. (1983) Marais, Halles, le créneau gay, *Samouraï,* 8 (June), pp. 34–38.

KAIN, R. (1981) Conservation planning in France: policy and practice in the Marais, Paris, in: R. J. P. KAIN (Ed.) *Planning for Conservation*, pp. 199–233. London: Mansell.

KERNE-VIANNOY, J. DE (2001) Édito, *e-m@le magazine,* 6(30 November–13 December), p. 3.

KNOPP, L. (1987) Social theory, social movements and public policy: recent accomplishments of the gay and lesbian movements in Minneapolis, Minnesota, *International Journal of Urban and Regional Research,* 11, pp. 243–261.

KNOPP, L. (1990) Some theoretical implications of gay involvement in an urban land market, *Political Geography Quarterly,* 9, pp. 337–352.

KRIEG, P.-C. (1996) Éditorial, *Paris Centre: Le Journal du IVème,* 93(October/November), p. 3.

LAFORGERIE, J.-F. (1997) Argent rose, argent roi, *Ex Aequo,* 6(April), pp. 24–29.

LAFORGERIE, J.-F. (1998) Dehors, dedans: mon ghetto, *Ex Aequo,* 14(January), pp. 16–22.

LAFORGERIE, J.-F. (2003) Le Marais est-il en crise?, *Illico,* 78(6 June), pp. 30–33.

LAMIEN, E. (Ed.) (1998) Dossier: en être ou ne pas en être, *Ex Aequo,* 14(January), pp. 26–36.

LAURIA, M. and KNOPP, L. (1985) Toward an analysis of the role of gay communities in the urban renaissance, *Urban Geography,* 6, pp. 152–169.

LE BITOUX, J. (1997) Marcher dans le gai Marais, *Revue h,* 1(July), pp. 47–51.

LE CLÈRE, M. (Ed.) (1985) *Paris de la préhistoire à nos jours*. Saint-Jean-d'Angély: Éditions Bordessoules.

LE DOUCE, A. (1983) Maurice McGrath, un patron gay, *Samouraï,* 8(June), pp. 39–41.

LE MOËL, M. (1997) Désaffection et dégradation du Marais au XVIIIe et XIXe siècle, *Cahiers du Centre de Recherches et d'Études sur Paris et l'Ile-de-France,* 59, pp. 65–78.

Le Parisien (2001) Joël a 'inventé' le Marais en 1978, 18 April, pp. 12–13.

LEVINE, M. P. (1979) Gay Ghetto, *Journal of Homosexuality* 4(4), pp. 363–377.

LEWIS, M. (1994) A sociological pub crawl around gay Newcastle, in: S. WHITTLE (Ed.) *The Margins of the City: Gay Men's Urban Lives*, pp. 85–100. Aldershot: Arena.

LOCKARD, D. (1985) The lesbian community: an anthropological approach, *Journal of Homosexuality,* 11, pp. 83–95.

LUZE, H. DE (1996) *Une Morale ondulatoire: enquête sur les sauvages parisiens de l'archipel du IVe arrondissement et plus particulièrement de l'île du Marais*. Paris: Loris Talmart.

MADESCLAIRE, T. (1995) Le ghetto gay, en être ou pas?, *Illico,* 57(August), p. 48–55.

MARTEL, F. (1997) Gare au ghetto gay!, *L'Express,* 2398(19–25 June), p. 132.

MARTEL, F. (1999) *The Pink and the Black: Homosexuals in France since 1968*. Stanford, CA: Stanford University Press.

MÉREAUX, J. (2001) *Le "Marais": l'espace homosexuel comme métaphore du groupe: éléments pour une socio-anthropologie d'une culture territorialisée*. Unpublished Mémoire de Diplôme d'Études Approfondies, Université de Paris X-Nanterre.

MICHEL, C. (1988) *Les Halles: la renaissance d'un quartier 1966–1988*. Paris: Mason.

MINC, A. (2003) *Épîtres à nos nouveaux maîtres*. Paris: Grasset.

NEUVILLE, P. (1995) Le Grand Bernard, *Le Frondeur,* 18 December, p. 6.

NOIN, D. and WHITE, P. (1997) *Paris*. Chichester: John Wiley & Sons.

Paris Centre: Le Journal du IVème (1997) 'Gays' dans le IVème, 95(April/May), p. 7.

PATTISON, T. (1983) The stages of gentrification: the case of Bay Village, in: P. L. CLAY and R. M. HOLLISTER (Eds) *Neighborhood Policy and Planning*, pp. 77–92. Lexington, MA: Lexington Books.

PITTE, J.-R. (1997) L'avenir du Marais, *Cahiers du Centre de Recherches et d'Études sur Paris et l'Ile-de-France,* 59, pp. 49–54.

POLLAK, M. (1985) Male homosexuality—or happiness in the ghetto, in: P. ARIÈS and A. BÉJIN (Eds) *Western Sexuality: Practice and Precept in Past and Present Times*, pp. 40–61. Oxford: Basil Blackwell.

PRIGENT, A. (1980) *La réhabilitation du Marais de Paris*. Paris: École des Hautes Études en Sciences Sociales.

PRIMO, T. (1998) Bernard Bousset: de la mémoire avant toute chose, *e-m@le magazine,* 17(29 January), pp. 18–19.

QUILLEY, S. (1997) Constructing Manchester's "new urban village": gay space in the entrepreneurial city, in: G. B. INGRAM, A.-M. BOUTHILLIER and Y. RETTER, (Eds) *Queers in Space: Communities, Public Places, Sites of Resistance*, pp. 275–292. Seattle, WA: Bay Press.

RAZEMON, O. and GALCERAN, S. (1996a) La bataille du Marais, *Ex Aequo,* 2(December), pp. 16–19.

RAZEMON, O. and GALCERAN, S. (1996b) Marais: la guérilla, *Illico,* 73(December), pp. 6–10.

REDOUTEY, E. (2002) Géographie de l'homosexualité à Paris, 1984–2000, *Revue Urbanisme,* 325(July–August), pp. 59–63.

RÉMÈS, E. (1996) Les policiers harcèlent les bars gays du Marais, *Libération*, 18–19 May, p. 15.

RETTER, Y. (1997) Lesbian spaces in Los Angeles, 1970–90, in G. B. INGRAM, A.-M. BOUTHILLIER and Y. RETTER (Eds) *Queers in Space: Communities, Public Places, Sites of Resistance*, pp. 325–337. Seattle, WA: Bay Press.

REVEL, R. (2001) Une clientèle très courtisée, *L'Express*, 21–27 June, p. 90.

ROLAND-HENRY, O. (1983) Le Central, *5 sur 5,* 4(December), p. 7.

ROUSSEL, Y. (1995) Le mouvement homosexuel français face aux stratégies identitaires, *Les Temps Modernes,* 50(582), pp. 85–108.

ROUY, P. (1990) Bars gais: nuisances policières, *Gai Pied Hebdo,* 449/450(20 December), p. 15.

ROYER, A. (1998) Christophe Girard, *e-m@le magazine,* 54(19 November), p. 10.

ROYER A. (2000) Pour moi, donc, j'aime le Marais, *e-m@le magazine,* 56 (16 November), p. 11.

SIBALIS, M. (1999) Paris, in: D. HIGGS (Ed.) *Queer Sites: Gay Urban Histories since 1600*, pp. 10–37. London: Routledge.

SIBALIS, M. (2001) Les espaces des homosexuels dans le Paris d'avant Haussmann, in: K. BOWIE (Ed.) *La Modernité avant Haussmann, Formes de l'espace urbain à Paris 1801–1853*, pp. 231–241. Paris: Éditions Recherches.

SIBALIS, M. (2003) Ghetto, in: L.-G. TIN (Ed.) *Dictionnaire de l'homophobie*, pp. 194–196. Paris: Presses Universitaires de France.

STRYKER, S. (2002) How the Castro became San Francisco's gay neighborhood, in: W. LEYLAND (Ed.) *Out in the Castro: Desire, Promise, Activism*, pp. 29–34. San Francisco, CA: Leyland Publications.

STUNGO, A. (1972) The Malraux Act 1962–72, *Journal of the Royal Town Planning Institute,* 58(8), pp. 357–362.

ULRICH, C. (1996) Le Marais, quartier général du lobby homosexuel, *L'Événement du jeudi*, 20–26 June, pp. 28–29.

VANIER, L. (1991) Sexe sur Seine, *Gai Pied Hebdo,* 482(21 March), pp. 56–58.

WHARTON, S. (1997) Financial (self)-identification: the pink economy in France, in: S. PERRY and M. CROSS (Eds) *Voices of France: Social, Political and Cultural Identity*, pp. 172–186. London: Pinter.

WINCHESTER, H. P. M. and WHITE, P. E. (1988) The location of marginalised groups in the inner city, *Environment and Planning D,* 6, pp. 37–54.

Where Love Dares (Not) Speak Its Name: The Expression of Homosexuality in Singapore

Kean Fan Lim

> Freedom is a *relation*—a power relation (Bauman, 1997, p. 27).

> There is plenty of essentialism to go round. Boundary fetishism has long been, and in many circles continues to be, the norm (Pieterse, 2001, p. 224).

> [T]o change life ... we must first change space (Lefebvre, 1991, p. 190)

1. Introduction

On 23 July 2001, three undercover policemen entered a sauna called 'One Seven' in Singapore and arrested two men for having oral sex. Significantly, the police appeared to know beforehand that gay men frequented the sauna and there was thus a possibility that homosexuals could be having sex.[1] The two men were initially charged for performing "gross indecency" under Section 377 (A) of the Penal Code,[2] before ultimately convicted and fined S$600 (US$360) each under Section 20 of the Miscellaneous Offences (Public Order and Nuisance) Act (Cap 184, 1997 Ed)[3] (see Au, 2001a). Intriguingly, why did anyone know that those two men were going to have sexual contact before it happened? In what is reportedly the fourth most 'globalised' and 'open' city in the world

Kean Fan Lim is in the Department of Geography, National University of Singapore, 1 Arts Link, Singapore 117570, Republic of Singapore. Fax: +65-67773091. E-mail: g0305818@nus.edu.sg. *The author would like to thank T. C. Chang and Lisa Law (from the National University of Singapore and the University of St Andrew's respectively) for their valuable comments throughout this study. The author is very indebted as well to his respondents for taking time to share with him their opinions and experiences. The author's gratitude also goes to the five anonymous referees of this paper for providing very constructive comments. The words of encouragement and support from Fiona F. Yang have been very important to the author who cannot thank her enough. Of course, all responsibility for any errors and misinterpretations remain solely the author's.*

today (*The Business Times*, 2003), are homosexuals still under state surveillance?

In Singapore, a south-east Asian city-state of 4 million people,[4] the socio-spatial regulation of homosexuals has had a long history (Leong, 1997; Heng, 2001; Ng, 2003). Heterosexual expression is dominant—almost 'compulsory'—and, as in other places, "desire of lesbians and gay men and bisexual desire for both men and women are often seen to be abnormal, unnatural or deviant" (Blunt and Wills, 2000, p. 128). Despite the cosmopolitan nature of this city-state, Singaporeans are still disallowed from even speaking in public places about homosexual-related issues. For instance, while the government launched the Singapore 21 initiative in 1997 to stimulate active and polyvocal citizenry, with the promulgation that "Every Singaporean Matters" (S21 speech, Minister Teo Chee Hean, 1999, n.p.), homosexuals were and still are not allowed to form societies or hold public forums.

Despite this negative socio-legal atmosphere, however, homosexuals are growing more visible and vocal, and there has been a steady growth in places where homosexuals congregate and express their sexuality. Ammon (2002), for instance, calls Singapore's Chinatown district the "emerging gay 'ghetto'" and "'new' gay pulse", as this is where several clubs and saunas serving a largely gay/lesbian clientele are located.[5] The successful organisation of the annual 'Nation' party on Sentosa island since August 2001 by Fridae.com, a homosexual on-line organisation, is unprecedented and has been heralded as a milestone in the 'coming out' of the homosexual community. The state's censorship of homosexual themes in artistic performances, whilst remaining strict, has also been loosened in the past decade. Perhaps the most surprising indication of a change in the state's stance came when Prime Minister Goh Chok Tong recently announced that his government now allows homosexual employees to take up even 'sensitive' positions. In fact, Goh's unprecedented announcement acknowledged that homosexuals are not sick, but just different: "We are born this way and they are born that way, but they are like you and me" (*Time*, 2003a).

This apparent mutation in attitude towards homosexuals in turn raises several interesting questions. Will heteronormative social space in Singapore become increasingly destabilised with this rise in the socio-spatial expression of homosexuality? How are homosexuals' senses of place *changing* with the emergence of more homosexual-friendly places? More significantly, is the government *genuinely* sincere about proffering homosexuals more freedom for expressing their sexuality? Or does it merely boil down to economic considerations (as has mainly been the case in Singapore's post-independence developmental trajectory), with the potential contributions of the 'pink economy' high on the statesmen's planning agenda?

With the above questions in perspective, this paper provides an analysis on the socio-spatial expression of homosexuality in Singapore. My objectives are twofold. First, I explore how social space is *constructed* and reinforced as heterosexual, paying particular attention to the moral-spatial dialectic. This focus on the production of heterosexual space is important, as Chouinard and Grant (1996, p. 184) accentuate, for examining homosexual spaces *per se* does not address "the heterosexual and hostile environments in which lesbians and gay men spend most of their time". Heterosexual space, as shall be argued, is *not* ontological—it is *produced*. This production, however, does not remain unchanged over time; it is continuously challenged. I therefore also interrogate some issues regarding the *forms* of expressions homosexuals undertake to negotiate heteronormative social space. While I acknowledge that practices within so-called queer spaces are important counterpoints to the heteronormative order, this paper examines the overt spatial expression of homosexuality. As it is, homosexuality continues to be expressed in this city-state, but not only in what architect William Lim (2001a) notes as marginal 'spaces of indeterminacy', but also increasingly through overt practices in the public domain.

The discussion is primarily informed by interviews conducted in 2002/03 with three prominent figures—namely, Mr Alex Au (webmaster of the prominent Singaporean homosexual website Yawning Bread); Dr Stuart Koe (Chief Executive Officer of Fridae.com and a key organiser of the 'Nation' party); and James Gomez, a well-known Singaporean political activist. Additional interviews were also conducted with ten homosexuals (three lesbians, seven gays) and their responses are supplemented by analyses of personal accounts posted by homosexuals in cyberspace, newspapers and in unpublished academic writings. Textual analysis was done on journal, newspaper and magazine articles reporting on homosexual issues in Singapore. I also examined three competing sets of statistics on social attitudes towards homosexuality. Here, there is insufficient space for me to comment extensively on the virtues and drawbacks of the methodological approaches adopted in this study (see Kong, 1998; Elwood and Martin, 2000; Williams, 2000). In short, I make *no* claim to represent objectively the opinions of the entire homosexual community in Singapore and individual accounts are not extrapolated to be reflective of the homosexual community's experiences. More importantly, I am well aware that my analytical 'way of seeing' is necessarily subjective and conditioned by my position as a heterosexual, male researcher (for issues on research positionality, see McDowell, 1992; Merrifield, 1995; Binnie, 1997; Rose, 1997; Binnie and Valentine, 1999), but it would surely be essentialist and biased to perceive that just because of my gender and sexuality, this paper will necessarily provide a less cogent and critical account of homosexual-related issues.

After this prolegomenary section, this paper is structured along the following lines. In the subsequent section, I will review the existing literature on homosexuality in Singapore and establish the theoretical framework that undergirds this study. I will then proceed into the discussion proper in sections 3–5 before concisely discussing some theoretical issues in the concluding section.

2. The Spatiality–Sociality–Sexuality Nexus

2.1 On Sexuality and the City

There has always been an inextricable relationship between spatiality, sociality and sexuality (Castells, 1983; Bell, 1991; Massey, 1994; Duncan, 1996; Myslik, 1996; Pile, 1996a; Binnie and Valentine, 2000; Blunt and Wills, 2000; Brown, 2000; Pritchard and Morgan, 2000; Rushbrook, 2002; Valentine, 1993, 1996, 2002; Brown and Knopp, 2003; Kitchin and Lysaght, 2003). As Knopp articulates

> Cities and sexualities *both shape and are shaped by* the dynamics of human social life. They reflect the ways in which social life is organised, the ways in which it is represented, perceived and understood, and the ways in which various groups cope with and react to these conditions (Knopp, 1995, p. 149; emphasis added).

Indeed, while issues relating to homosexuals stretch beyond urban areas (see, for instance, Philips *et al*, 2000), "It is in the modern city that the suppressed groups have formed inclusionary communities" (Baeten, 2002, p. 113) and this trend is more likely to persist as the world's population gets increasingly urbanised. Cities are also key nodes within which major info-communication infrastructure is embedded, thus enabling civic engagement not only within the local community, but also across different geographical scales. This is especially so in Singapore's context, where 68.4 per cent of households own a computer and the home Internet penetration rate stands at 59.4 per cent (Singapore Infocomm Development Authority, 2002). And with cities getting more entrepreneurial and place marketing being arguably more important than ever, planners have been seeking out new spaces that mark their cities as cosmopolitan and culturally tolerant, and this is why, as Rushbrook (2002, p. 188) articulates, queer space is one of several cultural spaces that serve as "a marker of cosmopolitanism, tolerance and diversity for the urban tourist". And as I shall argue, this could be the primary reason behind the emergence of 'visible' queer spaces and activities in Singapore.

Studies on homosexual spaces in Singapore are presently limited, but definitely increasing. On the spatio-temporal front, Heng (2001) provides a detailed outline of the historical emergence, surveillance and closure of homosexual entertainment spots in Singapore. Research has also been conducted on the 'courtship spaces' of homosexual couples (Tay, 1997) and, more extensively, into the lived experiences of gay men within territories of the home, the workplace and public spaces (Chng, 1999). Ng (1999) and Soh (2003) respectively illuminate Singaporean gay men's and lesbians' expression of homosexuality in cyberspace, a site where they can live their lives in conditions of their own choosing, while Tan (1998) depicts the lives of gay Christians in Singapore and how they grapple with their sexuality *vis-à-vis* a relatively hostile religious landscape. Perhaps the most important addition to the literature is a collection of essays in *People Like Us: Sexual Minorities in Singapore* (edited by Lo and Huang, 2003). While this collection's themes are not specifically geographical, it does provide an incisive understanding of some socio-cultural and legal issues facing homosexuals. Building on these works, this paper examines the production of heteronormative social space in Singapore and how homosexuals negotiate the overt expression of their sexuality in such a milieu.

2.2 Conceptual Parameters

The theoretical sub-stratum of this study draws from Henri Lefebvre's (1991) spatial conceptual triad. Specifically, Lefebvre focuses on how "social and spatial relations are dialectically interreactive, interdependent" (Soja, 1997, p. 248). Hence, to comprehend the spatial expression (or closeting) of homosexuality, there is a need to understand how heterosexual hegemony is spatially (re)produced (Valentine, 1993, p. 114).

To explicate this socio-spatial interconnection, Lefebvre offers a triad of concepts—namely, 'representation of place', 'representational space' and 'spatial practice' (see Lefebvre, 1991). As Donald (1997, p. 182) enunciates, "The living space of the city exists as representation and projection and experience as much as it exists as bricks and mortar or concrete and steel". Since the imbrications of these three concepts make them more analytically useful when employed as an integrated triad (Merrifield, 1993; Allen and Pryke, 1994; Liggett, 1995), they will not be taken as mutually exclusive entities in the impending discussion. However, attention is specifically paid to the *relation* between spatial practices and representations of space. While representational spaces have often been interpreted as primary and creative sites of resistances to dominant representations of space, spatial practices may also be overt acts of resistances (rather than subjugation) to heteronormativity in Singapore.

Representation(s) of space, governmentality and the panoptic effect. 'Representation(s) of space' refers to the infusion of dominant conceptual understandings (ideological, linguistic and symbolic) of space into the spatial practices and lived experiences of people. *Who* gets to 'represent' space and formulate what is spatially (un)desirable is contingent on the power of the dominant person/group. This paper looks at the Singapore government as the dominant group and its representation of Singapore's moral geography as naturally *hetero*sexual. The analytical focus is on the role of the state government because Singapore is a city-state, wherein the urban and national scales are actually conflated. However, I am also *au courant* that power relations exist more locally (also noted as "compound spaces"; see below) within Singaporean urban/national space, and the dominant person/groups may not be concatenated to the state. As Foucault expatiates

Power relations are extremely widespread

> in human relationships. Now this does not mean that political power is everywhere, but that there is in human relationships a whole range of power relations that may come into play among individuals, within families, in pedagogical relationships, political life *et cetera* (Foucault, 1984/1996, p. 434).

Representing space discursively is insufficient, however, if dominant groups are to ensure that the demeanours of subordinated groups fall within palatable parameters. There is a need for *governmental* and *disciplinary* technologies to control potential 'deviants'. To some degree, Foucault's notion of *governmentality* is useful to apprehend the strategies that seep subtly throughout society to shape the behaviours of homosexuals (Danaher *et al.*, 2000). Rather than top–down direct oppression, the governance of subjects is through disseminating expert discourses of social and/or medical pathology about the 'deviancy' of homosexuality, to the extent that homosexuals' self-regulation and self-censorship are effected (which in turn will also affect heterosexuals' dispositions; see below). Hence, we have to take into account not only 'techniques of domination' but also 'techniques of the self' (Danaher *et al.*, 2000). In Singapore, statistics depicting societal opinion have been used to legitimise socio-spatial policies prohibiting discussion of homosexuality in public spaces and sex education in secondary schools and junior colleges provides nebulous definitions of 'homosexuals' and 'homosexuality', mainly emphasising the downsides of this sexuality. As will be discussed, these are strategies of governmentality that reinforce the 'otherness' of homosexuals in Singapore.

However, as Smith (2000, p. 285) warns, an excessive focus on governmentality "dislocates law from the sites and practices of power". Rather, sovereign law is alive and well, and the aforementioned expert discourses also work to support and naturalise legislation (see also Fitzsimons, 1999). Law, in Moran's (2001, pp. 107 and 109) words, is a "cartographic technology" that determines places entailing "incessant and regular observation". The role that law plays is therefore crucial in analysing how the dominant application of *disciplinary* technologies effects homosexuals' 'techniques of the self' and consequently maintains a heterosexual sense-of-place. Such disciplinary technologies are best conceptualised by Foucault's (1979) 'panoptic effect'. Originally, Foucault's analysis of the panopticon (following Bentham) is contingent on the precondition of spatial confinement, since it is only when spaces are confined that observation, judgement and punishment can be easily facilitated. Extending this analysis, Hannah (1997) calls such spatial control 'compound discipline', wherein disciplinary power is exerted within a compound space. In addition, Hannah's definition of 'compound space' is expanded to include places without stringent barriers to entry and exit. However, Hannah's claim that the public and private status of compound space will affect disciplinary effectiveness does not hold in Singapore's context, because even private spaces can be scrutinised, intruded upon and 'disciplined'. Then again, this does not mean that homosexuals may be subjugated and will rigidly conform to heterosexual norms and practices. Resistances to dominant representations of space are always in effect (Pile, 1996b; Thrift, 1996) and, as will be elucidated, resistant acts by homosexuals may occur in both representational spaces *and* through spatial practices in the public domain.

'Spacing off' in representational space. Representational space is the directly *lived* space of everyday life (Lefebvre, 1991, p. 33). These spaces often remain *un*seen and yet they are where the impacts of power relations are at their most ocular (Duncan, 1996). Following de Lauretis, representational spaces are interpreted as

> spaces in the margins of hegemonic discourses, social spaces carved in the interstices of institutions and in the chinks and cracks of the power-knowledge apparatus

(de Lauretis; quoted in Soja, 1999, pp. 272–273).

The representational spaces of oppressed groups, however incomprehensible and even abhorred they may be by majority communities, are very meaningful and important because of their symbolic and liberating potential. As Baeten puts it

> The city's dangerous places may form an inexhaustible source of inspiration for middle-class versions of the urban dystopia, but they can equally be an inclusionary safe haven for society's Others (Baeten, 2002, p. 113).

For instance, while architectural designs are almost always planned without homosexual couples in mind, homosexuals turn to creating cosy, intimate and sensual interior spaces, spaces where their sexuality can be *celebrated away from* the discriminatory glares of the 'straight' public (Betsky, 1997).

Attaining a true understanding of the lived, "usually invisible" (Betsky, 1997, p. 12) spaces of homosexuals is difficult, however, if not impossible—space, indeed, hides consequences from us (Berger, 1972). It is theoretically possible that representational spaces are subjected to discipline and punishment too, if the authorities so desire. Instances of representational spaces in this paper include the One Seven sauna, cyberspace and discotheques/pubs where lesbian and gay parties were held.

Spatial practices: just about ensuring continuity and cohesion? According to Lefebvre (1991, p. 38), spatial practices refer to how people orient themselves geographically through their perceptions of space, such as the ways they move around places and where they want to go or avoid. In Soja's words (1999, p. 265), spatial practices are "the directly experienced world of empirically measurable and mappable phenomena", which essentially means the spatially *observable*. Through socio-spatial practices, societal continuity and cohesion are maintained (Lefebvre, 1991, p. 33). This is so because in socio-spatial practices

> There is an element of performance involved, whereby specific practices attempt to construct and maintain a particular sense of place, and in so doing limit alternative interpretations (Allen and Pryke, 1994, p. 454).

For some theorists, the 'performance' of sexual identity and its spatial manifestation are seen as the *effect* of dominant discourses (Butler, 1990, 1993). Subjects thus face a "compulsion to repeat" or resignify dominant discourses (Butler, 1990, p. 145), which consequently affects their spatial practices. To understand in greater depth the intentionality behind homosexuals' spatial practices, I follow Bourdieu's (1989, 1992) work on the development of habitus, which McNay (1999, p. 99) refers as "the subtle inculcation of power relations upon the bodies and dispositions of individuals". The spatial practices of homosexuals are therefore contingent on unequal power relations and what are *seen* may not be their absolute lived experiences and desires, but rather 'techniques of the self' that reinforce heteronormativity. Thus

> It is vital that we conceptualise performers as in some sense produced by power, and not...[as] virtuoso, theatrical, anterior agents at one remove from power's social script (Gregson and Rose, 2000, p. 441).

Specifically, I will argue that heteronormativity also shapes the habitus of heterosexuals and may work to internalise/fortify discriminatory attitudes towards homosexuals, thus leading the latter to perform the closet in their quotidian spatial practices. However, heterosexuals' habitus may *not just* be reinforced—but *also altered*—by accumulated experiences in a 'field' that contains different geographical contexts, especially in contexts where overt expressions of homosexuality are increasingly commonplace. Such expressions need and should not be belligerent, of course, but smaller manifestations in public spaces that traverse against heterosexual hegemony—expressions that may be transposed from representational spaces onto public space. Indeed, while the 'coming out' process is extremely difficult for many homosexuals (Rhoads, 1994; Markowe, 1996; Whisman, 1996), and many homosexuals' spatial practices may thereupon be self-censored and self-restrained by hegemonic discourses in order to maintain societal 'cohesion', I argue that spatial practices may *also* be subversive and contradictory acts to the heteronormative order, and it is through such practices that the heterosexuals' habitus and homosexuals' 'techniques of the self' may be mutated. The discussion shall focus on two forms of these practices—namely, the organisation of the annual 'Nation' party by the homosexual community over the past three years, and the expression of homosexuality through artistic media.

3. Constructing Heteronormativity in Singapore

3.1 Casting the Homosexual as Alien

The Singapore government has always been at pains to portray the national space as immanently heterosexual. Even today, homosexuality is classified as an illness in the Singapore military and is recorded in a gay soldier's medical file (for elaborate discussions, see C.-S. Lim, 2002a, 2002b). Consensual sex between men is still illegal in Singapore, a colonial-era law that the government obdurately clings to.[6] As Brown (2000, p. 91) notes, "through their rules and regulations, states sanction the presence of gays and lesbians in their national territories". Notably, lesbianism is totally *un*recognised by the law, which can be construed as an even more extreme form of oppression

> Oppression works through the production of a domain of unthinkability and unnameability. Lesbianism is not explicitly prohibited in part because it has not even made its way into the thinkable, the imaginable, that grid of cultural intelligibility that regulates the real and the nameable (Butler, 1991, p. 20)

At the 1993 United Nations World Human Rights Conference in Vienna, the then Foreign Minister of Singapore Wong Kan Seng commented that "Homosexual rights are a Western issue, and are not relevant to this conference" (quoted in Au, 2001b). This statement effectively legitimises the obviation of homosexuals from Article 19 of the United Nations Universal Declaration of Human Rights, which states that

> Everyone has the right to freedom of opinion and expression: this right includes freedom to hold opinions without interference and to seek, receive and impart information and ideas through any media and regardless of frontiers (http://www.un.org/Overview/rights.html).

Clearly, as Wong's words imply, 'everyone' in the Singaporean context excludes homosexuals, because only rights that are *non*-Western *and* heterosexual are worthy of consideration. The supreme irony is therefore hard to miss when Minister David Lim (1999a) proclaimed that "The moving spirit of Singapore 21 is that everyone must act ... Every citizen ... must also put in effort to build [society]". But as gay activist Alex Au questions

> How do we expect gay Singaporeans to feel passionate about Singapore if they perceive that they suffer discrimination, legal and social, in this country? (*Agence French Press*, 2000).

After failing to register the informal homosexual group People Like Us (PLU) as a society with the Registrar of Societies in November 1996, Au applied to the Singapore Public Entertainment Licensing Unit (Pelu) in May 2000 to hold a public forum named "Gays and lesbians within Singapore 21" (*The Straits Times*, 2000a). This is an attempt to foster dialogue on the place of homosexuals within the S21 framework. The proposal can be interpreted as a socio-spatial practice that was to be transposed from representational space to the visible public domain. However, that an application was required before the forum can be held in public space *underscores* homosexuals' exclusion from the 'public'. Not surprisingly, the application was rejected, with the official reason premised strongly on heterosexual norms

> There are major implications in endorsing and encouraging homosexuality openly, and promoting homosexual behaviour to become an accepted social norm. One major implication is the effect on our young. If more Singaporeans end up embracing this sexual orientation openly, the foundation of the strong family, which is the core building block of Singapore 21, would be weakened. This is why I feel it is better that we exercise great caution, be conservative, and stick to the basic concept of family and family values as much as we can, for as long as we can (Minister Lim Swee Say, in *The Straits Times*, 2000c).

Lim's response is debatable in at least two aspects. First, he immediately construed the forum as 'endorsing' and 'encouraging' homosexuality 'openly'. Homosexuality is thus interpreted as *ominous* to the dominant heterosexual space in Singapore and the forum may also cause more people to 'embrace' this sexuality. But is it really? Can homosexuality be 'encouraged'? Can sexuality be attained and jettisoned according to one's whim and fancy? As Stuart Koe, CEO of the homosexual website Fridae.com, contends

> To answer the question—is homosexuality 'chosen', I ask you in return, 'is heterosexuality chosen?' Why would anyone *choose* a sexuality that will forever condemn one to alienation, discrimination, ridicule, even hatred? ... Whilst the *expression* of one's sexuality may be tempered or controlled, the *basis* for that sexuality is quite natural, I can assure you.
>
> This means that homosexuality cannot be 'promoted'. You cannot make someone who is not homosexual into one by exposing them to homosexual issues or imagery. It would, however, be very civilised of our society to not only be tolerant of the diversity of sexualities of our constituents, but also allow each individual to develop his/her own views by allowing free access to information and by outlawing discrimination of any kind (personal communication; Stuart Koe, original emphasis).

And as two members of the public remark

> I am heterosexual. I would have attended the forum [proposed by Au] because I feel the homosexual population in Singapore has been treated unfairly over the years. But I would not have felt compelled to become gay as a result of my attendance. After all, my years of contact with gay friends have not had that effect on me (Siew Kum Hong, *The Straits Times*, 2000b).
>
> I am willing to discuss the role of gays within Singapore 21. But I do not think that in doing so, I am advocating being gay, because it has nothing to do with the merits and demerits of being gay ... I am unconvinced that present societal norms are as anti-homosexual as believed (Siew, *The Straits Times*, 2000d).
>
> Are homosexuals the result of nature or abusive nurture? I really can't say but I can honestly say it doesn't matter to me because this much I know: I have homosexual friends who are kind, compassionate, honest and intelligent, just as I know of heterosexuals who are dishonest and unkind—and just as capable of sleeping around (Lim Suat Hong, *The Straits Times*, 2003d).

These comments clearly problematise Lim's assertion that Singaporeans should "exercise great caution" against this perceived threat [of homosexuality] "for as long as we can", which he believes can be achieved if Singaporeans "stick to the basic concept of family and family values". But just what *is* the basic concept of the family? And how do we define family 'values'? As Sack (1997, p. 64) contends, "home is a relative concept" and what constitutes the 'family' and family 'values' varies from one familial space to another. Can we then generalise what the 'basic' concept of the family means? Are there *ecumenical* family 'values'? The answers are actually not clear-cut at all. As another member of the public counter-argued in a letter to *The Straits Times*

> A "better life in Singapore 21" to Mr. Lim [Swee Say] is to banish homosexual Singaporeans from public view. As long as they remain in private, we will leave them alone and leave our prejudiced views and penal code intact. That may be a "better life" for Mr. Lim himself and all who share his views, but it cannot be, in all honesty, a "better life" for society in the true sense of the word (quoted in Vincent's Lounge, 2000).

Indeed, a world that is heterosexual *in toto* may *not* necessarily equate to a better world replete with peace and love, as heterosexuals may also possess 'values' that may be very damaging to the family. As Anthony Yeo, a Clinical Director of a Counselling and Care Centre in Singapore, contends in a letter to *The Straits Times*

> The assertion that the gay lifestyle will erode moral values and expose the next generation to corrupting influences seems to suggest that the world we live in, that is predominantly heterosexual in orientation, is a perfect world, vulnerable to deadly influence if we permit the gay lifestyle to prevail. If we were to survey the kind of problems we experience daily, we would be familiar with physically and sexually abused children, females being raped and molested, people growing up emotionally and mentally disturbed, as well as a variety of deviant behaviours. Those in the mental-health profession would easily testify that the large majority of them come from homes where parents are heterosexual in orientation (*The Straits Times*, 2003d).

Nonetheless, just as homosexuality is prevented from being 'promoted', heterosexual love is *actively promoted* through a series of campaigns, the latest being "*Romancing Singapore*", kickstarted in 2003 and in 2004 turned into a year-long affair (see http://www.romancingsingapore.com) and several programmes engendered by social institutions such as the Ministry of Community Development and Sports (MCDS) and Social Development Unit (SDU). In this light, the government is evidently trying to *create and maintain* a heterosexual sense-of-place in Singapore. Such socio-spatial engineering may consequently influence homosexuals' 'techniques of the self' and leave them feeling out-of-place, for social space is perceived to be heterosexual in *nature* (more on this later). Homosexuals are apparently tolerated only *to the extent that* they remain in interstitial spaces, *in*visible to public eyes. Public forums about homosexuality are hence strictly proscribed. As Minister David Lim asks

> Does this mean that they cannot talk about homosexuality in Singapore? No. But it does mean that as a society, we are not yet ready to give recognition to this group of people, and the sexual mores they advocate (D. Lim, 1999a).

Yet, as the *Sunday Morning Post* (2000) reports, "[PLU] weren't planning a homosexual orgy. They simply wanted to talk".

This *idea* that homosexuality belongs to the private sphere until it is *fully* 'recognised' by the heterosexual majority clearly reinforces Mitchell's (2000, p. 177) contention that "actions in public ... must conform to the dictates of the 'normal' or hegemonic society". Au's application for a homosexual society and a public forum was hence a *potential transgression* of acceptable socio-spatial boundaries demarcated by the government, as the public exhibition of homosexuality could disturb and destabilise naturalised expressions of heterosexuality. Moreover, Au argues that

> The example of gay organisations in the West would not have been encouraging, since they are perceived to stand in opposition to the incumbent governments (well, actually they, as civil society groups, stand apart from their governments, not in oppo-

> sition to them, but to the [government], anyone apart from them is seen as opposed to them).
>
> It's entirely in line with the government's refusal to allow any alternative nuclei of influence, organisation or power to arise that they will persistently refuse to allow any *organised* gay voice to be heard. They won't even allow any media in Singapore very much independence, let alone lobby groups (personal communication; A. Au; original emphasis).

The Singapore authorities may also fear that allowing homosexuals 'access' to public space would send out a 'wrong' message to society. As lawyer Simon Tay (*Asiaweek*, 2000) notes, "To allow a [homosexual] society or a public meeting can be likened to ending the ban on Playboy … It's a question of symbols, of what is officially allowed". To maintain a low profile, some homosexuals also prefer to keep extant socio-spatial boundaries untouched; remaining in the closet is preferred instead (see Butler, 2001). The public expression and debate of/about homosexuality may be deemed as an inappropriate form of resistance, for

> Doing so may threaten the 'social fabric' and force the mainstream public to decide on the issue of homosexual rights in Singapore. I don't believe this is necessary in the big picture—the gay community is given sufficient leeway currently to evolve and grow without any untowards or negative pressures from the government. It may instead backfire if a referendum is sought. (personal communication, S. Koe).

In light of Prime Minister Goh's apparent relaxed stance on homosexuals, Koe may be justified in feeling that the homosexual community is "given sufficient leeway". But then, it is telling that PLU's subsequent application to register as a society in February 2004 was rejected *again* by the Registrar of Societies in April 2004 (*The Straits Times*, 6 April 2004). The fact that the socio-legal prohibitions faced by homosexuals remain *un*changed may persistently cast homosexuals as 'outsiders' and possibly even troublemakers if they grow more vocal, as

> What makes certain people 'strangers' and therefore vexing, unnerving, off-putting and otherwise a 'problem', is—let us repeat—their tendency to befog and eclipse boundary lines which *ought to be clearly seen* (Bauman, 1997, p. 25, emphasis added).

But would there not be fewer conflicts and misunderstandings if such boundaries are *obliterated*? Is the "sufficient leeway" that Koe talks about really *sufficient*? For Au, it certainly is not

> There's always underground freedom. The point must be what happens overground. The issue is why aren't we given our due—given legal equality, political recognition and social recognition above ground (Reuters News Agency, 2001).

While moving 'overground' does not necessarily entail a 'referendum', some argue that homosexuals should not be deprived of fundamental rights, which raises the issue of *sexual citizenship* (Weeks, 1998; Isin and Wood, 1999; Bell and Binnie, 2000; Richardson, 1998, 2000). What exactly counts as a Singapore 'citizen'? What 'equality' are Singaporeans entitled to? Can differences in sexuality be recognised? As two Singaporean men accentuate

> Just how long can sex and partying last, before we discover that we cannot do many, many other things that straight people can do? For example, marry legally the person we love, adopt kids and even hold a high-flying government job with an open conscience that you're gay? After everybody's got his share of *** and dancing, it's these basic rights that count in defining human dignity and self-respect (Sal, online account, undated).
>
> Its [*sic*] been thousands of years since the start of civilisation and homosexuality has been around ever since. If only people ever take time to know more about homosexuality. Maybe there won't be so much

> bigotry around. After all, people are afraid of unfamiliar territory (Edmund Wee, Letter to the BBC, 24 February 1998).

So is it due to the fear of "unfamiliar territory" that homosexuals are still portrayed as the 'other'? Of course, some may construe Prime Minister Goh's relaxed attitude towards homosexuals as a signal of increased tolerance, but tolerance is fundamentally still *different* from *acceptance* (Wilson, 1993), as the rejection of PLU's recent application to form a society only serves to accentuate. As it is, homosexuals are still bereft of rights that are taken-for-granted by heterosexuals, which Sal has strongly mentioned above. Moreover, it is evident that Goh's recent comments on homosexuals is not at all a recantation of the state's official stance, as he qualified in the same interview with *Time* that extant legislation against homosexuals will not change, because "It's more than just the criminal code. It's actually the values of the people. The heartlanders [i.e. the less cosmopolitan Singaporeans] are still conservative" (*Time*, 2003a). This essentially means that constructing a heteronormative sense-of-place remains necessary because 'society' thinks so, but this argument is peculiar, given the Singapore government's poor record of public consultation with 'society' (see, for instance, Kong, 1993). Nonetheless, with the assistance of statistics, the government may continue to naturalise heteronormativity, and it is this issue that I shall now elaborate.

3.2 Statistics, Societal Opinion and Social Justice

To justify heteronormativity and the regulation of homosexuals, the government has always argued that it is merely acting on behalf of 'society'. What can or cannot be spoken and done is circumscribed by so-called out-of-bounds (OB) markers that are putatively determined by what 'society' demands

> OB markers ... reflect what society as a whole believes in. What response the government makes depends on what we judge society to be ready to accept or tolerate, and what we believe society is able to bear (D. Lim, 1999a).

But as Minister David Lim (1999a) is also well aware, "because the OB markers are not clear, people are afraid to speak their minds". Since drawing OB markers is unconstitutional (Gomez, 2000), the government has often relied on the use of statistics that purportedly reflect societal opinion to legitimise such markers. Axiomatically then, this reference to OB markers smartly deflects the basis and consequences of decisions to 'societal' demands. As Sack puts it,

> Social monitoring of what takes place is justified *in the general sense* that society has given license to and thereby condoned or encouraged the creation of [oppressive] places (Sack, 1999, p. 36; emphasis added).

In the 1990s, the 1992 Censorship Review Committee Report has been a primary substratum upon which the government portrays public disapproval of homosexuality. It was probably predicated on this when Minister Teo Chee Hean argued that "We would not encourage or facilitate some groups—for example, gay groups, because there is a broad consensus among Singaporeans today that we should not do so" (*The Straits Times*, 1999b). But was there *ever* a "broad consensus"? As Siew Kum Hong contends, the 1992 Report's

> methodology is questionable, because respondents were asked sensitive questions in face-to-face interviews. Given public perception that Singaporean society is anti-homosexual, the respondents would have felt compelled to give answers that conformed to the perceived norms, and not those that they believed in (*The Straits Times*, 2000d).

And regarding Goh's recent remarks that the "heartlanders are still conservative", how do we know? And even if they are, how does it matter to social justice?

Recently, the Ministry of Community Development and Sports (MCDS, 2002) released the first monograph from its survey project, 'Social Attitudes of Singaporeans'. Of all the 1481 survey respondents, 85 per cent found 'homosexual behaviour' unacceptable. Whether this figure may be used to justify the government's future policies towards homosexuals remains speculative. At first glance, it may appear that the society remains anti-homosexual, but there is a need to interrogate critically how the survey *represents* homosexuality—the emphasis is not on homosexual *orientation*, but on homosexual 'behaviour', which strongly suggests sexual contact. Thus, the respondents are already indirectly *disciplined* by a survey that connotes homosexuality as sexual in character. However, sexual orientation should not be conflated with sexual 'behaviour', which heterosexuals *also* engage in. If the government wants to infer from its survey about society's stance towards homosexuality, then the term 'orientation' should replace 'behaviour'. Given the differences in meanings between these two terms, it is quite impossible that the word 'behaviour' has been employed inadvertently; the word could rather have been used deliberately to elicit a negative response.

Notwithstanding this, it was found that among Singaporeans aged below 30, 29 per cent found homosexual 'behaviour' acceptable, compared with merely 12 per cent for those aged above 30. Of Singaporeans with post-secondary education, 26 per cent found 'homosexual behaviour' acceptable, compared with 10 per cent for those with secondary-and-below education. While the SAS statistics still evince an overwhelming disapproval of homosexual 'behaviour', the study also suggests that younger and more-educated Singaporeans are growing more accepting of homosexuals and even of their 'behaviour'. By comparison, in a different recent study conducted by Vivien Lim (2000), 54 per cent of her 413 respondents (average age 20) felt they would not feel uncomfortable with homosexual friends, while 28 per cent felt that they would 'feel comfortable' when a person of the same-sex made 'advances' towards them. Still, Lim concludes that Singaporean youths possess "rigid gender associated beliefs" and are "still quite conservative in their attitudes" towards homosexuality.

However, another study on 591 people conducted by PLU in April–May 2000 found that 46 per cent of the street-side respondents and 74 per cent of those interviewed on-line 'felt that they would be able to accept a gay brother or sister, if not immediately, then after a while' (see http://www.geocities.com/WestHollywood/3878/gls21p02.htm). It was also found that 73 per cent of street-side respondents and 83 per cent of on-line respondents 'agree or strongly agree' that companies should not discriminate against homosexual employees. The PLU stresses that many of its research assistants are 'straight', which can be seen as "a harbinger of a more broad-minded society". And in sum, the PLU argues that the survey findings are "leading indicators" of Singapore society's opinions in time to come.

Which of the above studies is more reliable then? If younger Singaporeans increasingly find homosexuality (and by implication the practices of homosexuals) acceptable, then is it still plausible to generalise about what a whole 'society' thinks? This is certainly tricky. If the government wants to use statistics that depict societal opinion as the bedrock of its decisions, this suggests that public opinion is the ultimate arbiter of policy-making in Singapore. But is that so? Moreover, if the issue is more about social justice, not about popularity (Au, 1999a), then is societal opinion actually important? As Chua Mui Hoong comments

> Legal strictures against homosexual behaviour aim to 'protect' the majority from being offended by such behaviour. But should the state determine private morals? Should minority interests be set aside for the majority? (*The Straits Times*, 2003b).

If so, we may add, how can we say that social justice has been extended to *all* individuals in Singapore?

In the words of Taylor (quoted in Petrovic, 1999, p. 203), justice requires that "we all recognise the equal value of different cultures; that we not only let them survive, but acknowledge their worth". If homosexuals have their distinct ways of lives that are innocuous to anyone, then would it not be *unconscionable* and arrogant for 'us' to regulate their demeanours just because 'they' are not like 'us'? It is also ironic that the term 'society' is used to defend regulations of homosexuals when the S21 manifesto emphasises that every Singaporean matters, as it appears that homosexuals' views do not matter much, if at all. As Au stresses

> I'm uncomfortable with the suggestion ... that there is indeed a 'Singaporean society' with a viewpoint that we can locate. I don't see much consensus in this society on many social issues—perhaps because debate has been suppressed for so long that no understanding and consensus could develop. ... [The] opinion-leaders in Singapore tend to be uncomfortable with having different sections of society looking in different directions, and instead wanting a homogeneous national model (personal communication, A. Au).

We should hence be aware that statistical data are not 'innocent', 'objective' and 'natural'. As Hackings argues

> Statistics has helped determine the form of laws about society and the character of social facts ... the collection of statistics...itself part of the technology of power in a modern state (quoted in Brown, 2000, pp. 93–94).

However, to infer from statistics that 'society' is *always* unready publicly to discuss homosexuality is a self-fulfilling prophecy—if society is disallowed from gaining awareness about homosexuality, then misconceptions and prejudices may *remain internalised* in heterosexuals' habitus. Homosexuality will thus remain an 'unfamiliar territory'. As Au questions, "When is a society ever ready to deal with homosexuality, unless it first goes through a process of being forced to confront it?" (personal communication).

4. Law as Cartographic Technology: Does Panoptical Surveillance Exist?

4.1 Some Gay Men's and Lesbians' Perceptions

On 11 December 1998, Singapore's Senior Minister (SM) Lee Kuan Yew went on the Cable News Network (CNN) and responded to a question from a homosexual man saying that "What we are doing as a government is to leave people to live their own lives so long as they don't impinge on other people. I mean, we don't harass anybody" (*The Sunday Times*, 1998). While SM Lee's comments paint a picture of a benign government that acknowledges the existence of homosexuals, they also disguise the fact that the entrapment of gay men continues to be an on-going issue. The government also appears to be closely monitoring lesbian parties and, in some instances, has tried to regulate these parties. This *perception* that a panoptical system of surveillance exists is certainly not unfounded. As the International Gay and Lesbian Association reveals:

> Since the late 1980s police swoops on homosexual haunts have been routine. There are no official figures for the number of arrests, but between 1990 and 1994 newspapers reported 67 convictions arising from police undercover activities. This is likely to be a minute fraction of the total. Of the 67 cases, 50 were for "molest" (s.354) (typical punishment from 1993: 2–6 months in prison plus caning, usually three strokes), 11 for soliciting (s.19) (fines of $200 to $500) and 6 for obscene acts (s.294A) ($200 to $800) (IGLA; 2000).

On the reason why these arrests and convictions took place, Heng (2001, p. 87) contends that "There seemed to be an agenda to make examples of them". And indeed, such 'examples' may shape homosexuals' 'techniques of the self', with the message that an overt expression of their sexuality will only bring trouble. For instance, through interviews with 31 gay men, Chng (1999, p. 66) argues that "gay men often have to survey and modify their spatial behaviour in public so as not to reveal their sexual identity and risk indicting themselves as part of the denizens of society". However, with the increase of numerous outlets catering to a largely homosexual clientele since the late 1980s and Prime Minister Goh's recent announcement, it does *seem* that the government is really 'loosening' its control of homosexuals (as long as they remain in private, as aforementioned). As one gay man sees it

> We are seeing more and more of these [queer] places, and I think the government allows these places to remain because homosexuals can hang out within certain spaces. It would probably be more problematic controlling the gay community if homosexuals are allowed express their sexuality on the open streets (personal communication, Andy, pseudonym).

While Andy might view the growth in queer spaces positively, this growth could in fact be a subtle form of governmentality to keep homosexuals 'in check'; homosexuals are at least 'somewhere' within compound spaces, which facilitates ease of monitoring. But then, the marked decrease in entrapments of gay men over the past few years has probably reduced their sense of intimidation, which explains why many gay men expressed consternation at the entrapment incident in the One Seven sauna in 2001. Whether this arrest was attributed to surveillance or other unknown reasons is actually *not* the issue. What matters rather is that the law can be activated against gay men anytime. Thus, the government may appear to be 'closing one eye' on homosexuals' socio-spatial practices in recent years, but that eye can open any time, as the One Seven incident demonstrates, and when it will happen again is anyone's guess (or fear)

> Singapore has no privacy laws, and the police can practically enter any premises at their whim and fancy. But why enter as undercover cops and thus instill fear in this community? Must these men in blue be so draconian as to even regulate zealously private sexual activities between two consenting adults? (Sal, on-line account, 2001).

> Naturally, the arrests have affected a lot of homosexual men. However, I wouldn't say that it has heightened a fear of surveillance—that is always there—but it's given them a real example they can refer to, to justify their fears. They can't lightly dismiss their fears as unfounded when a real event in recent memory can be cited, and is cited every now and then (personal communication, A. Au).

At this juncture, it is important to make a distinction between the sense of intimidation experienced by lesbians and gay men. Since the law does not apply to lesbian sexual contact, the socio-spatial practices of lesbians may thus be less influenced by a fear of surveillance. As these two lesbians note

> Well, I don't go to the sauna, so I have no fear of 'someone watching me'. My behaviour is still as usual. I often go to lesbian parties or bars, but there are a lot of people there who are not gay or lesbian. Thus, there is no way to really confirm for sure whether those people are gay, lesbian or bisexual. Anyway, if we are not doing anything illegal, there will be no fear (personal communication, Linda, pseudonym).

> The surveillance does apply more to gay men. I don't feel like I am being watched as such. But … I do curb the public display of my sexuality. I think baby butches, butches in general get a harder time at the clubs. The fact that they look more 'out' does probably make them more vulnerable to police questioning when clubs are raided. I know that I sometimes avoid such clubs (where there may be too many underaged butch girls hanging out)—prefer to avoid trouble (personal communication, Arina, pseudonym).

While lesbians may experience a reduced sense of intimidation, the two accounts suggest, again, that if homosexuality is unexpressed in public spaces, but rather enmeshed with heteronormative expression, it is highly unlikely that they would be subject to 'harassment'. Their 'techniques of the self', and possibly that of many others, may have been shaped to perform the closet 'to avoid trouble', a point that is also accentuated in Soh's (2003) work on lesbians' perception of space in Singapore. Nevertheless, this does not mean that authorities are not paying attention to places where lesbians congregate. For instance, a bar in Boat Quay (a central part of the Singapore city centre) received a police notice to change the gender mix of its weekly lesbian party in 1994, although the party was advertised as a 'private' event (Heng, 2001, p. 91). On 12 March 1999, Mr Peh Chin Hua, a Member of Parliament, lambasted lesbians for the spatial expression of their sexuality, even when it was again *within* private premises

> In November last year, a disco in Singapore specially organised for-women-only type wild parties every Thursday night. Patrons were normally young girls aged 18 or 19. One half dresses as they are—as women, while the other half dress up as men. They bind their breasts, wear T-shirts, slacks, men's shoes and very short hair with glossy hairgel—just like young boys. If they don't open their mouths to speak, bystanders would even think they were real boys. Even worse, the girls would hug each other and kiss; or one 'type' would sit on the lap of the other 'type'.
>
> One cannot believe that these kinds of things happen [in] Singapore. How can this phenomenon not worry parents and teachers? If this goes on, I believe Education Minister Teo Chee Hean, principals and teachers would feel that their hard work in the past at educating the young have gone to waste and would feel sad about this. To prevent our young from going astray, I sincerely hope the Home Affairs Ministry would—just as in the past—come down hard on discos and bars that do these 'strange' things, so that parents and teachers can feel reassured (Peh Chin Hua; translated from Mandarin by Au, 12 March 1999, n.p.).

Peh's tirade is significant in four ways. First, he clearly believes that females should script their socio-spatial expressions according to predominant gender norms—i.e. that of a feminine-heterosexual woman. Any girls who wear "T-shirts, slacks, men's shoes and [have] very short hair with glossy hair-gel" are instantly perceived to have transgressed the boundaries of a 'normal' woman. Secondly, girls who "hug each other and kiss" or sit on one another's laps are considered to be "going astray". But is this not clearly a moralistic comment? Is Peh not imposing his personal values onto the judgement of the demeanours of others? Thirdly, discourses may have corporeal ramifications and Peh's call for the police to "come down hard" on representational spaces like pubs and discos drew an immediate response from the Minister of State for Law and Home Affairs, Mr Ho Peng Kee, who assured the media that 'immoral and indecent activities' in discos and lounges would be subject to 'regular and random spot checks' by the police (*The Straits Times*, 1999a). Fourthly, Peh's comments indicate that the "hard work in the past at educating the young" within educational institutions in Singapore has, implicitly or explicitly, denigrated homosexuality as a deviant sexuality that one should disassociate from. This point deserves further elaboration, for educational spaces should not 'educate' students to be discriminative towards others simply because of differences in sexuality (Petrovic, 1999).

4.2 Surveillance in Educational Spaces: Shaping Discriminatory Dispositions?

In Singapore's sex education syllabus, while students are taught that homosexuals are not strange or repulsive people, the message that is driven home is that homosexuality 'does not offer a complete, natural life experience'. Interpreted in a different light, being a homosexual necessarily means that one's life is *in*complete and *un*natural, although what becomes 'complete' is really very much a subjective, individual feeling. As *Time* reports

> In schools, sex-education courses focus almost exclusively on heterosexuality—the only mention of same-gender sex reminds students that it is against the law. Unsurprisingly, Singaporean society remains deeply conservative (*Time*, 2001).

Hence, if schools are where 'society' first forms its mindset about issues in life, then schools could well be the places where discriminatory dispositions towards homosexuality are first nurtured and internalised. Pedagogic actions are thus by no means about educating *per se*—such actions may also seek to conceal and, by extension, perpetuate extant societal divisions. In interviews with 25 lesbians in Singapore, Soh (2003) found that some respondents actually contemplated suicide because schools do not provide information about how to deal with homosexuality and do not provide an environment in which it feels 'alright' to discuss homosexual issues. Hence, the majority of her respondents chose to remain 'invisible' in schools because of "fear of gossip, lack of support and help as well as homophobia" (Soh, 2003, p. 34). The negative portrayal of homosexuality in educational spaces again exemplifies the role of *governmentality* in shaping heterosexuals' habitus towards sexual minorities; homosexuals must consequently either perform the closet or face possible expulsion. Consider this account

> The most negative experience was when I was in secondary school, I almost got thrown out of school because one of the teachers in the school found out I was a lesbian and soon all the teachers knew. They said things like: "homosexuality is abnormal, you are a good student, you must not be a lesbian". ... It was that attitude that was very damaging to me (playwright Madeleine Lim, 1997)

Within educational spaces, students may be disciplined even if they express their homosexuality *elsewhere*. In 1996, a 16-year-old boy who declared his homosexuality on his Internet homepage was disallowed from collecting his School Leaving Certificate and Testimonial unless he remove his homepage; he was also threatened by the school that the Singapore Broadcasting Authority (SBA) was going to 'take action' against him (*The New Paper*, 1996). While cyberspace is a representational space where homosexuals can 'be themselves' (Ng, 1999; Wakeford, 2002; Soh, 2003), this space may also be subjected to scrutiny and/or discipline. According to Singapore media regulations, any material/performance that 'promotes' homosexuality is strictly censored. In the above case, however, the boy was just proclaiming his sexuality, not 'promoting' it. As he says, "All I wanted to do was to make a statement that I'm proud of who and what I am" (*The New Paper*, 1996).

The SBA was alerted by the complaints nonetheless, as there was a possibility that the boy's homepage was really 'promoting' homosexuality. The boy's expression drew complaints most probably because he had disrupted the heterosexual sense-of-place usually associated with the compound spaces of schools. But this boy's expression also draws attention to how cyberspace functions as an 'alternative' space for homosexuals to construct and express their own sexual identities. As Soh (2003) shows, the Internet has: provided a significant amount of information for Singaporean lesbians seeking to understand more about their sexuality and yet unable to find enough resources in libraries and schools; become a site to organise outings and meet new lesbians; and, facilitated the discussion of personal problems through chatrooms and instant messaging services. Indeed, following Ester and Vinken, cyberspace may well become

> a realm where society's traditional senders (government, church, political parties, old media, etc) of information have no directive power and where they have to compete with all other senders and thus where

> they are confronted with the fact that the symbolic power of their message is no longer self-evident (Ester and Vinken, 2003, p. 668).

However, the boy's incident palpably shows that expressions in cyberspace have corporeal consequences (although nothing happened to him eventually) and it may also have raised the suspicions or fears of some homosexuals that their Internet expressions/interactions may be scrutinised. It is thus highly possible that many homosexuals would continue to live with a sense of unease if existing legislation remains. As political activist James Gomez contends

> You will always have this problem of surveillance and entrapment [of gay men] if you do not tackle the law head on. Is the sexual minority community willing to do it? … Sexual minority rights have to move to another level in Singapore. But there is no push. Only a lot of complaints. Progress over the Internet and some gay bars and saunas are not the answer. These can be shut down or policed under existing law. So the community needs to think about changing the law (personal communication, J. Gomez).

Indeed, the spatial practices of homosexuals have often come under the punishment and regulation of/by the authorities and other actors because the state and other associations possess socio-legal tools to control homosexuals. Such controls may consequently influence the socio-spatial expression of homosexuality in Singapore. However, spatial practices are increasingly not merely shaped by heterosexual 'norms'/regulations. Although it is not easy, as Gomez suggests, to change the laws, spatial practices *can* still be overt resistances to heteronormativity.

5. Where Love Dares To Speak Its Name: The Expression of Homosexuality Through Spatial Practices

5.1 The Annual 'Nation' Party: Celebrating Queerness in Singapore

In Singapore, overt resistances to the heterosexual order are indeed infrequent. However, as I shall show, spatial practices have also emerged as forms of resistances to heteronormativity. I will focus on: an annual 'Nation' party that was (and will continue to be) organised by a homosexual organisation, Fridae.com; and, the increasing artistic and intellectual expression of homosexual themes/issues to illustrate how the habitus of some homosexuals has *not* been shaped simply to perform the closet. In addition, I argue that it is only through overt spatial practices that 'continuity' and 'cohesion' of heteronormativity can be destablilised.

On 8 August 2001, the on-line homosexual organisation Fridae.com organised a coming-out party titled 'Nation'. This was the largest event organised by homosexuals (over 1500 guests attended in 2001, and 2500 in 2002) and was a very bold move in a city-state with a tough socio-legal climate towards homosexuals. As Dr Roy Chan, President of Action-For-AIDS (AFA), said, 'Nation'01' was "a stepping stone in community building in Singapore" (Fridae.com, 2002a). The rallying call for 'Nation'01' is certainly a brave proclamation of homosexuals' place in Singapore:

> This is the time to come out and celebrate our strength, individuality … in Singapore's first ever national pride party … You're proud. Of your country. Of your community. Of who you are (*Agence French Press*, 2001).

Would the emergence of 'Nation' spell the rise of more overt, and possibly confrontational, cultural politics? In this regard, Stuart Koe, one of the main party-organisers, was quick to emphasise that

> There's no political content … we're Singaporeans just like everyone else and we're here to celebrate just like everyone else … we're not about to demand any rights. For us, this is just another dance party (*Today*, 2001).

While Koe maintained that there is "no political content" in the party, his comments implicitly suggest that sexual citizenship is a key underlying theme of 'Nation', for "we're Singaporeans just like everyone else". As Koe explains further

> Nation is about the community *affirming* itself and its feelings about living in Singapore. We never intended to make this public but to make it discreet and to keep it within the ranks (*Today*, 2001; emphasis added).

That 'Nation' was staged again in 2002 and in 2003 (when a record 5000 people attended) without major societal controversy and governmental clamp-down is probably a major achievement for the homosexual community. As Koe stresses,

> Nation02 was extremely significant: by repeating the event, we are making a statement. We are for real, and we are here to stay, to be us, and to be a part of you. Our community is not going to hide behind a mask (Fridae.com, 2002b).

Again, a major impetus is about the 'affirmation' of a sense of community, that homosexuals need not "hide behind a mask" and perform heterosexual 'norms'. Furthermore, this 'affirmation' has transcended representational space into public space and the drive to be "proud" of "who you are" is clearly intended to underscore that homosexuality is *not* abnormal.

Contrary to the sense of intimidation that homosexuals may experience as a result of state regulation and discipline, the organisers of 'Nation' are unworried that there would be any possible entrapment during the party. As Koe notes

> It is highly unlikely that the Singapore government will 'entrap' anyone during the party. What will they be entrapping Singaporean citizens for? This is, after all, an event to celebrate National Day. The acceptance of Nation has already been widespread amongst the liberal straight, and foreign communities. It would be a shame if Singaporeans shunned the event out of a misinformed fear of what the government may do. Fridae.com, SGBoy, Taboo, Why Not, One Seven, etc. [websites, pubs and saunas] have all existed for many years in full view of the authorities for so many years. It would therefore be highly misinformed and paranoid to assume that Nation will fall prey when none of the other gay outlets have in all its years of existence (personal communication, S. Koe).

As in his earlier remarks, Koe's use of the term 'Singaporean' is clearly deliberate, for it refers to the entire population and *not just* the homosexual community. This is a subtle strategy of claiming that homosexuals deserve equitable rights *as* 'Singaporeans'. Indeed, that homosexuals are an *integral* part of the Singapore *nation* was accentuated in the 'Nation'02' party slogan "One People. One Nation. One Party". This slogan is palpably an allusion to the Singapore National Day Song "One People, One Nation, One Singapore". Moreover, to schedule 'Nation' just before National Day is also evidently to drive home the message that homosexuals are part of the 'One' people of Singapore. Certainly, it must be remembered that the play on words is also a strategy to circumvent existing laws against the 'promotion' of homosexuality. As X'Ho caustically comments

> Not billed as a gay party? Well, so much for 'transparency' then. Some things just can't be black-and-white. Can you imagine how 'unprepared' our *people* will be for news of such a scene in the press? So it's not the authorities' fault, hor (Ho, 2002; original emphasis).

Interestingly, as Koe notes, the authorities are also allowing other homosexual events/congregation places to operate, but for what reasons? Are the authorities really sincere about providing homosexuals with more outlets for expression? Or does it merely boil down to pure economic calculations? To X'Ho, the panoptical surveillance of homosexuals is still prevalent in Singapore, but the government allows homosexual outlets/expressions to be more overt because of homosexuals' contributions to the economy

> Question is—what called for this new 'liberalisation' of the gay-underground to operate with license above-ground? All you desperate outlet/avenue-hunting homos who think our society has gone more tolerant about your aberrant sexual preference, think again ... Methinks the real reason ... is ... more quintessentially Singapore: *money* ... The reason the authorities have allowed those Sodom-and-Gomorrah-sinful-despicable-depraved places to thrive is—they want raving homos to spend their *pink-dollar* here now (Ho, 2002).

Granted, not everyone will agree with X'Ho's contentions, but his point about the 'pink economy' is certainly worth elaborating. With cities growing more entrepreneurial (Hall and Hubbard, 1998), the 'cultural economy' is now perceived to be very strategic in enhancing a place's unique strengths (see Sayer, 1999; Clarke and Doel, 2000; du Gay and Pryke, 2002). Queer spaces and events now form significant *cultural capital* for urban economic development and place-imaging (Rushbrook, 2002). In Singapore's instance, Koe predicts that 'Nation'03' may generate some S$3–4 million (US$1.8–2.4 million) for the tourism industry (*The New Paper*, 2003). Homosexuals are also increasingly targeted by producers in Singapore, for it is perceived that they form a relatively affluent customer base. For instance, DandG Parfum's brand executive Audrey Lee feels that

> Gays appreciate the finer things in life, and thus fit into our desired image. They now form about 20 per cent of our customers and we're hoping to boost this to 50 per cent (*The Straits Times*, 2003f).

While not all homosexuals are rich, the fact that many of them do not form families means that they will have higher disposable incomes to expend on themselves and this is probably why many brands are becoming gay- and lesbian-friendly.

Recent research by Florida (2000, 2002) also reveals that there is a correlation between the tolerance of diversity in a city and its ability to attract creative global talents. As Florida says

> To some extent, homosexuality represents the last frontier of diversity in our society, and thus a place that welcomes the gay community welcomes all kinds of people (*The Straits Times*, 2002).

'Nation', for instance, has been likened to the Sydney Mardi Gras and has indeed raised Singapore's international profile as a 'fun' city

> There is no doubt that it has put Singapore on the map for partying in Asia—many consider Nation to be Asia's version of Mardi Gras—and many consider Singapore to now be the premier place to party (ahead of Hong Kong, Taiwan and even Bangkok) (personal communication, S. Koe)

> This is amazing, I've never been to anything like this in Asia. Whoever said we should skip Singapore because it's dull and sterile should experience this for himself [*sic*]. There was a lot of talk about Nation in the community back home [Hong Kong] and we are definitely not disappointed. We are really looking forward to Nation03! (Anthoni Chan, quoted in Fridae.com, 2002b).

> I've been texting my friends at home about what a ball I'm having here, and telling them that they must come next year! It's quite pricey, but I want to spoil myself a little. I had nothing but a good time here. And yup, I'll definitely be back next year (Michael Fisher, Australian tourist who came to Singapore just to attend Nation 2003, in an interview with *The Straits Times*, 2003).

Hence, is the government permitting the expression of homosexuality in Singapore just because it can help to augment Singapore's international profile and attractiveness to global talents? And is it also because queer spaces can help Singapore develop what Zukin (1995) calls the urban 'symbolic economy'? As Brian Lynch and Alex Au see it

> In a nation such as Singapore, where corporations often need to bring in people from other countries, it is frequently the case that the only people willing to take on such assignments, or who have the flexibility to do so (because of a lack of family or other commitments) are gays. Prime Minister Goh Chok Tong's recent statements are a recognition [*sic*] of the realities of life in an age of globalism (Lynch, letter to *The Straits Times*, 2003c).

> I see that the government has had to re-examine what parts of their dominant representation they need to defend, and what parts they can jettison, at little risk to their political hegemony. There are many factors at work here. Economics is certainly one of them. They need to find ways to make Singapore attractive to foreign talent, and to keep Singaporeans here. Yet, I think they're as cautious as ever regarding any threats to their hegemony. In order to avoid political challenge, the old adage of giving the people bread and circuses is as relevant as ever. It's just that in the past, they only gave bread and it was enough to keep the population quiescent. Now they have to allow some circuses too (personal communication, A. Au).

In this sense, the overt spatial expressions of homosexuality may be occurring, but that does not necessarily mean that homosexuals are accepted as part of 'mainstream' society; as aforementioned, they are merely *tolerated*. As Au (Reuters News Agency, 2001) stresses, "The further away you move from money towards speech, the more defined the restrictions coming your way". Moreover, as Ms Mavis Seow of property firm CB Richard Ellis believes, while retailers in Singapore are gunning for the Pink Dollar, "they have to be careful that they do not do so too 'loudly', otherwise straight people may begin to avoid these brands" (*The Straits Times*, 2003f). Undoubtedly then, socio-spatial boundaries still exist, and even brand names must be careful in their marketing campaigns not to become exclusively associated with homosexuality!

5.2 Artistic Expressions of Homosexuality

Still, as Koe (quoted in Ammon, 2002) puts it, "If you know the rules here you don't break them, but it doesn't mean you can't play right up to that limit, to be on the cutting edge". Another instance of such "cutting edge" form of overt resistance would be the artistic expression of homosexual-related themes/issues. Peterson (2003, p. 82)—for example, observes that, "By 1990 the general artistic climate in Singapore has begun to shift" and artistic expressions (especially theatre productions) have increased as censorship rules gradually eased. A notable production is *Asian Boys Vol. 1* (written by Singaporean playwright Alfian Sa'at), which deals with gay issues within the community (such as homophobia, racism) as well as the community's relationship with society. In the words of James Koh, *Asian Boys*

> marks an important step and direction for gay theatre in Singapore, at the very opportune moment where a more visible and burgeoning gay existence in Singapore is trying to find its precarious position in society (Koh, 2000).

Another notable play that explicitly deals with the fears and consequences of police surveillance of homosexuals, and the problems associated with organising public forums or congregation events, is *Mardi Gras*, which was performed in August 2003. Significantly, during the end of the performance, the actors made it known that the authorities have rejected the requests of homosexuals to 'out' themselves to the audience. This refusal clearly suggests that, while the authorities are prepared to allow plays of homosexual-related themes to go on stage, there is still a need to maintain socio-spatial boundaries and keep homosexuals relatively 'invisible'. Nonetheless, it still appears that artistic expression has generally become an important socio-spatial practice of being 'out' to society. As Sa'at observes

> What is more interesting to me as a sociological phenomenon is that there have been three separate readings at Chijmes, Substation and Borders [three different buildings in central Singapore] whereby there were these people who came up and read their own gay poetry. For me, it amounts to as much as a public coming out to complete strangers. It is interesting as it seems that homosexual expression has been so repressed to a point that if they can get outlets of self-expression such as these poetry reading sessions, they will do it. You have this absolutely silent minority, supposedly silent invisible minority that is using platforms like these. Interestingly, these people are actually quite young and they are not out to project themselves as flamboyant. These people are introverted writers and they find their right to self-expression through declaring, in a rather oblique way, their homosexuality to these kinds of audience (Sa'at, 1999).

Most significantly, the book *People Like Us* (2003) was finally published after several rejections from cautious publishers. This is the first book detailing the debates and issues regarding homosexual rights in Singapore and is sold in major bookstores. In addition, this book received wide press coverage and no report has been derogative (see Figure 1). The publication process of the book was not smooth-sailing, however, as "nobody wanted to touch it" until Select Publishing's owner Lena Lim accepted it for publishing. As Lim says

> We felt it was a very serious piece of work and it was a very worthy project ... I'm not advocating a lifestyle, but these are people who go about their daily lives and they deal with challenges all their lives (*The Straits Times*, 1 March 2003).

Again, how much of the relaxation of censorship regulations is tied to a genuine desire to incorporate homosexuals' voices in society? And how much of it is for economic objectives? As Singapore is gearing to be a "Global City for the Arts" (Chang, 2000; Yeoh and Chang, 2001; Chang and Lee, 2003), the authorities may be permitting more liberal expressions to portray an image of the artistic diversity generally associated with global cities. But as I have mentioned, the government is still strongly censoring artistic expressions that are deemed to be too disruptive of heteronormativity (see Au, 2000; *The Straits Times*, 2001; Peterson, 2003). Again, the issue may not merely be about money. As Law lecturer Kenneth Tan sees it

> There is a link between continued economic success and the PAP government's continued political legitimacy. But the moral majority is also the electoral majority, so the Government must perform a balancing act (*The Straits Times*, 2003a).

Quite simply then, while money is important, the reinforcement of heteronormativity is imperative

> Let me remind you that you're *not* exactly getting the delicious un-Hollywood gay cult-movies on the silver screen ... Try telling me that the authorities are finally tolerant of homos and I'll surely ask why Fassbinder's Querelle was missing from the Goethe Institute's program of the German film-director's career-retrospective ... I know the film's banned ... but many bans have been quietly "revoked", no? (Think True Romance—that Christian Slater film. Or check the new Rolling Stones compilation Forty Licks, aren't Get Off My Cloud and Wild Horses banned? Weren't they?). I mean, Querelle was Fassbinder's last film before he committed suicide. Doesn't that mean anything? Guess not, 'cos it's not about money or the economy (Ho, 2002; original emphasis).
>
> The point isn't money, or the making of money, but political power and their inviolability of it. It's not about what businesses

L16 LIFE! •

Sexual minorities here speak up

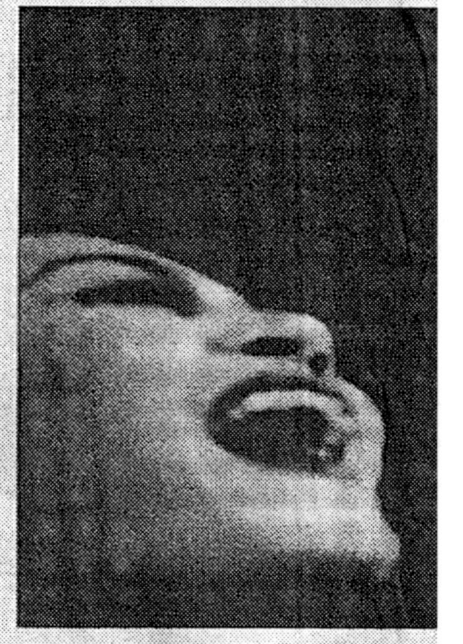

TRUTH OR TRASH?: My Sisters absolute rubbish, says Ms Leo

A transsexual who underwent an operation in Thailand is one of the 10 featured in one of two books, which look at communities of sexual minorities in Singapore

By **ONG SOR FERN**
ARTS CORRESPONDENT

TWO new books put the spotlight on the communities of sexual minorities in Singapore.

People Like Us, a collection of essays about the gay community in Singapore, was launched last Saturday at Select Books. It is co-edited by Joseph Lo and Huang Guoqin.

My Sisters — Their Stories, a coffee-table book focusing on the experiences of 10 transsexuals, is due to be published by Viscom Editions in April. It is written by Leona Lo.

People Like Us had a hard time finding a publisher before Select Books owner and publisher Lena Lim took it on, says Mr Lo, who is in his 30s.

"No one wanted to touch it with a 10-foot pole," says the project co-ordinator for a United Nations Development Programme's poverty alleviation project based in Tibet.

"I just take my hat off to Select who was willing to run with it. Without its institutional support, this book won't be possible."

Mrs Lim, however, is matter-of-fact about the affair: "We felt it was a very serious piece of work and it was a very worthy project."

The book's essays range from lighthearted discussions of gay culture to more academic essays on gay identity and rights.

She adds that it is similar to another Select book, Building Social Space In Singapore.

"It's about making space in society. I don't think our publishing it is in any way promoting the cause although you could take it that way. I think our society is mature enough to talk about these things honestly and with candour."

Ms Lo, 27, author of My Sisters, had an easier time finding a publisher.

She met Ms Sylvia Tan, publisher and owner of Viscom Editions, in 2001 to discuss a book project about musicals. But Ms Tan became more intrigued by Ms Lo's life story — the latter is a transsexual who underwent an operation in Thailand in 1997.

Ms Lo, who is in public relations, says: "I wanted to feature the stories of successful transsexuals. A lot of information out there is mediated through sensational films like Bugis Street which we felt was absolute rubbish."

But she ran into difficulties as some of the transsexuals she approached were reluctant to be "outed".

Of the 10 transsexuals featured, eight are Singaporean and two are Thai.

One of the Singaporeans is a well-known face who will be "familiar to those who watch local television".

The author says the book is aimed at "giving transsexuals a name and a voice".

She adds: "I'm not advocating a lifestyle. But these are people who go about their daily lives and they deal with challenges all their lives."

◆ *People Like Us, priced at $15, is available at Select Books. My Sisters — Their Stories, priced at $25, will be available at Select Books from April 5.*

Figure 1. 'Sexual minorities here speak up' (*The Straits Times*, 1 March 2003).

do, but about that they avoid doing. So long as businesses pose no threat to the government's political hegemony and their setting of the political agenda for Singapore, the government is not going to pay any special attention (personal communication, Alex Au).

Faced with existing stringent regulations, many homosexuals are still wary of confrontational challenges to heteronormativity.

They are quick to deny involvement in identity politics and many choose to establish their presence in the arts, entertainment and, especially, in business. As Koe says, a

> gradual and gentle integration [of gay events like 'Nation'] into the mainstream consciousness is one of the most effective means for Singaporeans to slowly learn that gay people in Singapore exist (personal communication, S. Koe).

But as Ammon notes, avoiding direct challenges to the heteronormative order is

> also a political process, [which reflects the] evolution of a community that has discerned wiser survival tactics against a very alert and reactive government (Ammon, 2002).

In short, these strategies of resistance accentuate the fact that homosexuals *can* be creative and anticipative subjects working to determine their own lives and enhance their dignity. Of course, structural constraints remain and resistances have to be gradual rather than revolutionary and mordacious (as happened in other parts of the world). In fact, a gradual move to widen societal awareness of homosexuality might work better to reshape the habitus of heterosexuals—heterosexuals' dispositions to conceive of homosexuals as the abnormal 'other' may be able to change, although this would definitely entail time and effort to materialise. At present, the degree to which spatial practices function as acts of resistance remains contingent on the state's approval, a caveat that is well summed up in a recent issue of *Time*

> Singapore is *not* an entirely liberated city—*yet*. The city's gays—mostly diligent, discreet workers and students—were relieved by [Prime Minister] Goh's recent announcement and by the generally supportive readers' letters published in the *Straits Times* (even if there were some hardened homosexuality-is-a-sin exceptions). But the *token* coming-out story that the newspaper also printed revealed just how many gays were *still* in closets, even if they do *risk* a visit to the clubs on weekends (*Time*, 2003b; emphasis added).

6. Conclusion

Adopting and expanding on Lefebvre's (1991) spatial triad, I have argued that the heterosexuality of quotidian space in Singapore is actively produced, although it is "always partial, in the process of becoming and unstable" (Valentine, 2002, p. 155). I have also discussed how the techniques of governmentality (the use of statistics, educational content) and direct juridical control may shape homosexuals' 'techniques of the self' in Singapore and may also reinforce heterosexuals' perceptions of homosexuals as the unnatural and undesirable 'Other'. Despite these tight socio-political constraints, however, it is hard to deny that homosexual expressions are increasingly 'visible' in this island city-state. Spatial practices thereby do not just maintain societal cohesion and continuity—they can in fact destabilise heteronormativity, leading in turn to the *dis*continuity of socially produced heterosexual space and, by extension, to a new and more inclusionary level of 'cohesion and continuity'. This suggests that a retreat into marginal representational spaces may not be the panacea for oppressed groups that are seeking to alter the *status quo ante* emplaced by predominant ideologies—it simply reinforces the idea that these groups belong *somewhere* in the margins, somewhere invisible. Changing space therefore require changing spatial practices.

Narrowly to construe spatial practices as repetitive acts that conform to the dominant order, and representational spaces as the primary potent sites of resistance, may work to reinforce the public–private binary. While it may be argued that extant research evinces that many homosexuals in Singapore choose mainly private spaces before actually revealing their sexual identities, often for fear of social opprobrium or loss of 'face' (Tay, 1997, Tan, 1998; Chng, 1999), does this then mean that Singaporean homosexuals are all confined to 'scripting' their sexual identities according to what is socially and morally 'acceptable' in public, while the expressions of their inner feelings and identities are only circumscribed to those private spaces that are not under the peril of socio-legal scrutiny? As this paper has shown, this need not be the case and the strong support of the 'Nation' party, the artistic expression of homosexual issues and the lively gay and lesbian entertainment scenes are just some indicators that homosexuals are increasingly unafraid of expressing their sexual identity in public, of leading lives *as* homosexuals. Here, I believe the crux of the issue centres on one important factor—human agency.

This courage to engage in spatial practices that do not 'fit' the dominant representation of space clearly accentuates the *conscious* agency of subjected individuals. It is through these overt socio-spatial expressions that greater awareness about minority groups like homosexuals can be effected. After all, the Minister David Lim (1999b) did stress that "Each citizen must contribute to our society's transformation, and not just by words, but also by actions". Moreover, it is interesting that gay issues have taken 'top spot' in a recent 'Remaking Singapore' e-forum hosted by MCDS, which suggests that Singaporeans are actually interested in discussing homosexual issues (*The New Paper*, 1 Octber 2001). To this, we may add that Singaporeans' habitus towards homosexual issues may be positively changed provided that there is room for discussion and greater awareness. As Sack (1999, p. 38) says, "Most evil is ultimately a lack of awareness ... the tightness of evil's grip is due to ignorance".

Of course, structural power is and remains important and I have shown how homosexuals are still boxed into an essentialised fixity, wherein the expression of homosexuality is regulated, stigmatised and, in some instances, even criminalised (sections 3 and 4). However, analysts should not undermine the creative ability of subjected beings (homosexuals in this instance) to swim against strong structural currents (heteronormativity; section 5). Indeed, the 'gradual' and non-confrontational forms of identity politics evince that negotiating structural power need not be through incendiary revolutions. As Alex Au (quoted in Ammon, 2002) says, "In the process of doing business, we are also exercising political muscle more than ever before".

Recently, it is interesting that a book, *Postmodern Singapore* (W. S. W. Lim, 2002; see also W. S. W. Lim, 2001b) actually celebrates Singapore as a 'post-modern' city. However, as many theorists see it, a 'post-modern' city is one that acknowledges syncretism, 'otherness' and 'difference' (Smith, 1993; Soja, 1996a, 1996b; Byrne, 1997; Dunn, 1998; Dear, 2000; Allmendinger, 2001; Relph, 2001; Sibley, 2001) and it may be too early to label the city-state as 'post-modern' in any sense given that we are not witnessing the evanescence of social engineering policies in Singapore. However, as this paper has evinced, the increasing room purveyed for the expression of homosexuality suggests that Singapore may indeed be growing into a city where multiple sexualities can not only co-exist, but also begin to respect one another—a city where love can really dare to speak its name.

Notes

1. In Singapore, just like anywhere else, it is difficult to define who exactly constitutes a 'homosexual'. The basis of a definition could be predicated on just sexual activity, or sexual orientation, or both. In this paper, I provide an inclusive definition of 'homosexuals' as men and women who feel attracted to the same sex, regardless of any sexual activity, and this include bisexuals and transvestites. Clearly then, 'homosexuals' do not constitute a homogeneous group. Gays and lesbians, for instance, may not share common perceptions and demeanours.
2. Singapore's Penal Code 377(A) states that

   > Any male person who, in public or private, commits or abets the commission of, or procures the commission by any male person of, any act of gross indecency with another male person, shall be punished with imprisonment for a term which may extend to 2 years.

3. Singapore's Section 20 of the Miscellaneous Offences (Public Order and Nuisance) Act (Cap 184, 1997 Ed) states that

   > Any person who is found guilty of any

riotous, disorderly or indecent behaviour in any public road or in any public place or place of public amusement or resort, or in the immediate vicinity of, or in, any court, public office, police station or place of worship, shall be guilty of an offence and shall be liable on conviction to a fine not exceeding $1,000 or to imprisonment for a term not exceeding one month and, in the case of a second or subsequent conviction, to a fine not exceeding $2,000 or to imprisonment for a term not exceeding 6 months.

4. According to the Singapore Department of Statistics' (2001) report on Singapore's 2000 Population Census, the resident population of Singapore is 3.26 million people, while non-residents comprise 0.75 million people. Of the resident population, 76.8 per cent are ethnic Chinese, 13.9 per cent Malays, 7.9 per cent Indians and 1.4 per cent Others (i.e. of other races). There are slightly more females than males, with the sex ratio being 998 males per 1000 females. Like many places world-wide, the Census does not provide any data on the 'homosexual' population in Singapore, so it is inconceivable even to determine the exact number of people who come under this classification term.
5. In Singapore, there is no specific number of 'gay and lesbian venues' as establishments do not advertise themselves as 'gay and lesbian' in any directory. There are also no clear boundaries demarcating what are gay and lesbian areas in Singapore.
6 Singapore's Penal Code 377 states that

Whosoever voluntarily has carnal intercourse against the order of nature with any man, woman or animal, shall be punished with imprisonment for life, or with imprisonment for a term which may extend to 10 years, and shall also be liable to a fine.

References

Agence French Press (2000) Ground-breaking survey shows Singaporeans accept homosexuality, 22 May.

Agence French Press (2001) Gay and Lesbian community parties into Singapore's National Day, 6 August.

ALLEN, J. and PRYKE, M. (1994) The production of service space, *Environment and Planning D*, 12, pp. 453–475.

ALLMENDINGER, P. (2001) *Planning in Postmodern Times*. London: Routledge.

AMMON, R. (2002) The new gay Singapore part three, March (http://www.traveland transcendence.com/g-sing02.html; accessed 3 January 2003).

Asiaweek (Hong Kong) (2000) Showing 'greater humanity' family values trump a tough stand on HIV, 9 June (http://www.asiaweek.com/asia week/magazine/2000/0609/nat.singapore.html; laccessed 2 December 2002).

AU, A. (1999a) Foam party in Parliament, March. (http://www.geocities.com/yawning_bread/yax-133.htm; accessed on 30 March 2004).

AU, A. (1999b) Don't ask, don't tell, May (http://www.geocities.com/yawning_bread/yax-142.htm; accessed on 19 February 2003).

AU, A. (2000) Do gays have a place in Singapore, June http://www.geocities.com/yawning_bread/imp-076.htm; accessed on 12 March 2003).

AU, A. (2001a) The arrests at One Seven and Section 20, November (http://www.yawning bread.org/index2.htm; accessed on 9 April 2004).

AU, A. (2001b) Society as ecology, August (http://www.geocities.com/WestHollywood/5738/yax-242.htm; accessed on 22 Jan 2003).

BAETEN, G. (2002) Hypochondriac geographies of the city and the new urban dystopia: coming to terms with the 'other' city, *City*, 6, pp. 103–115.

BAUMAN, Z. (1997) *Postmodernity and its Discontents*. Cambridge: Polity Press

BELL, D. J. (1991) Insignificant others: lesbian and gay geographies, *Area*, 23, pp. 323–329.

BELL, D. J. and BINNIE, J. (2000) *The Sexual Citizen: Queer Theory and Beyond*. Malden: Polity Press.

BERGER, J. (1972) *Ways of Seeing*. London: British Broadcasting Corporation; Harmondsworth: Penguin.

BETSKY, A. (1997) *Queer Space: Architecture and Same-sex Desire*. New York: William Morrow & Co.

BINNIE, J. (1997) Coming out of geography: towards a queer epistemology?, *Environment and Planning D*, 15, pp. 223–237.

BINNIE, J. and VALENTINE, G. (1999) Geographies of sexuality—a review of progress, *Progress in Human Geography*, 23, pp. 175–187.

BINNIE, J. and VALENTINE, G. (2000) Geographies of sexuality—a review of progress, in: T. SANDFORT *ET AL* (Eds) *Lesbian and Gay Studies: An Introductory, Interdisciplinary Approach*, pp. 132–145. London: Sage Publications.

BLUNT, A. and WILLS, J. (2000) *Dissident Geographies: An Introduction to Radical Ideas and Practice*. Essex: Prentice.

BOURDIEU, P. (1989) *The Logic of Practice*. Cambridge: Polity Press.

BOURDIEU, P. (1992) *An Invitation to Reflexive Sociology*. Cambridge: Polity Press.

BROWN, M. (2000) *Closet Space: Geographies of Metaphor from the Body to the Globe*. London: Routledge.

BROWN, M. and KNOPP, L. (2003) Queer cultural geographies—we're here! We're queer! We're over there, too!, in: K. ANDERSON, M. DOMOSH, S. PILE and N. THRIFT (Eds) *Handbook of Cultural Geography*, pp. 313–324. London: Sage Publications.

BUTLER, J. (1990) *Gender Trouble: Feminism and the Subversion of Identity*. London: Routledge.

BUTLER, J. (1991) Imitation and gender insubordination, in: D. FUSS (Ed.) *Inside/Out: Lesbian Theories, Gay Theories*, pp. 13–31. London: Routledge.

BUTLER, J. (1993) *Bodies that Matter: On the Discursive Limits of Sex*. London: Routledge.

BUTLER, J. (2001) Doing justice to someone: sex reassignment and allegories of transsexuality, *GLQ: A Journal of Lesbian and Gay Studies*, 7(4), pp. 621–636.

BYRNE, D. (1997) Chaotic places or complex places? Cities in a post-industrial era, in: S. WESTWOOD and J. WILLIAMS (Eds) *Imagining Cities: Scripts, Signs, Memories*, pp. 50–70. London: Routledge.

CASTELLS, M. (1983) *The City and the Grassroots: A Cross-cultural Theory of Urban Social Movements*. London: E. Arnold.

CHANG, T. C. (2000) Renaissance revisited: Singapore as a 'global city for the arts', *International Journal of Urban and Regional Research*, 24, pp. 818–831.

CHANG, T. C. and LEE, W. K. (2003) Renaissance city Singapore: a study of arts spaces, *Area*, 35, pp. 128–141.

CHNG, M. E. H. (1999) *All dressed up but no place to go? Gay men's space in Singapore*. Unpublished academic exercise, Department of Geography, National University of Singapore.

CHOUINARD, V. and GRANT, A. (1996) On being not even anywhere near 'the project': ways of putting ourselves in the picture, in: N. DUNCAN (Ed.) *Body Space: Destabilizing Geographies of Gender and Identity*, pp. 170–193. London: Routledge.

CLARKE, D. B. and DOEL, M. A. (2000) Cultivating ambivalence: the unhinging of culture and economy, in: I. COOK, D. CROUCH and S. NAYLOR (Eds) *Cultural Turns/Geographical Turns: Perspectives on Cultural Geography*, pp. 214–233. New York: Prentice Hall.

DANAHER, G., SCHIRATO, T. and WEBB, J. (2000) *Understanding Foucault*. London: Sage.

DEAR, M. J. (2000) *The Postmodern Urban Condition*. Maldon, MA: Blackwell.

DONALD, J. (1997) This, here, now: imagining the modern city, in: S. WESTWOOD and J. WILLIAMS (Eds) *Imagining Cities: Scripts, Signs, Memories*, pp. 181–201. London: Routledge.

DUNCAN, N. (1996) Renegotiating gender and sexuality in public and private spaces, in: N. DUNCAN (Ed.) *Body Space: Destabilizing Geographies of Gender and Identity*, pp. 127–145. London: Routledge.

DUNN, R. G. (1998) *Identity Crises: A Social Critique of Postmodernity*. Minneapolis, MN: University of Minnesota Press.

ELWOOD, S. A. and MARTIN, D. G. (2000) 'Placing' interviews: location and scales of power in qualitative research, *The Professional Geographer*, 52, pp. 649–657.

ESTER, P. and VINKEN, H. (2003) Debating civil society: on the fear for civic decline and hope for the internet alternative, *International Sociology*, 18, pp. 659–680.

FITZSIMONS, P. (1999) Michel Foucault: regimes of punishment and the question of liberty, *International Journal of the Sociology of Law*, 27, pp. 379–399.

FLORIDA, R. (2000) Forum: Pittburgh's prosperity depends on diversity, 15 October (http://www.post-gazette.com/forum/20001015edflorida8.asp; accessed on 17 March 2003).

FLORIDA, R. (2002) The economic geography of talent, *Annals of the Association of American Geographers*, 92, pp. 743–755.

FOUCAULT, M. (1979) *Discipline and Punish: The Birth of the Prison*. New York: Vintage Books.

FOUCAULT, M. (1984/1946) The ethics of the concern for self as a practice of freedom, in: S. LOTRINGER (Ed.) *Foucault Live (Interviews, 1961–1984)*, trans. by L. HOCHROTH and J. JOHNSON, p. 434. New York: Semiotext(e).

FRIDAE.COM (2002a) Singapore's Nation02 to be held on August 8, 25 June (http://www.fridae.com/magazine/en20020625_1_1.php; accessed on 11 March 2003).

FRIDAE.COM (2002b) Singapore celebrates gay pride, 14 August (http://www.q.co.za/2001/2002/08/14-singaporepride.html; accessed online on 12 March 2003).

GAY, P. DU and PYRKE, M. (Eds) (2002) *Cultural Economy: Cultural Analysis and Commercial Life*. London: Sage.

GOMEZ, J. (2000) *Self-censorship: Singapore's Shame*. Singapore: Think Centre.

GREGSON, N. and ROSE, G. (2000) Taking Butler elsewhere: performativities, spatialities and subjectivities, *Environment and Planning D*, 18, pp. 433–452.

HALL, T. and HUBBARD, P. (1998) *The Entrepreneurial City: Geographies of Politics, Regime, and Representation*. New York: Wiley.

HANNAH, M. G. (1997) Space and the structuring of disciplinary power: an interpretive review, *Geografiska Annaler*, 79B, pp. 171–180.

HENG, R. (2001) Tiptoe out of the closet: the

before and after of the increasingly visible gay community in Singapore, in: G. SULLIVAN and P. A. JACKSON (Eds) *Gay and Lesbian Asia: Culture, Identity, Community*, pp. 81–97. Binghampton: Harrington Park Press.

HO, X' (2002) Never mind the Sodom, here's the Moolah: remaking Singapore without question, *The Bigo Weekly Review*, 2 (http://www.bigOmags@bigo.com.sg/archive/ARxho files/ARxhonov02.html; accessed on 12 March 2003).

IGLA (2000) World legal survey: Singapore, June (http://www.ilga.org/Information/legal_survey/asia_pacific/singapore.htm; accessed on 19 February 2003).

INFOCOMM DEVELOPMENT AUTHORITY (Singapore) (2002) Annual survey on infocomm usage in households and by individuals for 2002, (http://www.ida.gov.sg/idaweb/doc/download/I2387/Annual_Survey_on_Infocomm_Usage_in_Hous eholds_2002.pdf; accessed on 31 March 2004).

ISIN, E. F. and WOOD, P. K. (1999) *Citizenship and Identity*. London: Sage.

KITCHIN, R. and LYSAGHT, K. (2003) Heterosexism and the geographies of everyday life in Belfast, Northern Ireland, *Environment and Planning A*, 35, pp. 489–510.

KNOPP, L. (1995) Sexuality and urban space: a framework for analysis, in: D. BELL and G. VALENTINE (Eds) *Mapping Desire: Geographies of Sexualities*, pp. 149–164. London: Routledge.

KOH, J. (2000) Asian Boys Vol. 1 by The Necessary Stage: Review (http://inkpot.com/theatre/asian2.html; accessed on 18 March 2003).

KONG, L. (1993) Ideological hegemony and the political symbolism of religious buildings in Singapore, *Environment and Planning D*, 11, pp. 23–45.

KONG, L. (1998) Refocusing on qualitative methods: problems and prospects for research in a specific Asian context, *Area*, 30, pp. 79–82.

LEFEBVRE, H. (1991) *The Production of Space*, trans. by D. Nicholson-Smith. Oxford: Blackwell.

LEONG, L. W. T. (1997) Singapore, in: D. J. WEST and R. GREEN (Eds) *Sociological Control of Homosexuality: A Multi-Nation Comparison*, pp. 127–144. New York: Plenum Press.

LIGGETT, H. (1995) City sights/sites of memories and dreams, in: H. LIGGETT and D. C. PERRY (Eds) *Spatial Practices: Critical Explorations in Social/Spatial Theory*, pp. 243–273. London: Sage.

LIM, C.-S. (2002a) Serving Singapore as a gay man—part 1, September (http://www.geocities.com/WestHollywood/5738/about_singapor2.htm; accessed on 12 March).

LIM, C.-S. (2002b) Serving Singapore as a gay man—part 2, October (http://www.geocities.com/WestHollywood/5738/about_singapor2.htm; accessed on 12 March).

LIM, D. T. E. (1999a) Speech at the NUS Political Association Forum, at Guild House, Kent Ridge at 7.00 pm, 6 August (http://www.singapore21.org.sg/speeches_060899.html; accessed on 12 April 2004).

LIM, D. T. E. (1999b) Speech at the Singapore 21 Conference "The Singapore 21 Vision: Moving Forward" at Singapore International Convention & Exhibition Centre (Suntec City), Level 2, 31 July (http://www.singapore21.org.sg/speeches_310799.html accessed on 12 April 2004).

LIM, M. (1997) Madeleine Lim: a Singaporean lesbian filmmaker, interview with Sintercom, 14 July (http://www.geocities.com/newsintercom/sp/interviews/madeleine.html; accessed on 19 March 2003)

LIM, V. (2000) Gender differences and attitudes towards homosexuality among Singapore youth, *The* Act, 23 (http://afa.org.sg/issue/issue23/frame_genderdiff.html; accessed on 18 March 2003).

LIM, W. S. W. (2001a) Spaces of indeterminacy. Revised version of paper first delivered at Conference on "Bridge the Gap" Fukuoka, Japan, July, organised by City of Kitakyushu and the Centre for Contemporary Art, CCA, Kitakyushu (http://www.btgjapan.org/app/app_002.html; accessed online on 19 December 2002).

LIM, W. S. W. (2001b) *Alternatives in Transition: The Postmodern, Glocality and Social Justice*. Singapore: Select Publishing.

LIM, W. S. W. (Ed.) (2002) *Postmodern Singapore*. Singapore: Select Publishing.

LO, J. and HUANG, G. (Eds)(2003) *People Like Us: Sexual Minorities in Singapore*. Singapore: Select Publishing.

MARKOWE, L. (1996) *Redefining the Self: Coming Out as Lesbian*. Cambridge, MA: Polity Press.

MASSEY, D. (1994) *Space, Place and Gender*. Minneapolis, MN: University of Minnesota Press.

MCDOWELL, L. (1992) Doing gender: feminism, feminists and research methods in human geography, *Transactions of the Institute of British Geographers*, 17, pp. 399–416.

MCDS (MINISTRY OF COMMUNITY DEVELOPMENT AND SPORTS) (2002) Survey of social attitudes of Singaporeans, conducted by D. Chan, September (http://www.mcds.gov.sg/MCDSFiles/download/MCDS%20Final.pdf; accessed on 19 March 2003).

MCNAY, L. (1999) Gender, habitus and the field: Pierre Bourdieu and the limits of reflexivity, *Theory, Culture and Society*, 16, pp. 95–117.

MERRIFIELD, A. (1993) Place and space: a Lefebvrian reconciliation, *Transactions of the Institute of British Geographers*, 18, pp. 516–531.

MERRIFIELD, A. (1995) Situated knowledge through exploration: reflections on Bunge's 'geographical expeditions', *Antipode*, 27, pp. 49–70.

MILLER, P. and ROSE, N. (1995) Producers, Identity and Democracy, *Theory and Society*, 24, pp. 427–467.

MITCHELL, D. (2000) *Cultural Geography: A Critical Introduction*. Malden, MA: Blackwell Publishers

MORAN, L. J. (2001) The gaze of law: technologies, bodies, representation, in: R. HOLLIDAY and J. HASSARD (Eds) *Contested Bodies*, pp. 107–116. London: Routledge.

MYSLIK, W. D. (1996) Renegotiating the social/sexual identities of places: gay communities as safe havens or sites of resistance?, in: N. DUNCAN (Ed.) *Body Space: Destabilizing Geographies of Gender and Identity*, pp. 156–169. London: Routledge.

NG, K. K. (1999) *The Rainbow Connection: The Internet and the Singapore Gay Community*. Singapore: KangCuBine Publishing.

NG, T. (2003) Law and homosexuals, in: J. LO and G. HUANG (Eds) *People Like Us: Sexual Minorities in Singapore*, pp. 17–20. Singapore: Select Publishing.

PETERSON, W. (2003) The queer stage in Singapore, in: J. LO and G. HUANG (Eds) *People Like Us: Sexual Minorities in Singapore*, pp. 78–96. Singapore: Select Publishing.

PETROVIC, J. E. (1999) Moral democratic education and homosexuality: censoring morality, *Journal of Moral Education*, 28, pp. 201–209.

PHILIPS, R., WATT, D. and SHUTTLETON, D. (Eds) (2000) *De-centring Sexualities: Politics and Representations Beyond the Metropolis*. London: Routledge.

PIETERSE, J. N. (2001) Hybridity, so what? The anti-hybridity backlash and the riddles of recognition, *Theory, Culture and Society*, 18, pp. 219–245.

PILE, S. (1996a) *The Body and the City*. London: Routledge.

PILE, S. (1996b) Introduction: opposition, political identities and spaces of resistances, in: S. PILE, M. KEITH (Eds) *Geographies of Resistance*, pp. 1–32. London: Routledge.

PRITCHARD, A. and MORGAN, N. J. (2000) Constructing tourism landscapes—gender, sexuality and space, *Tourism Geographies*, 2, pp. 119–139.

RELPH, E. (2001) The critical description of confused geographies, in: P. C. ADAMS, S. HOELSCHER, and K. E. TILL (Eds) *Textures of Place: Exploring Humanist Geographies*, pp. 150–166. Minneapolis, MN: University of Minnesota Press.

REUTERS NEWS AGENCY (2001) Gays find tacit acceptance but some seek more, 2 July (http://www.singapore-window.org/sw01/010702re.htm; accessed on 19 March 2003).

RHOADS, R. A. (1994) *Coming Out in College: The Struggle for a Queer Identity*. Westport, CT: Bergin & Garvey.

RICHARDSON, D. (1998) Sexuality and citizenship, *Sociology*, 32, pp. 83–100.

RICHARDSON, D. (2000) *Rethinking Sexuality*. London: Sage.

ROSE, G. (1997) Situating knowledges: positionality, reflexivities and other tactics, *Progress in Human Geography*, 21, pp. 305–320.

RUSHBROOK, D. (2002) Cities, queer space, and the cosmopolitan tourist, *GLQ: A Journal of Lesbian and Gay Studies*, 8, pp. 183–206.

SA'AT, A. (1999) *An Interview with Alfian Sa'at, A4ria Journal* 4, May (http://www.geocities.com/SunsetStrip/Studio/6728/issue_4.htm; accessed on 12 March 2003).

SACK, R. D. (1997) *Homo Geographicus*. Baltimore, MD: The Johns Hopkins University Press.

SACK, R. D. (1999) A sketch of a geographic theory of morality, *Annals of the Association of American Geographers*, 89, pp. 26–44.

SAL (2001) Is this my country? Is this my land?, July (http://www.geocities.com/WestHollywood/5738/guw-074.htm#foot1; accessed on 19 Feb 2003).

SAYER, A. (1999) Valuing culture and economy, in: L. RAY and A. SAYER (Eds) *Culture and Economy after the Cultural Turn*, pp. 53–75. London: Sage.

SIBLEY, D. (2001) The binary city, *Urban Studies*, 38, pp. 239–250.

SINGAPORE DEPARTMENT OF STATISTICS (2001) *Singapore population* (www.singstat.gov.sg/keystats/c2000/handbook.pdf; accessed 8 April 2004).

SMITH, C. (2000) The sovereign state v Foucault: law and disciplinary power, *The Sociological Review*, 48, pp. 283–306.

SMITH, N. (1993) Homeless/global: scaling places, in: J. BIRD, B. CURTIS, T. PUTNAM and L. TICKER (Eds) *Mapping the Futures: Local Cultures, Global Change*, pp. 87–119. London: Routledge.

SOH, G. C. P. (2003) *Cyberspace and sexuality: a geographical study*. Unpublished academic exercise, Department of Geography, National University of Singapore.

SOJA, E. (1996a) Margin/alia, in: A. MERRIFIELD and E. SWYNGEDOUW (Eds) *The Urbanization of Injustice*, pp. 180–199. London: Lawrence and Wishart.

SOJA, E. (1996b) *Thirdspace: Journeys to Los Angeles and Other Real-and-Imagined Places*. Oxford: Blackwell.

SOJA, E. (1997) The social-spatial dialectic, in: T.

BARNES and D. GREGORY (Eds) *Reading Human Geography: The Poetics and Politics of Inquiry*, pp. 244–256. London: Arnold.

SOJA, E. (1999) Thirdspace: expanding the scope of the geographical imagination, in: D. MASSEY, J. ALLEN and P. SARRE (Eds) *Human Geography Today*, pp. 260–278. Cambridge: Polity Press.

Sunday Morning Post (Hong Kong) (2000) Gays don't count in Singapore, 28 May.

TAN, S. S. H. (1998) *Modernity and identity: 'gay Christians' in Singapore*. Unpublished academic exercise, Department of Sociology, National University of Singapore.

TAY, R. C. H. (1997) *Your place or mine? A geographic study of courtship spaces in Singapore*. Unpublished academic exercise, Department of Geography, National University of Singapore.

TEO, C. H. (1999) Welcome Address at the Launch of the Singapore 21 Vision, Ngee Ann City Civic Plaza, 24 April (http://www.singapore21.org.sg/speeches_240499_2.html; accessed on 12 April 2004).

The Business Times (Singapore) (2003) S'pore slips to 4th place in globalization index, 9 Jan (http://business-times.asia1.com.sg/news/story/0,2276,69096,00.html?; accessed on 9 January 2003).

The New Paper (Singapore) (1996) Homepage SHOCK, 21 December.

The New Paper (Singapore) (2002) Gay issues take top spot, 1 October (http://newpaper.asia1.com.sg/top/story/0,4136,1348-1033487940,00.html; accessed on 20 February 2003).

The New Paper (Singapore) (2003) No Mardi Gras, says PM. Yet … gay poser on National Day, 6 July.

The Straits Times (Singapore) (1999a) Nightspots monitored closely, 13 March.

The Straits Times (Singapore) (1999b) Interview with Mr Teo Chee Hean, 27 April.

The Straits Times (Singapore) (2000a) Bid to open dialogue on gays in society, 6 May.

The Straits Times (Singapore) (2000b) How could forum have 'imposed' on others?, 2 June.

The Straits Times (Singapore) (2000c) All can be part of Singapore 21, 6 June.

The Straits Times (Singapore) (2000d) Let's face it, society is now more tolerant of gays, 12 June.

The Straits Times (Singapore) (2001) Censorship had its ups and downs, 12 October.

The Straits Times (Singapore) (2002) Making room for the three 'T's, 14 July.

The Straits Times (Singapore) (2003a) Government more open to employing gays Now, 4 July.

The Straits Times (Singapore) (2003b) Gum, gay and the goggle box: time to consider a U-turn, 12 July.

The Straits Times (Singapore) (2003c) Gay rights study heavily tainted, 15 July.

The Straits Times (Singapore) (2003d) Not All Heterosexuals feel the same as George Lim, 18 July.

The Straits Times (Singapore) (2003e) Keep an open mind and respect differing views, 18 July.

The Straits Times (Singapore) (2003f) Chasing the pink dollar, 17 August.

The Sunday Times (Singapore) (1998) Singaporeans too reliant on government? SM Disagrees, 13 December.

Time (US) (2001) Boys night out: we're here. We're queer. Get used to it. Can Singapore accept its gay community?, (http://www.time.com/time/asia/features/sex/sexgay.html; 19 March accessed on 16 October 2003).

Time (US) (2003a) The lion in winter, 30 June (http://www.time.com/time/asia/covers/501030707/sea_singapore.html; accessed on 17 August 2003).

Time (US) (2003b) Singapore: It's in to be Out, 18 August http://www.time.com/time/asia/magazine/printout/0,13675,501030818-474512,00.html; accessed on 13 October 2003).

THRIFT, N. (1996) The still point: resistance, expressive embodiment and dance, in: S. PILE and M. KEITH (Eds) *Geographies of Resistance*, pp. 124–151. London: Routledge.

Today (Singapore) (2001) Why all this fuss? Gays peeved as spotlight turns on N-Day bash, 7 August.

VALENTINE, G. (1993) 'Desperately seeking Susan': a geography of lesbian friendships, *Area*, 25, pp. 109–116.

VALENTINE, G. (1996) Renegotiating the 'heterosexual street': lesbian productions of space, in: N. DUNCAN (Ed.) *Body Space: Destabilizing Geographies of Gender and Identity*, pp. 146–155. London: Routledge.

VALENTINE, G. (2002) Queer bodies and the production of space, in: D. RICHARDSON and S. SEIDMAN (Eds) *Handbook of Lesbian and Gay Studies*, pp. 145–160. London: Sage.

VINCENT'S LOUNGE (2000) More on the public forum (http://www.geocities.com/WestHollywood/Stonewall/8218/GNews.html; accessed on 19 March 2003).

WAKEFORD, N. (2002) New technologies and 'cyber queer' research, in: D. RICHARDSON and S. SEIDMAN (Eds) *Handbook of Lesbian and Gay Studies*, pp. 115–144. London: Sage.

WEEKS, J. (1998) The sexual citizen, *Theory, Culture and Society*, 15, pp. 35–52.

WHISMAN, V. (1996) *Queer by Choice: Lesbians, Gay Men and the Politics of Identity*. New York: Routledge.

WILLIAMS, M. (2000) Interpretivism and Generalisation, *Sociology*, 34, pp. 209–224.

WILSON, A. R. (1993) Which equality? Toleration, difference or respect?, in: J. BRISTOW and A. R. WILSON (Eds) *Activating Theory: Lesbian, Gay, Bisexual Politics*, pp. 171–189. London: Lawrence and Wishart.

YEOH, B. S. A. and CHANG, T. C. (2001) Globalising Singapore: debating transnational flows in the city, *Urban Studies*, 38, pp. 1025–1044.

ZUKIN, S. (1995) *The Cultures of Cities*. Cambridge, MA: Blackwell.

Sexual Dissidence, Enterprise and Assimilation: Bedfellows in Urban Regeneration

Alan Collins

1. Introduction

There have been numerous contributions expounding in social theoretical terms the linkages between sexuality or vice and urbanism (see—for example, Park and Burgess, 1925; Park, 1952; Castells, 1983; Knopp, 1994,1995; and Mort, 1996). This paper has a much narrower focus and a different analytical framework. It adopts a pragmatic, principally economic approach, drawing heavily on some of the logic of the 'new economic geography' that has permeated contemporary economics. This study examines the 1980s and 1990s phenomenon of the emergence and development of the urban 'gay village' in various cities across England. The term 'gay village' possibly stems from the earlier gay agglomeration present in Greenwich Village, New York, and describes a visible physical clustering of gay enterprises and community within a city. These spatial entities have been the subject of sociological, cultural, historical and geographical scrutiny in North America, where some gay villages have been discernible since at least the 1970s (Chauncey, 1995; Stryker and van Buskirk, 1996; Grube, 1997; Boyd, 2003), in continental Europe (Sibalis, 1999) and in England (including, Quilley, 1997; Brown, 1997; Mort, 1998; Trumbach, 1999). In this paper, a principally economic case study of the UK's largest urban gay village in London is presented. Its evolution and development are considered with reference to the documented experiences of a number of other significant (in terms of number of gay enterprises) English urban gay villages. Retaining this wholly English focus provides an albeit imperfect means of controlling for some

Alan Collins is in the Department of Economics, Portsmouth Business School, University of Portsmouth, Richmond Building, Portland Street, Portsmouth, PO1 3DE, UK. Fax: 02392 844 037. E-mail: alan.Collins@port.ac.uk. *The author is grateful to anonymous referees and Ronan Paddison for constructive comments, to all those who agreed to be interviewed for relaying their valuable experiences and insights, to Bill Johnson for assistance with the map and also to Samuel Cameron, Stephen Drinkwater and Derek Leslie for advice on specific technical matters. The usual caveat applies.*

cross-cultural factors that may influence urban gay village development. In this limited way, an attempt is made to identify any commonalities and points of contrast in their evolution and development.

In attempting to discern the key features of the urban gay village phenomenon in England, the paper necessarily reflects on the reasons for their genesis in those particular urban locations and the economic development and urban planning issues that may arise with regard to the future dynamics of these areas and with regard to other nascent or potential gay villages in other English locations. Drawing on various bodies of evidence, ranging from field visits, local histories, public health orientated surveys, local authority documentation and a series of interviews with local planning officers across a number of English cities and gay business managers and entrepreneurs, arguments for the existence of a seemingly recurrent model of gay-led urban and economic regeneration are advanced.

The paper is organised in the following manner. In section 2, the theoretical arguments advanced in connection with gay agglomerations and gay villages are explored. The following section presents a reference case study of the gay village in London's Soho which extends into other parts of London's West End. In section 4, a simple model is advanced of urban gay village development. Concluding remarks are offered in the final section.

2. Genesis and Development of Urban Gay Villages: Theoretical Arguments

At a geographical scale of resolution somewhat broader than the sub-city level of urban gay villages, accounting for why certain locations have an unusually large gay male or lesbian community has been the subject of recent research attention. For example, in the context of the US, Murray (1996), amongst others, highlights the notion that a key driver of a gay male's residential choice is the nature of a city's social and political views towards gays. Accordingly, San Francisco, California, is perceived as a city where being gay is accepted and celebrated and thus is home to a disproportionately large gay population. Working from the perspective of pure rational choice economics and using data from the US census in 1990, Black *et al.* (2002) sought to advance a simple economic explanation, primarily to serve as the basis for a model to predict the spatial distribution of gay households across the US.[1] In the context of the San Francisco example, the explanation proceeds as follows. The city is physically one of the most attractive in the US with a pleasant mild climate, a splendid range of restaurants and a diverse and vibrant arts and entertainment community. The city thus forms a high-amenity location, where valued amenities are capitalised into the hedonic rent/property price gradient and/or wage gradient—i.e. in equilibrium, households pay for these amenities via higher property (rental) prices and maybe also lower wages. Since gay households face constraints that make bringing up children more costly for them than 'straight' (heterosexual) households, this reduces their lifetime demand for housing resources but frees up resources for reallocation to more amenity-focused consumption. Accordingly, Black *et al.* (2002) show that gay men may be observed to sort disproportionately into higher-amenity locations with further econometric evidence suggesting that measures of local amenities predict gay location choice more strongly than measures of 'gay friendliness'.[2]

In terms of an English focus, the thesis of Black *et al.* (2002) could be seen as contributing to a very plausible explanation as to why the resort city of Brighton and Hove in Sussex, on the south coast of England, with its urban gay village of Kemptown, has developed as a leading gay magnet. Yet, it is perhaps less useful in fully accounting for the *early* genesis of urban gay villages in—for example, Newcastle's 'pink triangle' (Pubs-Newcastle, 2003) or the Hurst Street area of Birmingham (Marketing Birmingham, 2002). Indeed, in the broader English context, it is perhaps worth considering how comprehensive and generalisable is such an explanation. In the UK context it is, arguably, too heavily reliant on high amenity-focused consumption. Remaining broadly consistent with the econometric work of Black *et al.* (2002), one might still reasonably argue that high levels of amenities are not an absolutely necessary pre-requisite for high lesbian/gay population densities, but their presence is certainly likely to help facilitate the 'take-off', or race, to a gay population critical mass. Once attained, this is then sufficient to provide a virtuous circle where high lesbian/gay household densities and/or gay commercial zones become self-sustaining. They may be self-sustaining because a gay population beyond the critical threshold (a large concentration of sexual like-mindedness) may be viewed as offering high amenity value in its own right, with club good characteristics (Buchanan, 1965). Arguably, the congestion limit for this club good could be extremely high. Indeed, if the gay population density rose to approach 100 per cent in the gay village, it is possible that many gay men might not actually encounter a region of diminishing returns at all.

Such a 'critical gay population size' focused perspective might thus be more useful in explaining—for example, the early genesis of the Birmingham Gay Village centred around Hurst Street, Birmingham, in England. In contrast to the Castro Street area in San Francisco, this gay village emerged in a city at the heart of an urban-industrial conurbation, with an unremarkable but temperate climate, and a decidedly inland, off-city-centre location. At the time, this location was certainly not renowned for its aesthetic qualities, or diverse range of amenities, of any obvious particular appeal to lesbians and gay men. The site of the gay village was a run-down small craft and industrial factory/warehouse district with some pubs and a few shabby shop premises. Perhaps more simply, as the main city of the UK Midlands conurbation, it had, and continues to have, a very large population. Accordingly, it must therefore contain a large number of lesbian and gay individuals/households. Arguably these individuals/households had many leisure/retail market needs that were relatively latent, or to varying degrees suppressed, prior to the full emergence of the gay village. Gradually over time and in the wake of a small existing service-sector-based commercial foothold, the benefits of relatively cheap premises and land values could be exploited with additional injections of lesbian/gay-focused entrepreneurial effort. This has resulted in these leisure/retail market needs being met more directly by the 21 service-sector enterprises[3] that currently give shape and focus to their urban gay village, which also claims to be the third-largest gay tourist destination in England (Marketing Birmingham, 2002). Inevitably, this widening service-sector base and growing gay community profile has attracted other, particularly Midlands-based, lesbians and gay men (households and entrepreneurs) to live and work in the city, or at least within a travel-to-leisure, or social commuting distance of it.

It is also a salient feature to note that, while the econometric findings of Black *et al.* (2002) provide a strong and convincing body of evidence for why gay men now live or aspire to live in San Francisco, they do not help to explain why the location of San Francisco developed as a gay magnet in the first place, compared with some other 'high amenity' US cities that did not. Black *et al.* (2002) dismiss this as an 'historical accident' and sought to distinguish *the* economic argument from historical and sociological perspectives on individual gay residential choices. Indeed, they cite the work of D'Emilio (1989) in describing the triggering, causal role of discharged gay US Navy sailors based in the San Francisco area deciding to remain living there.

In contrast to Black *et al.* (2002), this study finds and considers *an* economic argument within these historical accidents, or series of historical accidents. Further, this study also suggests that the economic dimension or economic significance of these historical accidents helps the better tracing of a path in the evolution of gay villages that may be observed in the contemporary urban landscape of England. That there are significant economic forces linked to such historical accidents has been an emerging theme in the 'new economic geography' that has entered the corpus of contemporary economics. For example, in a consideration of why Santa Clara County in California has become a favoured location for many firms in the microelectronics and software industries, Krugman notes that

> like most such agglomerations, Silicon Valley owes its existence to small historical accidents that, occurring at the right time, set in motion a cumulative process of self-reinforcing growth (Krugman, 1997, p. 239).

Switching the focus back to gay residential choice, one might thus contend that an initial cluster or agglomeration of gay households and gay amenities will naturally attract more and more gay men. This cumulative self-reinforcing growth (inward migration) inevitably helps further to shape the emergence of a spatial structure featuring urban gay villages or districts.

Considering the development of urban gay villages from the vantage-point of the new economic geography, it is relatively easy to translate its key pillars as identified by Krugman (1997), of path dependence, self-organisation and discontinuous change, into circumstances which would lead and shape the development of gay districts in a particular location. Following the pure logic of Schelling's segregation model (Schelling, 1971, 1978), it may be observed that in terms of path-dependence, individual microeconomic or microsocial decisions such as the opening of a gay-friendly bar or a gay nightclub in a particular city district, for whatever reason (such as low rents, the opportunity of vacant commercial premises in near-central locations, entrepreneur's personal taste), can direct and help to shape the future course of broader macrobehavioural outcomes. Such outcomes might relate to the future residential locational clustering of gay men already in the city and gay men migrating to the city. In terms of self-organisation, lesbians and gay men are not required to cluster and settle around gay agglomerations of services. Yet such relatively large-scale observed ordering can be conceived of as the spontaneous overall outcome of a large number of free-market atomistic interactions among gay and heterosexual households. While one might suspect fewer constraints to residential mobility among gay men/households, such segregation could conceivably also arise, in part, from the preferences of some heterosexual households to move to locations of more uniformly heterosexual character.

Turning to discontinuous change, it is reasonable to posit that, with the growth of new urban sub-centres and with a constant or increasing percentage of lesbian/gay male households, there must exist a critical mass (or in the language of Garreau, 1991, 'a point of spontaneous combustion') at a particular population/city size that is sufficient to foster and sustain higher-order gay amenities such as gay nightclubs, gay hotels, gay gyms and gay sauna/health clubs. This causes 'secondary explosions' as businesses form or move to the location (gay plumbers, gay carpenters, gay cleaners, gay accountants, gay law practices, etc.) to serve the gay businesses and the large gay population already there. Such critical mass economic behaviour could be mathematically formalised in terms of the standard economic base-multiplier model of traditional regional science or regional economic analysis, but augmented with increasing returns to scale (Pred, 1977).

Evidence for such 'secondary explosions' manifest in terms of the existence of gay community-orientated enterprises, may be readily found in the classified pages of the national and local gay press (*Pink Paper*, *Gay Times*, *Boyz*, *QX*, *Gscene*, etc.). It might seem strange to a world of heterosexual consumers why gay businesses or households might prefer the services of an explicitly gay plumber to one who is not explicitly gay. In part, it may be gay community camaraderie, but there are also more functional possible reasons relating to deliberately minimising the likelihood of encountering homophobic behaviour, especially when inside gay-owned homes/businesses. Further, it also assists in minimising the effort associated with the pressure to 'de-gay' homes or premises to

Table 1. Experiences of discrimination, abuse and violence against gay men in England (percentages) ($N = 14\ 632$)

Experience in the past 12 months	No	Yes
Experienced discrimination while using bars and restaurants	93.3	6.7
Experienced discrimination while shopping	95.1	4.9
Experienced discrimination while dealing with tradespeople and business services	95.4	4.6
Experienced discrimination with workmates and colleagues	86.6	13.4
Experienced discrimination with strangers in public	74.2	25.8
Experienced verbal abuse because of sexuality	65.7	34.3
Experienced physical attacks because of sexuality	92.9	7.1

Source: Sigma Research (2003).

make heterosexual persons more at ease. For some geographers, these phenomena have been described as examples of the intrusion of heteropatriarchy, which they have observed as characterising most social space, even gay social space (Johnston and Valentine, 1995; Bell, 1995; Kirby and Hay, 1997).

Evidence that there may be some rational, self-defensive basis for such gay community orientated business preferences, by gay men in the UK, can be found in the National Gay Men's Sex Survey 2002 (Sigma Research, 2003). Selected results are presented in Table 1. They suggest that the scale of discrimination in ordinary commercial environments is not trivial, although lower when they typically relate to such situations where it would be difficult to discern homosexuality in the course of a relatively swift transaction. This might become more apparent in the home or business environment of a gay man wary of a stranger's disposition to gay men (recalling that 25.8 per cent of gay men had experienced discrimination by strangers in public in the past 12 months). It is certainly more apparent for gay men where others have much more time to discern sexuality, such as in the work environment (where 13.4 per cent of the sample had experienced discrimination in the past 12 months). The figures are also relatively high for verbal abuse and physical assault across England. Of course, these figures do mask variations between small county towns with isolated gay pubs or club nights and larger urban gay villages in more cosmopolitan cities, where assaults still do take place, but where there can be a greater sense of safety in numbers.

It is worth emphasising that the founding gay amenities and the subsequent 'secondary explosions' of gay community orientated businesses are typically elements in a widening service-sector base. It is these services that act as a magnet to gay households, since they provide a means for helping to sustain the defining features of an 'out' gay lifestyle. The movement of (young) gay men to an often-distant and large urban metropolis offers greater scope for the more explicit exploration of their sexuality with virtually complete anonymity if desired. In this environment, they are typically remote from the social controls exerted by family and home communities (Cameron and Collins, 2000). This is important since part of the motivation for such migration can be linked to the nature and extent of social disapproval from their domestic roots. Such migrants can easily discern and participate in the social and recreational opportunities afforded by the commercial centres of these gay village areas.

For a number of relatively mature urban gay village areas, they could be said to have thrived to such an extent that they have become the *chic* social and cultural centres of the city—the place to be seen, not just within a loyal and relatively recession-resistant gay customer base, but regardless of one's sexual preferences. Indeed, in the same manner as ethnically defined commercial enclaves or zones (for example, Chinatown, Manchester and London) provide all visitors and local residents with option value and direct benefits from a more diverse range of consumption opportunities (retailing, entertainment and gastronomy), gay villages clearly do likewise. As examples, one may point to the gay bars, clubs, shops and restaurants in London's Soho, Manchester's Canal Street and Brighton's Kemptown. Their overall customer base is not exclusively lesbian or gay. In the context of Canal Street, in particular, it has been reported and voiced in the gay community that the area is increasingly becoming 'too straight'. These concerns have become so common and rehearsed that they have even been featured in dramatic representations of gay life in Manchester, such as the Channel 4 television drama series *Queer as Folk*. Canal Street has now become well known as a safe zone for heterosexual women to socialise in, such that heterosexual men now also use this social space in pursuit of heterosexual women. Inevitably, this raises the amount of search effort required for gay men and women to find potential partners, since they increasingly now have to filter out more heterosexual individuals from the search pool, and also it may lead to some changes in the sexual ambience and sense of personal safety in the area (Whittle, 1994). In this way, assimilation of gay social space into the fashionable socio-sexual mainstream advances, but may feature, along the way, varying degrees of resentment from some quarters of the gay community that would prefer to retain more sexually exclusive social space.

3. Evolution of Soho Gay Village, London

While the 'new economic geography' can provide a rationale for the emergence and expansion of these gay agglomerations, it does not describe the physical process of urban change required to accommodate the presence and evolving identity of a gay village/district. Inevitably, the recent physical evolution of recognisable urban gay villages in the UK, as elsewhere, has become interwoven with debates raging over the process of urban regeneration, renaissance and gentrification. For the case study area under focus and various other principal English urban gay villages (in Manchester, Brighton and Birmingham), the open and visible emergence of their clustering of gay service-sector enterprises, even to the eyes of their heterosexual populations (not just the lesbian/gay *cognoscenti*), is uniformly a feature of the 1980s and 1990s. To provide a more nuanced account of Soho's route, to what is argued to be an ultimately convergent development path with the other urban gay villages, some brief prefacing consideration of its historical evolution is also presented.

Beginnings

Aside from its capital city status, London is a world city such that its resident and visitor population actually sustains a number of mature and nascent satellite gay clusters of service-sector enterprises in the north, south, east and west of London.[4] During the course of the 1970s and early 1980s, what could be said to be the principal gay centre of London, in terms of number of establishments, comprised a handful of fairly low-profile (in mainstream heterosexual society terms) pubs and clubs in the Earls Court area of West London. The pre-eminence of this small clustering was easily displaced during the course of the mid 1980s and early 1990s, such that central London's Soho now reigns as the largest gay cluster in the metropolis and the UK. It is currently based around Old Compton Street (see Figure 1) but significantly extends into surrounding streets and squares and other parts of the 'West End' of central London in walkable proximity, such as Covent Garden and the Strand.

Given the long history of gay cultural and sexual entertainment enterprises in London since at least the late 17th century (Norton, 1992; Trumbach, 1999), it would seem unsurprising if their were a *continuous* clustering or presence of gay enterprises in Soho till the present day. However, there is no such

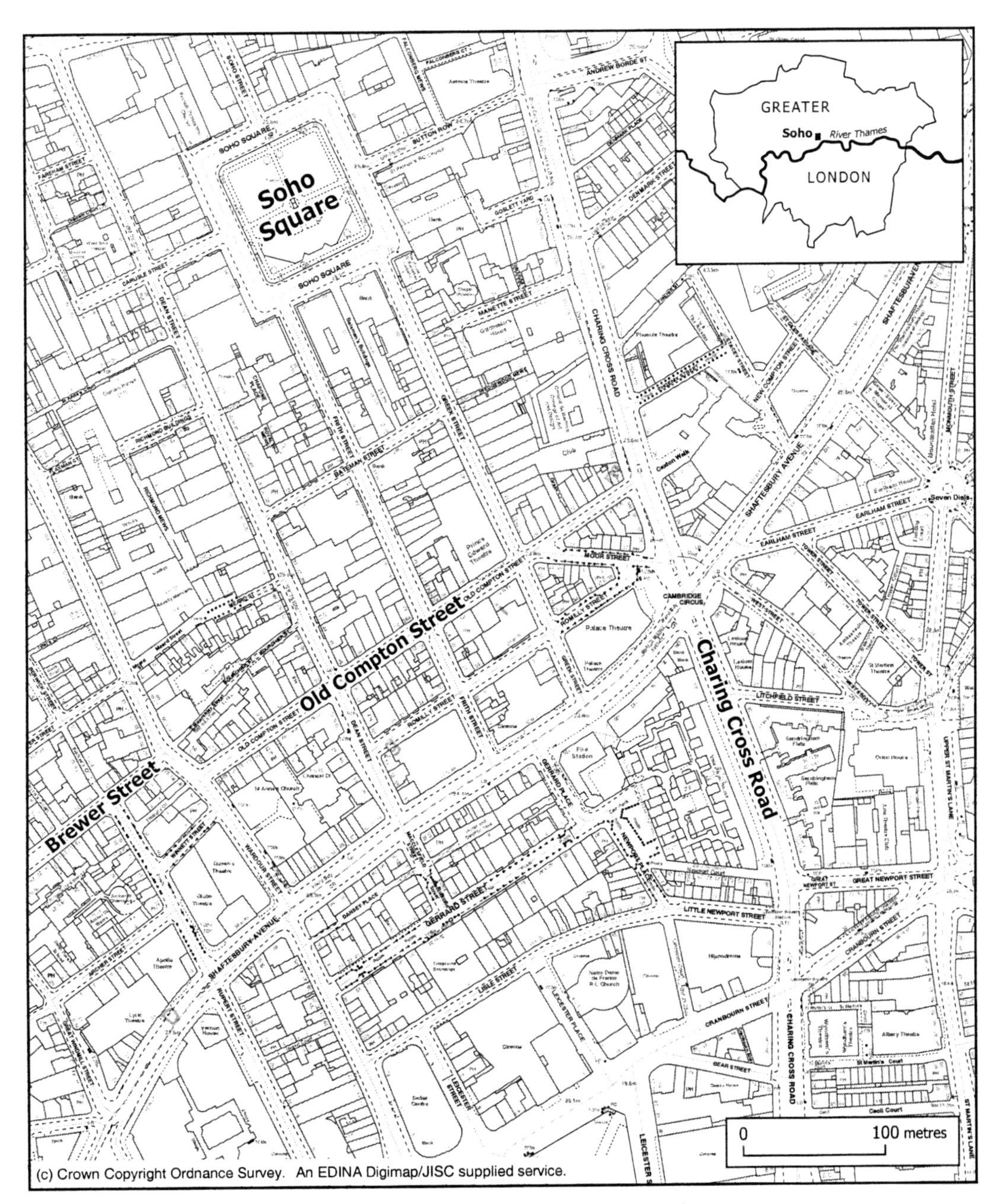

Figure 1. The Soho district, central London.

clear evidence for this. There may well have been gay enterprises such as the infamous ‘molly houses’ in the area in the 18th century, as there were many such establishments spread across London.[5] These were, however, closed down in various episodic moral panics and clamp-downs (Norton, 1992). The accepted wisdom that seems to emerge from these historical investigations is that there were probably more gay bars and gay brothels spread across London in the mid 18th century than existed in the 1950s.

The area of Soho was hunting ground at least from the middle ages through to the

17th century, used by royalty and other aristocrats.[6] Following a perceived need to alleviate central London overcrowding and provide safer, fairly central greenfield development land in the aftermath of the Great Fire of London in 1666, Soho was subjected to rapid urban development. The lay-out of streets and squares which, in the main, still survive, was devised by Gregory King, an eminent engraver, genealogist, statistician and urban planner. The area still had some *cachet* arising from its royal/aristocratic hunting associations. Hence, it became a fashionable residential district for the aristocracy and wealthy in the 1670s and 1680s (Summers, 1991; Hub Communications, 2003; Vinyl-Junkies, 2003). Given the area had available space, and some large properties to accommodate craft industries, in the early 18th century, Soho attracted considerable influxes of new settlers from continental Europe. Many of these were refugees escaping persecution, including Greeks (hence, Greek Street in Soho) and French Huguenots avoiding, respectively, Ottoman persecution and Louis XIV's reign. Predictably, these influxes of refugees and craft industry led the aristocrats and 'old money' wealthy, over time, to leave Soho to seek out and establish newly fashionable and exclusive residential areas, such as Mayfair (Summers 1991). Despite the emergence of some nearby slum housing, Soho became through the 18th and 19th centuries a thriving commercial centre renowned over various times for its doctors, lawyers' practices, tailoring, artists (Farson, 1993) and the first London showroom of the celebrated industrialist and ceramicist Josiah Wedgwood (Summers, 1991; Vinyl-Junkies, 2003).

Sex, Soho and Culture

Alongside the thriving legitimate industry also emerged the beginnings of the sex industry, which ultimately came to dominate the actual and perceived popular character of Soho from the late 1950s (Walford, 1987; Farson, 1993) to the mid to late 1980s. The earliest documented presence (in the early 18th century) relates to Hoopers Hotel (known for heterosexual brothel services) on Soho Square and also a continuous presence of primarily street-based sex workers, right through until the late 1950s (Farson, 1993; Vinyl-Junkies, 2003). The 1959 Street Offences Act, however, featured legislation to curb soliciting for prostitution and introduced serious penalties for so doing. Consequently, the already significant stock of off-street prostitution-related premises in nearby upper-floor flats began to increase further through the 1960s and 1970s. Alongside this came an increase in all the associated commercial sex paraphernalia of red lights, 'models' business cards near ground-level doorbells and telephone-booth postcard advertisements. Soho thus became the home to numerous 'clip joints'[7] and illegal brothels, whose ownership (often subject to bloody gang turf wars) or 'protection' typically lay with Italian and Maltese mafias and other local criminal gangs.[8] The area was generally perceived through countless media and popular cultural images as seedy, shabby and clouded in a haze of sexual and legal liminality. Riding on this sense of liminality (and concomitant lower property prices/rents) in the 1950s and 1960s, the area became a magnet for a number of artistic or bohemian individuals, who settled in Soho (Farson, 1993; Fryer, 1998). Hence, in due course, the area also saw the beginnings of more bohemian and *avant-garde* cultural enterprises emerging. (Examples include Ronnie Scott's jazz club (Fordham, 1996) and Peter Cooke's "Establishment Club" (a satirical comedy and revue club) (Thompson, 1997; Carpenter, 2000).) The area had underground cultural *cachet* which erupted via the popular media in the 1960s into a significant, but ultimately short-lived, musical, cultural and fashion wave, centred around enterprises in Carnaby Street in west Soho.

Further, following the media glare of London's Metropolitan Police Vice Squad and Obscene Publications Squad corruption charges (where many officers were found to be running vice and porn protection rackets), the revoking of many commercial licences in the early to mid 1980s crackdown on the sex and retail pornography industry and, alongside, the beginnings of more private and discreet sex-based e-commerce, there was a major contraction in the number of Soho commercial and residential premises used by this sector.[9] The area had declined further until, in the mid 1980s, it was a relatively run-down central urban area, tainted with a long-standing reputation for vice and crime and with its past glories from its role in the 1960s London cultural vanguard, fast fading in the public consciousness.

Village People, Enterprise and Village Community

Of crucial importance for the emergence of the Soho Gay Village, was the fact that there was *already* a small presence of gay pubs (or parts of pubs) in Soho itself and the 'West End' of central London, serving its Theatreland.[10] Thus, the development of this urban gay village was not actually taking place on a completely blank canvas. While Binnie (1995) seems to consider the Soho Gay Village to have actually emerged in 1991 with the opening of the café bar called *Village Soho*, this author disputes the historical definitiveness of that assertion. There were many earlier establishments existing in Soho and the West End in the 1980s (not exclusively centred on Old Compton Street as now). While not collectively then branded (commercially and informally) as a gay village, nevertheless, they were already then beginning to gel, through walkable proximity, into such a phenomenon.

From a very small existing commercial foothold, substantial expansion could be made, in large part, because it was conveniently helped by thc contraction of the sex and porn industry—a valuable historical accident in the Krugman (1997) sense. As more commercial premises became available from the mid 1980s and into the 1990s, more gay service-sector enterprises became manifest via a seemingly *ad hoc*, organic and gradually accelerating incremental process (as opposed to area-based planned development from the outset).

Gay entrepreneurs and households were conveniently present at that time as an under-exploited market niche, who were unlikely to be deterred by the pervading social stigma of living, operating businesses and providing a customer base, in an area whose very name at that time was dripping with connotations of sexual illicitness and danger. Partly, this could be explained in terms of the general backdrop of liminality that was already a feature of the lives of most lesbians and gay men (before and even after decriminalisation of homosexual activity between two consenting adult males over 21 years of age in the 1967 Homosexual Law Reform Act). This was especially so in the mid to late 1980s where there was a heightened sense of lesbians and gay men in particular, being further dispatched towards and beyond the margins of social and sexual acceptability. It is important to recall that, at the time, the Conservative government of the UK was pressing for more 'traditional family values' orientated policies, which militated against gay lifestyles; pressurising local government authorities (especially what were termed by press and right-leaning politicians as 'loony left London boroughs') to restrict or curtail policies and schemes that treated lesbian/gay households and lifestyles with equanimity (the introduction of Section 28 in the 1988 Local Government Act being one example); and, conveniently acquiescing in the wake of the tabloid press whipped up hysteria that recast the HIV/AIDS epidemic into the 'gay plague'.

The relevant local planning authority (Westminster City Council) has never had any special policy area status, formal or informal, of any kind for a 'gay village'. The sobriquet 'Soho Gay Village' does not feature in any Westminster planning/local economic development documentation. That said, there is some ambiguous reference in the Pre-inquiry Unitary Development Plan of the City of Westminster (dated 29 August 2002) to the fact that Soho is a "vibrant and cosmopolitan area" (p. 5). This may (and only may), in part, be suggestive of the presence of a large cluster of gay enterprises and its role as a focal point for the London gay community.

Westminster Council's seemingly more ambivalent stance in this direction is somewhat surprising, given the gay village's commercial vitality, tourism significance and the authority's explicit recognition of ethnically defined clusters. Based on detailed inspection of available public documentation, it is readily clear that the adjacent London 'Chinatown', centred on Gerrard Street has consistently attracted significant formal and explicit attention from the Conservative Party-led Westminster Council authority (for example, Westminster City Council, 2002, 2003). It has even appointed a Chinese Community Liaison Officer to oversee matters impacting on this community. Planning permission and change-of-use matters (for legitimate businesses), helping the development of the Soho Gay Village, however, do seem to have been considered wholly on their individual merits, but irrespective of their potential consistency with, and contribution to, any gay village identity.

These gay enterprises continue to provide services principally geared to supporting gay male social and recreational activities, including: gay pubs/bars (typically gay-run but not necessarily gay-owned, as breweries rightly appreciated the potential of the growing gay market), cafés, restaurants, fetish-wear retailers, gay accessory stores, nightclubs, gyms, saunas (in premises on the margins of Covent Garden), gay escorts (using rented central Soho apartments) and other gay-focused businesses and professional practices. They have emerged to trade in close proximity and shape the widening service-sector base that characterises the commercial foundations of the contemporary Soho Gay Village. In terms of scale, from a schedule of personal visits and using the address and postcode information in the *Gay to Z Directory* (Issue 12, 2003) (cross-referencing with other London free gay publications further to corroborate and discern explicit gay focus or ownership), it was calculated that, in August 2003, only taking into account explicitly gay nightclub venues, gay pubs/bars and gay cafés/restaurants in the Soho Gay Village (including some adjacent streets/squares), 56 such enterprises could be found in short walking distance of each other.

The increasing trading vitality and aesthetic improvements, linked to the development of an urban gay village in such a central London location, has of course meant that it has formed part of the spectrum of leisure options for all London residents and visitors. Witnessing its success with both gay and straight consumers, the Soho Gay Village has laid the foundations for the secondary growth of a large number of other bars/cafés/restaurants in its wake, that are not gay-run or gay-owned and which are targeted principally towards more homosexuality-tolerant, fashionable young singles, or just the mainstream heterosexual market.

While there has always been a sizeable, principally apartment dwelling, residential community in Soho, throughout the course of the 19th and 20th centuries, the construction of an increasingly stronger gay presence and identity did lead to a 1980s and 1990s influx of gay households in Soho. Understandably, this could not be sustainable unless gay households rarely moved and never sold their property to non-gay households, or the highest bidder irrespective of sexuality. As a fashionable central London location of potential interest to all, regardless of sexuality, and with several gated and secure new residential developments emerging to capitalise on its location and fashionable lifestyle desirability, property prices and rentals have increased markedly, even by London standards. This ensures income and wealth now ultimately conditions the dynamics of the contemporary residential character of Soho and presents constraints on the lesbian/gay residential population density. This is not to say that the Soho Gay Village is ultimately likely to be displaced as the principal gay cluster in London and the UK. It is not, since the presence of upwards of 56 gay enterprises demonstrates clearly it has never really relied on just Soho residents (estimated to comprise approximately 6000 individuals (Soho Society, 2001) as its significant customer base. It has long been reliant on the wider London gay population, since Londoners enjoy an extensive travel-to-leisure area facilitated by London's dense tube, train and bus network.

For some geographers, the urban renaissance of the Soho area, alongside the municipal and police actions to induce contraction of the sex industry, provides indicative evidence of a 'revanchist' (Smith, 1996) city agenda (Hubbard, 2004). By this is meant a process of urban change, whereby a White (heterosexual) middle-class 'conspiracy' supported by municipal political power and considerable capital resources, literally 'takes revenge' and reappropriates urban space back into the White middle-class (heterosexual) sphere of interest.[11] Even ignoring the role of e-commerce, as a booming substitute channel, in the spatial diminution of the commercial sex and porn industry, such an argument neglects the fact that the engine of this renaissance was not a direct municipally supported initiative and was not entirely resourced in its early genesis with wider middle-class society's capital. It is true that there was and continues to be, concerted action by Westminster Council and the Metropolitan Police to constrain through planning, licensing and law enforcement, the physical extent of the sex and porn industry (Hubbard, 2004). Yet for this author, its significant displacement by a large number of gay service-sector enterprises could not credibly be argued to be the preferred end-game outcome of such a White middle-class heterosexual conspiracy. The essence of this somewhat tortuous argument seems to be based on a view that the gay community were tacitly allowed to serve as the means of urban renaissance. Continuing in this conspiratorial vein, they would serve to provide revalorised and better-quality social and commercial inner urban space, for the ultimate benefit of White heterosexual middle-class society, who would eventually reacquire such space (by dint of their power and capital over time). It is true that the residential character of London's Soho is likely to have changed as prices/rents have risen. There may well be fewer lesbian/gay households now than in the late 1980s or early 1990s, as the wider desirability of the now-fashionable Soho area has added its force to the housing market. Yet, in terms of the commercial landscape, there seems to be a large, dynamic and sustainable body of gay service-sector enterprises that are likely to dominate the area for the foreseeable future.

In the light of the outcome that is the Soho Gay Village, perhaps the urban renaissance process could more accurately be described as the harnessing of the convenient presence of essentially class-blind, ethnicity-blind, but gay-sourced capital and entrepreneurial effort, of sufficient scale to be able to exploit the commercial space made available by the diminishing physical presence of the sex and porn industry. The feature of London's gay managed capital and entrepreneurial effort driving an urban renaissance process, at least in the early stages, seems to be more consistent with an interpretation of Lees' (2000) emancipatory city thesis. This suggests a particular path of urban regeneration and renaissance. In essence, it could be characterised as a process whereby people become united in the central area of a city and create opportunities for recreational/social interaction, ethnic and sexual tolerance and cultural diversity. This seems a neat encapsulation of the essential character of Soho Gay Village, since it resolutely presents itself as physically gay space (even in the midst of municipal disinterest), but also as part of an array of leisure consumption options, for the benefit of the wider urban liberal society of London and not just its lesbians and gay men.

4. Towards a Simple Model of the Evolution of Urban Gay Villages in England

In Quilley (1997) a brief insight into gay life in Manchester before the Gay Village prefaces a detailed politico-economic exposition of its emergence in the 1980s. In the Brighton Ourstory Project (BOP) (2001) a more straightforward historical narrative is unfolded. Yet what is clear is that, in common with London, what really constituted the physical presence of pre-gay village life in these places were a small number of pubs across the city. Brighton and Hove had gay pubs known by the lesbian and gay *cognoscenti*, (including military personnel stationed at the local garrison and also the naval port of Portsmouth along the coast) even in the 1930s (BOP 2001). Additionally in Brighton, given its seaside holiday orientation, a wider gay-orientated service-sector base could be identified far earlier than in most other English towns or cities. Even in pre-gay village days, a number of hotels or bed and breakfasts (BandBs) were known through an informal network (functioning by largely by word of mouth) to be places where lesbian and gay couples could stay with no 'awkward' questions (BOP 2001). Yet none of these enterprises was in a close physical clustering—perhaps because this would be more likely to draw attention to their then strictly illegal status in supporting homosexual activities and lifestyles. These enterprises were thinly spread across Brighton and Hove, but included some premises within the current Brighton gay village area of Kemptown, centred on St James Street (BOP, 2001).

Quilley (1997) also draws attention to the former status of the Canal Street environs, the centre of the Manchester Gay Village, as a marginal area, featuring red-light activities and 'skid row' or renewal-needy characteristics. Turning to Brighton, this south coast resort has long had echoes and associations with sexual and legal liminality, even since the Regency era. It was well known as a place for conducting illicit affairs and for its extensive smuggling operations. A pioneering early contribution to the image of sexual frisson and peccadilloes, connoted by weekends in Brighton, was made by the Prince Regent who 'kept' a number of mistresses there (Barlow, 1997; Fines, 2002). Brighton was also the home to a number of secret tunnels in the service of smuggling operations and had a reputation for hosting a sizeable criminal fraternity since at least the interwar period. The latter was given the oxygen of publicity by the publication of Graham Greene's 1938 novel and subsequent film adaptation, *Brighton Rock,* which actually further stimulated tourism demand (Fines 2002). Accordingly, as with Soho, all these characteristics help to paint a picture of these locations being enmeshed to varying extents in a mist of sexual and legal liminality, but discreetly and gradually presenting a more homosexually open presence. This was done principally through the semi-public meeting venues provided by 'pub' environments. This marginal area/twilight character coupled with an initial pub focus is also readily identifiable in the development of several other such gay villages in England, including those in Birmingham and Newcastle.

The reality of the decriminalisation of homosexual activity between consenting males in 1967 seemed to take some time to become established, with regard to the fuller range of commercial opportunities this would allow. During the mid to late 1970s in parts of London, and in Brighton, Manchester and Newcastle, small clusters of such pubs did emerge, although not necessarily fully visible to mainstream society in these locations. It was within these clusters that the beginnings of a widening of the service-sector base could eventually be discerned. Aside from Brighton's early 'gay-friendly' small hotels and BandBs, there emerged from the mid–late 1970s at all these locations a presence of one or two gay nightclubs, possibly a gay sauna and, for some, a gay-orientated sex/clothing accessories store. Cafés, gay tradespersons and professional practices explicitly operating in the locale, typically followed in the secondary wave of enterprises during the course of the 1980s and early 1990s and the drive to a readily identifiable gay village identity.

It is possible to note a key feature, apart from its ultimate scale, which does distinguish London's Soho from most other English gay villages. At no point during the course of its evolution was there any explicit municipal support given to the formation of a gay village identity. This neglect or underplaying of its identity markedly contrasts with the situation in several other UK cities with gay villages (Manchester, Birmingham, Brighton). Birmingham and Manchester both have explicit project elements in their development plans for what they explicitly refer to as 'the gay village' including, land redevelopment, installation of CCTV to increase the sense of safety in this space, pedestrianisation schemes and various other urban environmental amenity improvements. Newcastle has explicit references to investigating the development needs and potential of 'the gay village' in its City Centre Action Plan (Newcastle City Council, 2002a, p. 36) but has not hitherto fleshed out any detailed development plans featuring a gay village identity. The rapid development towards what is now a slightly more visible gay village (the 'pink triangle') in Newcastle, is thus, in large part, attributable to a proactive stance adopted by a small group of senior members of its planning department and the relevant licensing magistrates. In the light of a real background political will to foster equality (as now formally enunciated in The Newcastle Plan—A Community Strategy for the City (Newcastle City Council, 2002b), these parties met in 1999 with a view to removing obstacles to a gay clustering of service-sector establishments. The intention was primarily to improve the quality of social and recreational provision for a then underprovided for, lesbian/gay population. These problems were generated by an inadequate number of liquor licences in the city. The alleviation of this specific constraint, by municipal and regulatory authorities, directly facilitated and accelerated its expanding gay service-sector base. Further, Manchester and Brighton also maintain a number of municipal forums where lesbian and gay residents' issues can be discussed and they also feature a significant number of 'out' lesbian and gay local councillors.

For Quilley, the very development process culminating in Manchester's Gay Village was also intimately bound up with the changing agendas of the municipal urban left. He suggests that, from a policy centrepiece of resistance to Thatcherite policies, the idea of a gay village evolved to become harnessed explicitly within a property-led redevelopment strategy. In such a strategy

> the aesthetic of the gay scene has become articulated into a wider re-imaging of the city around the familiar theme of European style cafés, pedestrian streets, and arcades, as well as around a central role for leisure and cultural activities (Quilley, 1997, p. 275).

Remarkably then, even in the absence of a sympathetic or proactive municipal (political) advocate or patron, a similar physical outcome could also be said to characterise the contemporary commercial face of London's Soho Gay Village.

In distilling these experiences and findings, it seems reasonable to posit a simple developmental model of the evolution of urban gay villages in England. It is contended that the progression through these stages may be accelerated by direct municipal support (planning and/or licensing) and high amenity attributes in a location (following Black *et al.*, 2002), but neither of these elements is considered to be a sufficient and necessary condition. Of crucial importance, however, for a sustainable urban gay village is the eventual attainment of a gay critical mass population to serve as the customer base, within the travel-to-leisure area. Essentially, this developmental model describes a number of characteristic phases in the evolution of an urban gay village as presented in Table 2. Stage 1 describes the pre-conditions for an urban area typically apparent in a number of eventual gay village locations. The critical presence of licensed premises/pubs for the evolution of English gay villages is readily apparent in viewing the progression from stages 1 to 3. Clearly, an inadequate stock of liquor licences or a reluctance to issue supplementary licences could easily retard the pace of development to urban gay village status. Relatively few

Table 2. Stages in the development of urban gay villages in England

Stage 1: 'Pre-conditions'—urban area in decline: location of sexual and legal liminal activities and behaviour
Key features
1. Twilight/marginal area showing extensive physical urban decay
2. Presence of street-based and/or near off-street (predominantly heterosexual) prostitution
3. Significant stock of vacant commercial premises
4. Low property prices/rental values
5. Typical presence of at least one gay licenced public house

Stage 2: 'Emergence'—clustering of gay male social and recreational opportunities
Key features
1. Conversion of some other nearby licensed public houses into 'gay run' pubs
2. Increase in applications made for liquor licences to support conversion of some other existing commercial premises into gay nightclub or additional licensed public houses
3. Upgrading or renovation of existing gay pub(s) in the area
4. Substantial increase in gay male customer base and pub revenue stream

Stage 3: 'Expansion and diversification'—widening gay enterprise service-sector base
Key features
1. Conversion of some other existing commercial premises for gay service-sector enterprises: gay health clubs/saunas, gay retail lifestyle accessory stores, gay café-bars
2. Further applications for liquor licences and planning permission for additional gay nightclubs
3. Increasing gay household density in the existing stock of residential units in the gay village locale
4. Increasing physical visibility and public awareness of the urban gay village to mainstream society
5. Growth of gay tradespersons and professional practices operating in or near the gay village, or via its community media
6. Increasingly significant and sustained contribution to the gay service-sector enterprises' revenue streams from visiting gay tourists

Stage 4: 'Integration'—assimilation into the fashionable mainstream
Key features
1. Increasing presence of heterosexual custom in ostensibly gay pubs/bars
2. Conversion of some existing commercial premises for new mainstream society service-sector enterprises (bars, clubs, restaurants)
3. Influx of young urban professionals to the existing stock of residential units in the gay village environs
4. Outflow and suburbanisation of early gay residential colonisers
5. Increasing applications and construction of new-build (apartment) residential units in the gay village environs
6. Increasingly significant and sustained contribution to gay service-sector enterprises' revenue streams from the heterosexual community

such gay villages could currently be argued to have fully achieved progression through to assimilation into the fashionable mainstream as described in stage 4. One might include Manchester Gay Village, London's Soho and arguably Kemptown, Brighton, in this category, but this grouping seems likely to increase in number over time.

Looking to the future, as new urban gay villages do emerge over time, it is possible that, in market terms, real competition is generated amongst them for lesbian/gay custom and in terms of lesbian/gay residential attractiveness. In such a situation, it is plausible to envisage a situation where some urban gay villages will enter a declining phase and revert from villages back to lower-level districts or clusters. It may be that the result is a long-run equilibrium consisting of a relatively small group of large urban gay villages, towards which lesbians and gay men gravitate (subject to income and job constraints) and a larger number of smaller gay districts and clusters.

5. Concluding Remarks

This paper has explored the generic concept of the urban gay village in England, in terms of the reasons for their location and their evolutionary paths. A case study of the largest such urban gay village in England is presented and its characteristics and development path considered. This is unfolded in the light of the documented experiences of other significant English urban gay villages. Despite differing experiences of municipal support for such entities, a recurrent model of their development seems discernible. Typically from within a marginal or twilight area, a foothold, comprising the presence of one or more gay-run licensed premises (pubs), may attract additional pubs and eventually a wider range of gay service-sector enterprises, exploiting the existence of adjacent or nearby vacant and low-value space.

Further economic case studies elsewhere, such as in other nascent urban gay villages in England and in other countries, might usefully be undertaken to confirm the validity and level of generalisation possible with this simple model. It may offer an initial analytical framework, or can provide some broad-level insight, to inform local economic development and planning strategy. It seems that the harnessing of the gay village concept in these domains is likely to be feasible, with a shorter and more definitive time-scale, where there is a genuine willingness by municipal authorities to support lesbian and gay communities and where this is maintained alongside the practical desire to realise more swiftly their potential, in enhancing the commercial vitality and physical regeneration of particular urban areas. The experience of London's Soho Gay Village shows, however, given the appropriate favourable pre-conditions and the absence of actively repressive constraints (for example, ample liquor licences, fair and objective assessments in planning permission requests) such entities may evolve anyway, even in the face of municipal political ambivalence.

Notes

1. Black *et al.* (2002) in part account for differences between lesbian and gay male residential choice by virtue of the fact that they observe lesbian households are more likely to have children, so that (they argue) many lesbians are more likely to prefer smaller towns.
2. Albeit with some systematic error, to permit identification of lesbian and gay households, the 1990 US census long form asked questions which can reveal single never-married households without children (by age and gender) and explicit questions relating to same-sex unmarried partner status for other residents in the home. It is this latter category of 'gay couples' that Black *et al.* (2002) focus on in their study. In the past, the US census form only gave the choice of husband and wife for partner status. From the individual and aggregate data in the 2001 UK census, one can readily identify single never-married households without children by gender through various age-bands and same-sex households (i.e. not same-sex unmarried partners). That said, this might become possible, to some extent, with the soon to be released household micro data from the UK 2001 census. The current difference between the US and UK individual and aggregate census data generates, in statistical/econometric terms, some serious problems, since the definition of same-sex households provides a very 'noisy signal' (and probably a substantial overestimate) of lesbian/gay household status. This is clearly the case because same-sex households need not be lesbian or gay at all. Further, there are also mixed sexual orientation households which must go unrecognised (in fact, are systematically completely neglected) in both the US and UK census. In principle, it may be possible that future UK-based work could corroborate the proportions (or at least provide a lower bound estimate) of lesbian/gay households or partly lesbian/gay households, from the UK Census cohort of same-sex households. This could be done by grossing up the proportions from some other large-scale survey that actually does consider sexual identity or preferences—for example, the UK 2000–2001 National Survey of Sexual Attitudes and Lifestyles soon to be available from the ESRC Data Archive at the University of Essex, England.

3. Author's count from *B1 Community* (Issue No. 1, May 2003)—a free news and listings publication for the lesbian and gay community in Birmingham, England.
4. These include clusters based around Clapham Common High Street (south), Earls Court/Brompton Road (west), Camden Town and Islington (north) Commercial Road and Bethnal Green (east).
5. 'Molly houses' were a kind of combined gay bar, cabaret, social club and brothel featuring some cross-dressing behaviour.
6. The name Soho comes from the then-popular hunting cry 'so-ho' to rally the hounds when in pursuit of quarry.
7. The term 'clip joint' refers to places where men were fleeced out of their money by being enticed to pay large sums for drinks and often ultimately non-existent sexual entertainment by women working at the establishment. Those unwilling to pay had to face the prospect and visible threat of confrontation with the clip joint's hired thugs.
8. Recent investigations suggest that Albanian gangs currently dominate the more scaled-down and volatile (in terms of entry/exit rates) brothel scene. It has been suggested that this has been achieved by either employing or enslaving eastern European women to undercut the sex tariff prices of already-resident sex workers (Coward, 2003). This volatility is reported to be largely a consequence of the more sustained and aggressive action by the local authority and police to search out and close such brothels in Soho. Yet often, once closed down they were found to re-open quickly in different premises, staying in business there until detected again. That said, the motivation and action of the authorities are the subject of considerable contention by some sex workers (Silverman, 2003).
9. One estimate considers the industry contracted to a sixth of its former hey-day size in the early 1980s (Vinyl-Junkies, 2003).
10. These included *The Golden Lion* in Dean Street, Soho, which was apparently popular with gay servicemen and, recalling an even earlier era, *The Salisbury* in Upper St Martins Lane, where it is reputed that Oscar Wilde and his *entourage* visited from time to time.
11. Castells (1983) and Lauria and Knopp (1985) in the context of gentrified *residential* areas in the US, however, assert that there may be observed a homosexualisation of such areas by dominant interests and gays, *both* comprising mainly White middle-class men. In the course of this particular study, no obvious class bias was detected in the commercial presence, orientation and customer base of the Soho Gay Village.

References

BARLOW, A. (1997) *The Prince and his Pleasures*. Brighton: The Royal Pavillions Libraries and Museums.

BELL, D. (1995) Perverse dynamics, sexual citizenship, and the transformation of intimacy, in: D. BELL and G. VALENTINE (Eds) *Mapping Desire: Geographies of Sexualities*, pp. 304–317. London: Routledge.

BINNIE, J. (1995) Trading places: consumption, sexuality and the production of queer space, in: D. BELL and G. VALENTINE (Eds) *Mapping Desire: Geographies of Sexualities*, pp. 182–199. London: Routledge.

BLACK, D., GATES, G., SANDERS, S. and TAYLOR, L. (2002) Why do gay men live in San Francisco? *Journal of Urban Economics,* 51(1), pp. 54–76.

BOYD, N. A. (2003) *Wide Open Town: A History of Queer San Francisco to 1965*. Berkeley, CA: University of California Press.

BRIGHTON OURSTORY PROJECT (2001) *A history of Brighton's lesbian and gay community* (http://brightonourstory.co.uk; accessed August 2003).

BROWN, J. (1997) *The lesbian and gay communities in the built environment – a planning obligation.* Working Paper No. 41, Centre for Research into European Urban Environments, Department of Town and Country Planning, University of Newcastle upon Tyne.

BUCHANAN, J. (1965) An economic theory of clubs, *Economica* 32, pp. 1–14.

CAMERON, S. and COLLINS, A. (2000) *Playing the Love Market: Dating, Romance and the Real World.* London: Free Association Press.

CARPENTER, H. (2000) *That Was Satire That Was: The Satire Boom of the 1960s*. London: Victor Gollancz.

CASTELLS, M (1983) *The City and the Grassroots*. Berkeley, CA: University of California Press.

CHAUNCEY, G. (1995) *Gay New York: Gender, Urban Culture and the Making of the Gay Male World, 1890–1940*. New York: Basic Books.

COWARD, R. (2003) Slaves in Soho, *The Guardian*, 26 March (http://guardian.co.uk/comment/story/0,3604,921977,00.html; accessed 15 August 2003).

D'EMILIO, J. (1989) Gay politics and community in San Francisco since World War II, in: M. B. DUBERMAN (Ed.) *Hidden from History: Reclaiming the Gay and Lesbian Past*, pp. 456–473. New York: NAL Books.

FARSON, D. (1993) *Soho in the Fifties*. London: Pimlico.

FINES, K. (2002) *A History of Brighton and Hove*. Chichester: Phillimore.

FORDHAM, J. (1996) *Jazz Man: The Amazing Story of Ronnie Scott and his Club*. London: Trafalgar Square.

FRYER, J. (1998) *Soho in the Fifties and Sixties*. London: National Portrait Gallery Publications.

GARREAU, J. (1991) *Edge City: Life on the New Frontier*. New York: Doubleday.

GRUBE, J. (1997) 'No more shit': the struggle for democratic gay space in Toronto, in: G. B. INGRAM, A.-M. BOUTHILLETTE and Y. RETTER (Eds) *Queers in Space: Communities, Public Spaces, Sites of Resistance*, pp. 127–145. Seattle: Bay Press.

HUBBARD, P. (2004) Cleansing the metropolis? Urban prostitution and the politics of zero tolerance, *Urban Studies*, 41(9), pp. 1687–1702.

HUB COMMUNICATIONS (2003) *Soho* (http://www.hubcom.com/channel/soho/history.htm; accessed 15 August 2003).

JOHNSTON, L. and VALENTINE, G. (1995) Wherever I lay my girlfriend, that's my home: the performance and surveillance of lesbian identities in domestic environments, in: D. BELL and G. VALENTINE (Eds) *Mapping Desire: Geographies of Sexualities*, pp. 99–113. London: Routledge.

KIRBY, S. and HAY, I. (1997) (Hetero)sexing Space: Gay Men and 'Straight Space' in Adelaide, South Australia. *Professional Geographer*, 49(3), pp. 295–305.

KNOPP, L. (1994) Social justice, sexuality and the city, *Urban Geography*, 15(7), pp. 644–660.

KNOPP, L. (1995) Sexuality and urban space: a framework for analysis, In: D. BELL and G. VALENTINE (Eds) *Mapping Desire: Geographies of Sexualities*, pp. 149–161. London and New York: Routledge.

KRUGMAN, P. (1997) How the economy organizes itself in space: a survey of the new economic geography, in: W. B. ARTHUR, S. DURLAUF and D. A. LANE (Eds) *The Economy as an Evolving Complex System II: Proceedings*, pp. 223–237. Boulder, CO: Westview Press..

LAURIA, M. and KNOPP, L. (1985) Towards an analysis of the role of gay communities in the urban renaissance, *Urban Geography*, 6, pp. 152–169.

LEES, L. (2000) A re-appraisal of gentrification: towards a geography of gentrification, *Progress in Human Geography*, 24(3), pp. 389–408.

MARKETING BIRMINGHAM (2002) *Gay Birmingham* (http://www.birmingham.org.uk/nightlife/gay_birmingham.php; accessed 15 August 2003).

MORT, F. (1996) *Cultures of Consumption: Commerce, Masculinities and Social Space*. London: Routledge.

MORT, F. (1998) Cityscapes: consumption, masculinities and the mapping of London since 1950, *Urban Studies*, 35(5/6), pp. 889–907.

MURRAY, S. (1996) *American Gay*. Chicago, IL: Chicago University Press.

NEWCASTLE CITY COUNCIL (2002a) *City centre action plan* (http://www.newcastle.gov.uk).

NEWCASTLE CITY COUNCIL (2002b) The Newcastle plan: a community strategy for the city (http://www.newcastle.gov.uk).

NORTON, R. (1992) *Mother Clap's Molly House: The Gay Subculture in England 1700–1830*. London: Gay Men's Press.

PARK, R. E. (1952) *Human Communities: The City and Human Ecology*. Glencoe, IL: The Free Press.

PARK, R. E. and BURGESS, E. W. (1925) *The City*. Chicago, IL: Chicago University Press.

PRED, A. (1977) *City-systems in Advanced Economies*. London: Hutchinson.

PUBSNEWCASTLE (2003) *Pink triangle/gay village* (http://www.weir86.freeserve.co.uk/GayVillage.html; accessed 15 August 2003).

QUILLEY, S. (1997) Constructing Manchester's 'new urban village': gay space in the entrepreneurial city, in: G. B. INGRAM, A.-M. BOUTHILLETTE and Y. RETTER (Eds) *Queers in Space: Communities, Public Spaces, Sites of Resistance*, pp. 275–292. Seattle: Bay Press.

SCHELLING, T. C. (1971) Dynamic models of segregation, *Journal of Mathematical Sociology*, 1(2), pp. 143–186.

SCHELLING, T. C. (1978) *Micromotives and Macrobehaviour*. New York: W.W. Norton & Co.

SIBALIS, M. (1999) Paris, in: D. HIGGS (Ed.) *Queer Sites: Gay Urban Histories since 1600*, pp. 10–37. London: Routledge.

SIGMA RESEARCH (2003) *Vital Statistics 2002: English Strategic Health Authorities Data Report: Findings from the National Gay Men's Sex Survey 2002*. London: Sigma Research (http://www.sigmaresearch.org.uk).

SILVERMAN, J. (2003) Sex workers say 'let us stay'. *BBC News*, 18 February (http://news.bbc.co.uk/hi/uk/2754019.stm; accessed 15 August 2003)

SMITH, N. (1996) *The New Urban Frontier: Gentrification and the Revanchist City*. London: Routledge.

SOHO SOCIETY (2001) *Soho present* (http://www.thesohosociety.org.uk; accessed 15 August 2003).

STRYKER, S. and BUSKIRK, J. VAN (1996) *Gay by the Bay: A History of Queer Culture in the San Francisco Bay Area*. San Francisco, CA: Chronicle Books.

SUMMERS, J. (1991) *Soho*. London: Bloomsbury.

THOMPSON, H. (1997) *Peter Cook: A Biography*. London: Hodder & Stoughton.

TRUMBACH, R. (1999) London, in: D. HIGGS *Queer Sites: Gay Urban Histories since 1600*, pp. 89–111. London: Routledge.

VINYL-JUNKIES (2003) *Soho* (http://www.vinyl-junkies.com/soho.html; accessed 15 August 2003).

WALFORD, E. (1987) *Old London: Strand to Soho*. London: Alderman Press.

WESTMINSTER CITY COUNCIL (2002) *Draft Pre-inquiry unitary development plan*, 29 August, City of Westminster Council, London (http://www.westminster.gov.uk).

WESTMINSTER CITY COUNCIL (2003) *Draft action plan for Chinatown May 2003*. City of Westminster Council, London. (http://www.westminster.gov.uk).

WHITTLE, S. (1994) Consuming differences: the collaboration of the gay body with the cultural state, in: S. WHITTLE (ed.) *The Margins of the City: Gay Men's Urban Lives*, pp. 27–41. Aldershot: Ashgate Publishing.

Authenticating Queer Space: Citizenship, Urbanism and Governance

David Bell and Jon Binnie

Introduction

This paper seeks to make an intervention into a number of debates about contemporary cities and city-spaces. In particular, we want to explore the interrelationships between the global cities thesis, discussions of new forms of urban citizen and citizen-spaces, changing forms of urban governance and the production and consumption of 'gay space' in cities. Our key focus is the impact of the 'new urban order' on sexualised spaces in cities. These spaces, we want to suggest, are caught between imperatives of commodification and ideas of authenticity.

The paper begins with a brief discussion of the contested notion of 'global cities' as a broad theoretical backdrop to thinking about urban sexual citizenship and about the new kinds of identities that global cities produce, host and attract.[1] It then moves on to introduce related ideas about consumer citizenship and to explore how the logic of consumer citizenship refracts the way cities are marketed to attract particular kinds of residents, visitors and capital. It then explores how sexual 'others' are conscripted into the process of urban transformation and, by turn, how this city rebranding becomes part of the sexual citizenship agenda. Key to our argument is that this interweaving of urban governance and sexual citizenship agendas produces particular kinds of sexual spaces, at the exclusion of other kinds. This observation brings up the vexed issue of authenticity; seen as some kind of essential retreat or defence from commodification and spectacle, the discourse of authenticity is one that we need to engage with carefully in discussions of sexualised spaces.

The paper moves on to consider the extent to which the idea of sexual citizenship has

David Bell is in the Faculty of Arts, Media and Design, Staffordshire University, Flaxman Building, College Road, Stoke on Trent, ST4 2DE, UK. Fax: 01782 294 760. E-mail: d.bell@staffs.ac.uk. Jon Binnie is in the Department of Environmental and Geographical Sciences, Manchester Metropolitan University, John Dalton Building, Chester Street, Manchester, M1 5GD, UK. Fax: 0161 247 6318. E-mail: j.binnie@mmu.ac.uk.

been woven into the tournament of urban entrepreneurialism—for example, through the discourse of pride—and how this affects sexualised spaces, as more and more urban space (and more and more types of urban space) becomes ever-more commodified. We argue that this process can be read as an instance of 'the new homonormativity', producing a global repertoire of themed gay villages, as cities throughout the world weave commodified gay space into their promotional campaigns. The paper ends with a discussion of a recent exemplar of this process, in the form of Richard Florida's index of the 'creative class'—an index which favourably weights gay spaces in ranking the creativity and creative appeal of US cities.

Global Cities and (Sexual) Citizenship

A number of writers argue that globalisation is leading to the development of 'world cities' or 'global cities'—a new urban species, where new forms of politics and identity are birthed (for example, Friedmann, 1986; Sassen, 1991). In *The Sexual Citizen* (Bell and Binnie, 2000), we drew attention to the relationship of sexual citizenship to cities and discussed the impact of globalisation on sexual politics; in this paper, we want to extend that analysis by exploring the relationship between cities, the transformation of citizenship and urban sexual cultures and communities.

Cities are increasingly seen as key sites in struggles over citizenship, as they are argued to be the prime site where difference is encountered (see, for example, Young, 1990). However, how we conceive of and theorise 'the city' and city-spaces will have consequences for how we think through citizenship. Even so-called global cities are marked by local specificities—too often ignored in grand theories of urban change. This is not to say that we do not agree with the point that globalisation intersects with urban space and people to produce new cities and citizens—just to reiterate a point made by Nigel Thrift (2000), who cautions against mythologising 'The City'. In part, work in what we might call queer urban history is of tremendous use here, in detailing the ways in which the particularities of cities interact with broader processes of structural transformation (see Higgs, 1999, for examples).

How we conceptualise cities has huge consequences for how we imagine citizenship. If you accept the global cities thesis, then these cities are now dwarfing other economic spaces and territories, and challenging the nation-state as a container of social relations—and as the hearth of citizenship. Michael Peter Smith (2001), however, counters the 'global cities' approach, arguing that it can sometimes provide an economically deterministic view of urban change which he suggests needs more careful contextualisation, in order to see the full complexity of influences that restructure cities, spaces and identities. The disjunctive global flows famously sketched by Appadurai (1990) produce an uneven landscape; moreover, these flows interact with pre-existing urban forms and urban lives—these are not erased, but reworked. When our attention turns to the spaces of sexual citizenship, we can see how these interactions work on the ground, whether in Le Marais, the 'gay quarter' of Paris (Sibalis, 1999), the South of Market district of San Francisco (Rubin, 1998) or New Park in Taipei (Martin, 2000).

A cornerstone of much work on global cities is the emergence there of new identities and class groups—another factor that has important implications for thinking citizenship (including sexual citizenship). Isin and Wood argue that a new cosmopolitan class of professionals and managers is forging an identification with a cosmopolitan imaginary centred on consumption practices: "the new professional-managerial groups have become less concerned about their national interests and turned their back on the nation-state: they display cosmopolitan tendencies" (Isin and Wood, 1999, pp. 100–101). They go on to suggest that members of new professions may be more loyal to their profession than the state or city, naming them 'professional-citizens'. Isin (1999) develops this argument, focusing on the role of cultural capital in accessing new forms of citizenship and on the function of networking and conferencing in the solidifying of transnational professional-citizen identities. Almost as an afterthought, Isin admits that other new political identities are being forged in the global city; equally based on the logics of networks and transnational flows, forms of 'insurgent citizenship' are promoting new rights claims in this context. Although attempting to force an absolute separation between the new professional class and these insurgent identities (which include immigrants, greens and sexual minorities) and seeing these two 'classes' as oppositional, Isin is nevertheless right to point to the need to see global cities as complex sites where all kinds of new identities and practices play out.

Consumer Citizenship and the City

The consumer citizen is a figure centre-staged in new debates on world cities and the practices of cosmopolitanism. However, we want to argue that the focus on the consumer citizen must be complemented by a concern with notions of production, especially the production of space. Nikolas Rose (1990, cited in Isin and Wood, 1999) has argued that consumption is now central to how citizenship is defined. Here, the construction of our identities, and the management and disciplining of the self, occurs through choices we make as consumers. 'Freedom' and 'power' are thus increasingly (even exclusively) articulated through the market—a familiar story in the shifting meaning of citizenship. An exclusive focus on consumption is too short-sighted, however,—we need to pay more attention to labour in the production of consumption spaces and experiences (see Adkins and Lury, 1999). Moreover, we need to ask the key question: what is consumed by the 'new' consumer citizen? John Urry argues that

> Global rights might also include the right to be able to buy across the globe the products, services and icons of other cultures and to be able to locate them within one's own culture (Urry, 2000, p. 70).

If Urry is right, then we also need to be attuned to the power relations at work here—who's consuming whom?—and also mindful of the labour of production that enables these consumption practices. Further questions arise. Who can participate in consumer citizenship? Who is the Other of the consumer citizen? Who is excluded from consumption spaces and practices?

Jasbir Puar's (2002) recent work on 'gay tourism' makes clear just how important these questions are for thinking sexual citizenship. The focus on the global city in much work on queer tourism is, however, problematic in both overstating the mobility of queers and understating the agency of the state (and especially the local state) (see Brenner, 1999). In the conclusion to *Globalizing Cities*, Marcuse and van Kempen (2000, p. 262) rightly argue that globalisation means a change in terms of what the state does, rather than a decline in its power—it moves from being an agent of redistribution to a promoter of enterprise. The changing role of the state—specifically, the local state—increasingly rests on making cities more desirable places for what Richard Florida terms the 'creative class'—the focus of a later section of this paper. The role of the state here is to foster spaces of/for consumption, to act 'entrepreneurially'. Florida cites the work of Edward Glaeser to argue that

> The future of most cities depends on their being desirable places for consumers to live. As consumers become richer and firms become more mobile, location choices are based as much on their advantages for workers as on their advantages for firms (Florida, 2002, p. 259).

As we shall see later, sexual 'others' are among the groups seen in this formulation as marking cities as 'desirable'—a paradoxical rebranding for groups more used to being labelled as 'undesirables'. Sexual 'others' are also clearly constructed as agents of consumer citizenship—not least given the pink economy discourse, plus the discussions of global gay tourism, the globalisation of Pride/Mardi Gras mega-events and so on (see below). At the same time, other sexual 'others' are constructed as exotic objects to be consumed, perhaps most vividly in forms of sex tourism. Considering cities as sites for modalities of consumer citizenship is therefore very important for unpacking contemporary urban sexual citizenship.

In a further attempt to unpack the current modalities of citizenship relevant to our argument here, Nikolas Rose (2000) has written about the relationships between urban governance, consumption and health, arguing that the management of health has become a key part of the agenda of active citizenship. As a result of this, "health is not a state to be striven for only when one falls ill, it is something to be maintained by what we do at every movement of our everyday lives" (Rose, 2000, p. 101). Stress clinics and gyms are depicted by Rose as sites where bodily discipline is recast as an element of civic duty and active citizenship: "The imperative of health thus becomes a signifier of a wider—civic, governmental—obligation of citizenship in a responsible community" (Rose, 2000, p. 101). Health promotion as an arm of urban governance is mirrored, moreover, in the purification of space. In particular, spaces deemed as potentially 'unhealthy' (such as cruising grounds demonised as sites of unsafe and deviant sex) are policed under the guise of 'health promotion' (Dangerous Bedfellows, 1996). As part of the rewriting of urban consumer citizenship, this has profound implications for sexualised spaces, as the next section shows.

Queer Theory/Urban Theory

In this section, we want to examine how queer politics and theory relate to urban politics and theory. Work at the intersection of queer and urban has taken a number of forms. There has been an increasing focus on cities as sites of/for queer consumption—for example, by tourists (for example, Markwell, 2002; Puar, 2002), a growing interest in the spatial formations of sexual practice (for example, Ingram *et al.*, 1997; Leap, 1999) and a continuation of earlier work focused on gay gentrification (for example, Knopp, 1995). In addition, a number of authors, such as Berlant and Warner (1998) and Hubbard (2001), have focused on 'counterpublic' sexualised spaces such as cruising grounds and red-light zones. In connecting sexual politics to the politics of space, the main argument has been to link rights-claims to contests over space: to establish forms of queer territoriality as the base for political work—a developmental model of gay space unchanged since Harvey Milk's campaign of the late 1970s (if not before).

However, the tactic of claiming space has also received criticism. Shane Phelan, for example, writes that

> Making our communities into armed camps is not good politics; rather than shoring our borders to prevent infection, we must work on infecting the body politic with the dangerous virus of irreverent democracy (Phelan, 2001, p. 132).

Phelan's remarks have a clear spatial connotation. Many 'gay' consumption spaces are bounded communities, where processes of exclusion operate, for instance on the basis of race and gender. Phelan suggests that such bunkering-in is inappropriate to the task of reimagining sexual citizenship. However, boundaries can be seen as necessary, to keep 'unwanted others' out. But—as has been witnessed in case studies of such spaces—what sometimes happens is that the boundary of 'unwantedness' gets redrawn, so that in opening up to (non-gay-identified) consumers, the spaces push out what we might call the 'queer unwanted' (Binnie, 2004). The new publicity of more mainstream manifestations of gay consumer cultures—thoroughfares, street cafés, trendy bars, themed gay villages—has driven the less-assimilated queers underground, back into subterranean, back-street bars and cruising grounds (see Califia, 1994; Dangerous Bedfellows, 1996).[2] As Michael Warner writes, in the context of the recent desexualising of New York via Mayor Guiliani's zoning amendment bill

> All over New York … a pall hangs over the public life of queers. Much more is at

> stake here than the replacement of one neighbourhood by another, or the temporary crackdowns of a Republican mayor. As in other U.S. cities, sex publics in New York that have been built up over several decades—by the gay movement, by AIDS activism, and by countercultures of many different kinds—are now endangered by a new politics of privatisation (Warner, 1999, p. 153).

A similar story is recounted by Gayle Rubin (1998) in her discussion of the disappearance of spaces of leathersex in the South of Market district of San Francisco as a result of gentrification, which has not only erased the established spaces of the leather community, but also planned out the 'edge zones' of the neighbourhood that have been of central significance to the evolution of local sexual subcultures. Of course, this driving underground refers to particular kinds of sexual cultures and spaces, and has, as Warner and Rubin both note, been supported by segments of the gay community. Moreover, its accomplishment has not been total—queer counterpublics have emerged as a consequence of this process, but these are often ephemeral, contested zones, easily erased by forces of purification.

As Shane Phelan (2001) also argues, queer politics have become marginalised with (even by) the mainstreaming of gay politics—an argument easily transferable to a geographical register. In *Striptease Culture*, Brian McNair (2002) suggests that the mainstreaming of gay culture is a significant development within British popular culture. This means that gay images and icons have become much more prominent within the public sphere—especially, perhaps, the public sphere of television. But the greater visibility of gay culture has produced mixed effects, not all of them positive. Joshua Gamson's (2002) work on TV talk-shows vividly and usefully highlights how greater gay visibility on televison has brought with it tensions over 'appropriate' forms of sexual identity—another example of the casting-out of the 'queer unwanted', seen here as potentially damaging an assimilationist agenda of respectability (which has all-too-clear class and space dimensions to it). The production of what Lisa Duggan (2002) names 'the new homonormativity' works to exclude 'undesirable' forms of sexual expression, including their expression in space—for example, by reducing the 'gay public sphere' to consumption spaces and gentrified neighbourhoods only.

Gamson's argument is also useful for rethinking issues of the public and the private in relation to sexual citizenship. In his historical account of US gay culture, George Chauncey (1996) argues that 'privacy could only be had in public' for gay men in early 20th-century US cities—in the sense that public spaces such as parks and piers were key spaces where men could be intimate and private with one another. The converse may now be the case—now that we see the return of the dimly lit back-alley bar as a counter to the assimilation of other gay consumption spaces into the urban fabric. This is the opposite of the evolutionary model of gay liberation (critiqued by Hoad, 2000), which suggests that we are on a trajectory or path to liberation. In this context, Phelan (2001) points towards the recloseting of the butch lesbian, arguing that certain queer identities have become almost pathologised within lesbian and gay culture—and likewise the spaces where those identities are performed (for examples of this pathologisation by gay writers, see among others Bawer, 1993; Jeffries, 2003). As Gamson's work on talk-shows illustrates, the availability of forms of public space is contingent, meaning that aspects of queer culture are rendered invisible and are denied access to the same public space being claimed as a right by newly empowered sexual citizens. We can observe identical processes occurring in other public spheres and with the same effects. As we now go on to show, the issue of public visibility is itself intensely problematic, especially when visibility is used as a marker of cosmopolitanism.

Sexual Strangers in the City

In some writings on urban cosmopolitanism, the possibilities of encountering difference in cities are romanticised. For example, Iris Marion Young (1990) sees the coming-together of strangers as a definitional feature of cosmopolitan urbanism, writing that 'eroticism' lies in the consumption of difference

> We walk through sections of the city that we experience as having unique characters which are not ours, where people from diverse places mingle and then go home (Young, 1990, p. 239).

This reading suggests that straights may wish to consume (queer) difference 'and then go home', where home is a space set apart from the whirl of difference on the streets. However, some forms of difference are inevitably excluded from this 'cosmopoliticisation' because they are too different, too strange. Moreover, just as Young (1990, p. 239) states that "City life also instantiates difference as the erotic, in the wide sense of an attraction to the other", Samuel Delany (1999) stresses the role of contact between strangers as the core of urban experience. But where Young's eroticism amounts to little more than voyeurism, for Delany contact means spontaneous encounters between strangers, where a different modality of eroticism occurs

> Contact is also the intercourse—physical and conversational—that blooms in and as 'casual sex' in public rest rooms, sex movies, public parks, singles bars, and sex clubs, on street corners with heavy hustling traffic, and in the adjoining motels or the apartments of one or another participant, from which nonsexual friendships and/or acquaintances lasting for decades or a lifetime may spring (Delany, 1999, p. 123).

This contact is random and cannot be promoted or produced by capital or the state. Public sex sites such as public toilets, bathhouses and cruising areas are here given a distinct significance and Delany writes that these are inherently democratic spaces and encounters—in that they resist commodification and purification. And in an essay on Delany's book, Rofes (2001) also notes that this contact enables identification and opportunities for communication and friendship, notably across class and racial boundaries.

However, there is perhaps an ever-present danger in romanticising contact between strangers in these spaces as somehow an 'authentic' alternative to the more commodified or marketised aspects of gay culture. Which forms of contact are seen as authentic and to be valued, and which are seen as inauthentic and without value? And how can the devalued forms of contact be legislated against? This raises further questions about the nature of representation, and the dangers inherent in these matters, as authors reproduce this dichotomy by caricaturing some forms of contact as being more valuable or more worthy than others (see, for example, Delany, 1999). Finding answers to these questions is particularly difficult as gay cultures have routinely been represented as lacking authenticity, with gay performances of masculinity and femininity, for example, being cast as bad, inauthentic copies of 'original' straight identities (Butler, 1990; Bell *et al.*, 1994).

While the question of authenticity in the representation of urban encounters between sexual dissidents is problematic, so too is the question of the (often-troubling) presence of straights within spaces defined as gay. These tensions have been studied in research on the conflicts associated with the increased use of the consumption spaces of Manchester's gay village by straight women (Binnie and Skeggs, 2004). Here, the authenticity of the space of the village as a *gay* space is used as a commodity to attract these women, as 'the 'gayness' of the village is promoted to the wider community as a non-threatening authentic commodity' (Binnie and Skeggs, 2004, pp. 56–57). However, the presence of straight-identified women in this space is contested, particularly by some gay men who fear a heterosexual invasion of 'their' space.

Studying the narratives used by straight women to rationalise and legitimate their presence within gay-defined spaces, the question of an authentic or 'correct' form of contact with sexual dissidents was clearly an emotive one. Some women in the study legitimated their presence within this space on the basis of their knowledge or cultural capital, in terms of knowing 'how to behave' in the presence of gay men. However, in order to have their authentic experience of gay space, the women were concerned that other straight women should be excluded for lacking the requisite capital and because too many straight women in the village would dilute its gayness—which was the original quality and property that attracted them in the first place (see Binnie and Skeggs, 2004, for a full discussion).

The notion of authenticity is often counterposed with the inauthenticity of 'the spectacle'. We should be wary, however, of dismissive uses of terms such as spectacle, for they carry a value judgement about what forms of cultural production and practice are authentic and which are 'merely' spectacle. Marxist writers on the post-modern city often imply that urban spectacles are inauthentic cultural expressions (for an overview, see Thomas, 2002, ch. 6). How, then, are we to understand the production of lesbian and gay pride festivals in terms of the discussion of authenticity versus spectacle? Spectacles are too easily dismissed as providing only inauthentic and commodified encounters with difference, but this oversimplifies the uses and meanings of 'spectacle' and any discussion of 'spectacular' events has to explore in more detail the production and consumption of spectacle alongside the effects of that production and consumption, especially in relation to the inclusion and exclusion of forms of otherness. Moreover, given the long and problematic relationship between homosexuality and capitalism, it would be wrong totally to dismiss gay events as mere spectacle and commodified pleasure (see Bell and Binnie, 2000).

So, we can see two imperatives structuring the production of 'queer' urban spaces and events, both of which bring to the surface the problematic issues of authenticity and spectacle. On the one hand, there is the promotion of gay spaces and events as part of broader urban entrepreneurialism agendas. On the other, there is the process of purifying space and the concomitant eradication of strangeness and danger that inevitably results from these strategies of boosterism. There are of course clear connections between these two tendencies—the production of safe space is part of the process of promoting gay space. Like Rose's discussion of health promotion as urban governance, the production of safety here becomes an element of regulation. But it is important to note that 'spectacular' events, such as Mardi Gras in Manchester or in Sydney, are also significant in emphasising the role of queer cultures within the narratives particular cities tell about themselves, so their promotion can be seen to bring benefits in terms of sexual citizenship.[3] This is, however, a complex and uneasy relationship on both sides, as cities negotiate different narrative threads in their overall promotional arsenals and as queer cultures ambivalently submit to, or embrace, or reject, their partial incorporation into that arsenal.

Queer Place Promotion, Neo-liberalism and Homonormativity

Debates on entrepreneurialism, city promotion and urban governance have centred on promotional campaigns based on a neo-liberal ideological framework. The main focus of these debates has been on entrepreneurial governance and the socially exclusionary nature of these policies. Kevin Ward (2000) has argued that such campaigns perform an ideological function in cementing neo-liberalism within urban governance. Such strategies reflect the changing role of the local state—the emphasis is on promoting entrepreneurialism as opposed to actively addressing redistribution. It is argued that the vigorous promotion of city-centre living based on the development of up-market consumption spaces has produced very real exclusions of the urban poor from the city. This context is crucial to an understanding of the politics of gay space in contemporary cities.

It is also crucial in any discussion of queer place promotion and marketing to reflect on the relationship between discourses of gay pride and civic pride. Elspeth Probyn has explored the discourse of pride in the context of the 1998 Gay Games in Amsterdam, arguing that the celebration of pride at the event "operates as a necessity, an ontology of gay life that cannot admit its other" (Probyn, 2000, p. 19). In the spectacle of the Games, shame cannot be explicitly acknowledged and this includes *sexual* shame: for Probyn the Games promote an entirely asexual and apolitical version of pride. The imperative to be proud, to display pride and to wed gay pride to civic pride (which involves making the city proud of its gays and the gays proud of their city) can be intensely problematic and can lead to the exclusion of 'shameful' aspects of gay and urban cultures.

Matching gay pride to civic pride means that cities have to respond positively to gay culture in order to maintain their competitive edge. However, as has been observed in relation to other urban theming models, there is a 'me-too-ism' that means that every city that considers itself a player must have the requisite features—ethnic quarters, hi-tech corridors, festivals, gay villages (Bell and Jayne, 2004). In a bid to be unique, cities have in fact become more alike—but there is no alternative, as not having those features means not even being in the race. This means that more and more cities have developed their own versions of themed spaces, including gay villages, leading to a normalisation of the presence of certain types of gay space in cities. This can be read in two ways: any aspiring competitive city must have a themed gay space and the only type of gay space that an aspiring competitive city can have is a themed one. The globalisation of the gay village model and the globalisation of entrepreneurial urban governance, moreover, are reproducing a narrow range of cloned spaces world-wide (Binnie, 2004)—itself the object of local resistance, as in the case of Le Marais in Paris, where the move towards developing a 'gay ghetto' is seen by some as a further instance in the American colonisation of French culture (Sibalis, 1999). There is also a tension over the extent to which city imagineers focus on local gay culture and a question about whether this might alienate other potential visitors and investors (Binnie, 1995).

Queer place promotion also needs to be considered in the context of debates around the pink economy; we have to pay attention to the politics behind the discursive construction of the pink economy myth, which has itself been woven into urban competitiveness. We also have to remember the relationship between production and consumption: lesbians and gay men are workers, not simply consumers—and any discussion of queer place promotion and the marketing of cities must remember this. The hype about gay spending power made gay culture 'sexy' in a commercial sense, while simultaneously desexualising it, and the creation of new gay consumption spaces rests on a labour force who may be priced out of participating in those spaces as anything other than bartenders or go-go dancers. Moreover, the requirement to perform for customers—familiar to all interactive service encounters but perhaps especially hyped-up in gay commercial spaces—places emphasis on appropriate bodies, clothes and behaviour, bringing yet more possibilities for exclusion.

Gay global events such as Mardi Gras festivals and the Gay Games, meanwhile, have assumed a greater profile within lesbian and gay communities. For instance, in his discussion of the Sydney Gay Games held in 2002, Waitt (2003, p. 171) notes that the event that year attracted 12 000 athletes, compared with the 1350 that participated in the first Gay Games in San Francisco in 1982. The dramatic growth in scale of such events has in turn led to increasing attention from the wider media. The high media profile of these events has, however, inevitably sparked discussion within lesbian and gay communities about whether such events represent a mainstreaming, commodification and depoliticisation of these communities (see Waitt, 2003, for a discussion of these debates in Sydney). Moreover, in her essay on the Amsterdam Gay Games, Probyn (2000, p. 18) notes that "the key tenets of the [Gay] Games movement are 'inclusion', 'participation', 'the achievement of one's personal best' ". Such a discourse works to deny multiple exclusions and barriers to participation—for example, on the grounds of cost: there is a hefty price tag attached to being proud in the context of events like this.

Regeneration and Purification

Gays are often cast as model citizens of the urban renaissance, contributing towards the gentrification of commodifiable cosmopolitan residential and commercial areas. Gays are attracted to neighbourhoods as pioneers, so the script runs, then others are attracted in because of the cultural capital gained from living in boho areas. Both gays and straights, though, have been behind moves to push out sleaze from gentrifying neighbourhoods. For many assimilationist gays, for example, gay male sex zones are seen as an embarrassment that must be cleaned up. Eric Rofes (2001) argues that conservative gay health officials and community leaders have often been among the most vocal critics of sex zones; these spaces are seen as the province of the least desirable and least assimilable elements of male–male sexual culture—certainly not spaces of pride.

Both gay male sex areas and the desexualised gentrifying gay districts are subject to the logic of the ecology of gentrification. As gay districts become gentrified, they become more desirable for wider gentrification and colonisation by trendy (and less trendy) straights. These new residents of the area often will not tolerate noise and disturbance from nightclubs and late-opening café bars and therefore act to limit commercial venues. As areas become more 'respectable', so gay commercial spaces are forced out. Of course, at one level this is an inevitable part of the gentrification script: 'pioneering' gentrifiers move on as their life-courses evolve and as they seek further to exploit the housing market, and the hike in house values as neighbourhoods become identifiably boho prices out some of the very people that lend parts of cities their 'boho-ness' or edginess in the first place (Bridge, 2003; O'Connor, 2001). However, the important point to make here is that these long-term gentrification effects work to chase out 'unwanted' activities, with profound implications for the sustainability of sexual spaces and cultures, as for other 'alternative' uses of city-spaces (O'Connor, 2001).

Gay male sex areas are even more vulnerable to crusades, such as that undertaken in New York by Mayor Giuliani. Giuliani's zoning law represents a concerted attack on public sex through making it impossible for businesses such as adult bookstores, porn cinemas and saunas to operate profitably; as Berlant and Warner (1998, p. 562) argue, "The law aims to restrict any counterpublic sexual culture by regulating its economic conditions". So while some aspects of gay culture are readily colonised by capital, others have to be rendered uneconomic, uncommercial (see also Warner, 1999).

In terms of policy-making and attempts to promote or develop gay villages, what is significant about these spaces is that they have historically grown 'organically' without much in the way in conscious place promotion or investment (at least until very recently). They have tended to be located in parts of the city that were seen as beyond the control and active policy-making reach of the state. These spaces are of course regulated—policed and subject to planning controls—so they were subject to forms of state intervention in a negative sense; more recently, however, active state promotion of gay spaces in cities has brought them into the entrepreneurial, neo-liberal frame.

The recent development of Manchester's gay village, for example, took place within a wider discourse around the 'Manchester model' of urban regeneration—the Manchester script that has to be ascribed to. In this way, the Manchester model of becoming an erstwhile cosmopolitan and European city has become enabling and empowering in helping to shape a sense of entitlement by many users of the gay village that they do have a stake and a voice. The key to the 'success' of the gay village, however, has been the production of a desexualised consumption space where an asexual non-threatening (especially to women) gay identity can be enacted. Gayness is here used as a resource to attract women as consumers into the space, as discussed earlier (Binnie and Skeggs, 2004).

Another issue to consider here is the way in which sexuality can be configured within debates on urban governance. Davina Cooper's (1994) elegant exploration of the 'New Urban Left' in 1980s Britain focused on the failure of 'progressive' initiatives promoted by the local state at a particular moment in the political history of the UK. The landscape of urban politics (and, indeed, sexual politics) has now radically changed (on Manchester's shift from municipal socialism to entrepreneurial governance, see Quilley, 2002). What we now need to do is to link together the literature on governance and regeneration with the material on queer space; the neo-liberalisation of urban politics and sexual politics has rewritten the terms by which urban sexual citizenship operates.

Rose (2000) provides a way into rethinking citizenship which is useful to this task, through his discussion of the 'games' of citizenship. Arguing that citizenship is no longer primarily realised in relation to the state, Rose examines how citizenship relations are instead embodied in everyday material practices. In particular, he is concerned with examining active citizenship and the demands and responsibilities of the citizen associated with neo-liberal 'Third Way' politics: "games of citizenship today entail acts of free but responsible choice in a variety of private, corporate, and quasi-public practices, from working to shopping" (Rose, 2000, p. 108). In this context, gay consumers of gay villages are positioned as producers of spectacle for straight observers, their duties being to conform to an accepted and 'respectable' notion of gay identity. What is being promoted is a very safe form of 'exotic difference' in order to attract mobile capital, most particularly in the form of international tourism.

Debates on sexual citizenship have been characterised by the conflicts between those in favour of strategies of assimilation versus those who insist on the rejection of heteronormative values. As Lisa Duggan (2002) argues, the assimilationist agenda accords to the sexual politics of neo-liberalism, or what she calls the 'new homonormativity'. Tracking the neo-liberal agenda in US gay politics, Duggan shows how this new homonormativity is gaining ground, particularly as it chimes with the broader neo-liberal political landscape. Its aim, she writes, is to produce

> a politics that does not contest dominant heteronormative assumptions and institutions but upholds and sustains them while promising the possibility of a demobilized gay constituency and a privatized, depoliticized gay culture anchored in domesticity and consumption (Duggan, 2002, p. 179).

In terms of the production of space, this formulation maps perfectly onto the reshaping of gay villages and neighbourhoods under the entrepreneurial governance agenda.

From the Pink Economy to the Creative Class

Economic restructuring has chased out queer counter-publics and spaces of public sex have become even more marginal as a result of entrepreneurial governance. Gentrification has meant the eating-up of derelict land and marginal spaces in the city, making nice neighbourhoods for 'guppies' but erasing cruising grounds. As Dereka Rushbrook (2002) argues, a variety of urban transformations are currently reshaping 'gay space'. These include commercialisation, gentrification, entrepreneurial governance and the growth of cosmopolitan tourism. In each of these processes, gay-friendliness has come to be used as a form of cultural capital deployed by powerful groups and by cities themselves as they jockey for position on the global urban hierarchy. Gays are now seen as strange attractors of global venture capital. So, the arguments about the effects of gay gentrification get replayed over the pink economy (Badgett, 2001) and now come back to us in the guise of the creative class.

The hype around Richard Florida's *The Rise of the Creative Class* (2002) powerfully demonstrates much about the current state of the economics of sexual citizenship. The obligations of sexual citizenship are the sacrifices we have to make. *The Rise of the Creative Class* is significant for highlighting, albeit problematically, the role that gays play in urban regeneration. Florida calculates indexes for creativity and constructs a 'gay index' to measure the level of significance of a gay community within each US city in his study. Florida makes some interesting assertions

> To some extent, homosexuality represents the last frontier of diversity in our society, and thus a place that welcomes the gay community welcomes all kinds of people (Florida, 2002, p. 256).

He argues that there is strong correlation between the 'gay index' and regions where hi-tech industries are located, suggesting that where gays go, geeks follow (another interesting revaluing of previously marginalised identities). In indexing his new creative class—a class not unlike Isin's professional-citizens—Florida makes a telling observation

> The Gay Index was positively associated with the Creative Class ... but it was negatively associated with the Working Class. There is also a strong relationship between the concentration of gays in a metropolitan area and other measures of diversity, notably the per cent of foreign-born residents (Florida, 2002, p. 258).

Significantly, in a footnote on the census data on which Florida bases his discussion, he acknowledges that the data include only lesbians and gay men in same-sex partnerships. Thus lesbian and gay visibility (in terms of the national census) is based on being part of a couple or relationship: single gay men or lesbians do not figure in the equation. Once again, here is a statement of the 'appropriate' form of queer lifestyle—what if the 'gay index' was based on levels of arrest for public sex, or number of tea rooms or back-rooms? (A modified repeat of Florida's index for UK cities, conducted by the policy think-tank Demos, used the number of services provided for the lesbian and gay community in its equally flawed calculations.)

The gay index is therefore an index of respectability, of nicely gentrified neighbourhoods, and the only accepted presence in public space is the Mardi Gras festival, itself turned from a political demonstration into a celebration of 'difference' staged for tourists. As Rushbrook (2002, p. 195) says, "gay urban spectacles attract tourists and investment; sexually deviant, dangerous rather than risqué, landscapes do not". Hence, the new forms of 'gay space' in cities operate their own forceful exclusions—notably, the 'trash' and the 'pervs'. In a direct echo of Gamson's work on talk-shows, gay public space effects marginalisation even as it claims visibility. As Rushbrook writes, what we can observe here—replaying the effects of gay gentrification and of gay consumerism—is a "shift from a closed, introverted queer space to a more open appropriation of public space in which a blurring of boundaries is accompanied by a watering down of queerness" (Rushbrook, 2002, p. 198). In each phase, gays have been held up as pioneers—by doing-up run-down areas of town, by spending their way out of recession and by being at the forefront of the new creative economy. Like the 'positive representations' argument, there is something to be said for this—but we need to look much more closely at what is going on here: who counts in the creative class, who is invisibilised and what effects aligning homosexuality with the creative economy has for other queer spaces and identities? 'Cleaning up' city-spaces in order to attract transnational venture capital, or hi-tech geeks, or cosmopolitan tourists, has immense implications for classes not part of this new creative economy.

Summary

The aim of this paper has been to explore some key dimensions of the contemporary restructuring of urban space—often as a response to the heightened competitiveness promoted by the global cities discourse—and to track how these have influenced the changing shape of 'sexualised space' in cities. These effects are complex and paradoxical: the presence of gay communities and spaces has become part of the arsenal of entrepreneurial governance, giving sexual 'others' a central role in place promotion, as symbols of cosmopolitanism and creative appeal. Yet this incorporation has meant tightening regulation of the types of sexualised spaces in cities. This 'sexual restructuring' of cities, we argue, is a powerful component of the 'new homonormativity', a broader ideological project tied to the logic of assimilationist sexual citizenship.

Notes

1. There is not room in this paper to discuss the contested definition of 'sexual citizenship', nor its genealogies. For a detailed discussion, see Bell and Binnie (2000).
2. We do not mean to suggest here that all types of gay space are driven underground, nor that this move has impacted uniformly on the lives of all sexual citizens. Some participants in gay urban culture, for example, can move between 'mainstream' and 'underground' spaces as part of their leisure practices. However, the disappearance of certain forms of sexualised spaces in cities has profound impacts for the kinds of sexual cultures that can thrive there and those that are driven, if not to extinction, then to an increasingly precarious and marginal existence. Public sex, sadomasochism and fetishism, have been particularly affected by this process.
3. The use of the Mardi Gras 'brand' for gay mega-events has been argued to represent both a globalisation of particular forms of gay event and the depoliticisation of such events, especially where the name has replaced the earlier label 'Pride'. In addition, Mardi Gras festivals have been increasingly woven into place-promotion and tourism strategies, raising again the question of the 'spectacularisation' and commodification of gay culture (see Johnston, 2004). It is important to note, however, that while Manchester's annual event has been rebranded Mardi Gras, its origins and heritage, and its scale, format and meanings today, are still locally distinct.

References

Adkins, L. and Lury, C. (1999) The labour of identity: performing identities, performing economies, *Economy and Society,* 28, pp. 598–614.

Appadurai, A. (1990) Disjuncture and difference in the global economy, in: M. Featherstone (Ed.) *Global Culture: Nationalism, Globalization and Modernity*, pp. 295–310. London: Sage.

Badgett, M. V. L. (2001) *Money, Myths and Change: The Economic Lives of Lesbians and Gay Men.* Chicago, IL: University of Chicago Press.

Bawer, B. (1993) *A Place at the Table: The Gay Individual in American Society.* New York: Touchstone Books.

Bell, D. and Binnie, J. (2000) *The Sexual Citizen: Queer Politics and Beyond.* Cambridge: Polity Press.

Bell, D. and Jayne, M. (Eds) (2004) *City of Quarters: Urban Villages in the Contemporary City.* Aldershot: Ashgate.

Bell, D., Binnie, J., Cream, J. and Valentine, G. (1994) All hyped up and no place to go, *Gender, Place & Culture,* 1, pp. 34–47.

Berlant, L. and Warner, M. (1998) Sex in public, *Critical Inquiry,* 24, pp. 547–566.

Binnie, J. (1995) Trading places: consumption, sexuality and production of queer space, in: D. Bell and G. Valentine (Eds) *Mapping Desire: Geographies of Sexualities*, pp. 182–199. London: Routledge.

Binnie, J. (2004) *The Globalization of Sexuality.* London: Sage.

Binnie, J. and Skeggs, B. (2004) Cosmopolitan knowledge and the production and consumption of sexualized space: Manchester's gay village, *Sociological Review,* 52, pp. 39–61.

Brenner, N. (1999) Globalisation as reterritorialisation: the re-scaling of urban governance in the European Union, *Urban Studies,* 36(3), pp. 431–451.

Bridge, G. (2003) *Cosmopolitanism, provincialism and difference: circuits of cultural capital and gentrification.* Paper presented at the *Annual Meeting of the Association of American Geographers*, New Orleans, March.

BUTLER, J. (1990) *Gender Trouble: Feminism and the Subversion of Identity*. London: Routledge.
CALIFIA, P. (1994) *Public Sex: The Culture of Radical Sex*. Pittsburgh, PN: Cleis Press.
CHAUNCEY, G. (1996) 'Privacy could only be had in public': gay uses of the streets, in: J. SANDERS (Ed.) *Stud: Architectures of Masculinity*, pp. 244–267. New York: Princeton Architectural Press.
COOPER, D. (1994) *Sexing the City: Lesbian and Gay Politics within the Activist State*. London: Rivers Oram.
DANGEROUS BEDFELLOWS (Eds) (1996) *Policing Public Sex: Queer Politics and the Future of AIDS Activism*. Boston, MA: South End Press.
DELANY, S. (1999) *Times Square Red, Times Square Blue*. New York: New York University Press.
DUGGAN, L. (2002) The new homonormativity: the sexual politics of neoliberalism, in: R. CASTRONOVO and D. NELSON (Eds) *Materializing Democracy: Toward a Revitalized Cultural Politics*, pp. 175–194. Durham, NC: Duke University Press.
FLORIDA, R. (2002) *The Rise of the Creative Class: And How It's Transforming Work, Leisure, Community and Everyday Life*. New York: Basic Books.
FRIEDMANN, J. (1986) The world city hypothesis, *Development and Change*, 17, pp. 69–84.
GAMSON, J. (2002) Publicity traps: television talk shows and lesbian, gay, bisexual, and transgender visibility, in: C. WILLIAMS and A. STEIN (Eds) *Sexuality and Gender*, pp. 311–331. Oxford: Blackwell.
GLUCKMAN, A. and REED, B. (Eds) (1997) *Homo Economics: Capitalism, Community, and Lesbian and Gay Life*. London: Routledge.
HIGGS, D. (Ed.) (1999) *Queer Sites: Gay Urban Histories since 1600*. London: Routledge.
HOAD, N. (2000) Arrested development, or the queerness of savages: resisting evolutionary narratives of difference, *Postcolonial Studies*, 3, pp. 133–158.
HUBBARD, P. (2001) Sex zones: intimacy, citizenship and public space, *Sexualities*, 4, pp. 51–72.
INGRAM, G. B., BOUTHILLETTE, A.-M. and RETTER, Y. (Eds) (1997) *Queers in Space: Communities, Public Places, Sites of Resistance*. Seattle: Bay Press.
ISIN, E. (1999) Citizenship, class, and the global city, *Citizenship Studies*, 3, pp. 267–283.
ISIN, E. (2002) *Being Political: Genealogies of Citizenship*. Minneapolis, MN: University of Minnesota Press.
ISIN, E. and WOOD, P. (1999) *Citizenship and Identity*. London: Sage.
JEFFRIES, S. (2003) *Unpacking Queer Politics*. Cambridge: Polity Press.
JOHNSTON, L. (2004) *Mardi Gras as Tourist Spectacle*. London: Routledge.
KNOPP, L. (1995) Sexuality and urban space: a framework for analysis, in: D. BELL and G. VALENTINE (Eds) *Mapping Desire: Geographies of Sexualities*, pp. 149–161. London: Routledge.
LEAP, W. (Ed.) (1999) *Public Sex: Gay Space*. New York: Columbia University Press.
MARCUSE, P. and KEMPEN, R. VAN (2002) Conclusion; a changed spatial order, in: P. MARCUSE and R. VAN KEMPEN (Eds) *Globalizing Cities: A New Spatial Order?*, pp. 249–275. Oxford: Blackwell.
MARKWELL, K. (2002) Mardi Gras tourism and the construction of Sydney as an international gay and lesbian city, *GLQ: A Journal of Lesbian and Gay Studies*, 8, pp. 81–99.
MARTIN, F. (2000) From citizenship to queer counterpublic: reading Taipei's New Park, *Communal/Plural*, 8, pp. 81–94.
MCNAIR, B. (2002) *Striptease Culture: Sex, Media and the Democratisation of Desire*. London: Routledge.
O'CONNOR, J. (2001) *Creative industries and gentrification*. Paper presented at the *Annual Meeting of the Association of American Geographers*, New York, March.
PHELAN, S. (2001) *Sexual Strangers: Gays, Lesbians, and Dilemmas of Citizenship*. Philadelphia, PA: Temple University Press.
PROBYN, E. (2000) Sporting bodies: dynamics of shame and pride, *Body and Society*, 6, pp. 13–28.
PUAR, J. K. (2002) Circuits of queer mobility: tourism, travel and globalization, *GLQ: A Journal of Lesbian and Gay Studies*, 8, pp. 101–137.
QUILLEY, S. (2002) Entrepreneurial turns: municipal socialism and after, in: J. PECK and K. WARD (Eds) *City of Revolution: Restructuring Manchester*, pp. 76–94. Manchester: Manchester University Press.
ROFES, E. (2001) Imperial New York: destruction and disneyfication under Emperor Giuliani, *GLQ: A Journal of Lesbian and Gay Studies*, 7, pp. 101–109.
ROSE, N. (2000) Governing cities, governing citizens, in: E. ISIN (Ed.) *Democracy, Citizenship and the Global City*, pp. 95–109. London: Routledge.
RUBIN, G. (1998) The Miracle Mile: South of Market and gay leather, 1962–1997, in: J. BROOK, C. CARLSSON, and N. J. PETERS (Eds) *Reclaiming San Francisco: History, Politics, Culture*, pp. 247–272. San Francisco: City Lights.
RUSHBROOK, D. (2002) Cities, queer space, and the cosmopolitan tourist, *GLQ: A Journal of Lesbian and Gay Studies*, 8, pp. 183–206.

SASSEN, S. (1991) *The Global City*. Princeton, NJ: Princeton University Press.

SIBALIS, M. D. (1999) Paris, in: D. HIGGS (Ed.) *Queer Sites: Gay Urban Histories Since 1600*, pp. 10–37. London: Routledge.

SMITH, M. P. (2001) *Transnational Urbanism: Locating Globalization*. Oxford: Blackwell.

THOMAS, D. (2002) *The Transformation of Cities: Urban Theory and Urban Life*. Basingstoke: Palgrave.

THRIFT, N. (2000) 'Not a straight line but a curve', or cities are not mirrors of modernity, in: D. BELL and A. HADDOUR (Eds) *City Visions*, pp. 233–263. Harlow: Pearson.

URRY, J. (2000) Global flows and global citizenship, in: E. ISIN (Ed.) *Democracy, Citizenship and the Global City*, pp. 62–78. London: Routledge.

WAITT, G. (2003) Gay Games: performing 'community' out from the closet of the locker room, *Social and Cultural Geography*, 4, pp. 167–183.

WARD, K. (2000) From rentiers to rantiers: 'active entrepreneurs', 'structural speculators' and the politics of marketing the city, *Urban Studies*, 37, pp. 1101–1115.

WARNER, M. (1999) *The Trouble with Normal: Sex, Politics, and the Ethics of Queer Life*. Cambridge, MA: Harvard University Press.

YOUNG, I. M. (1990) *Justice and the Politics of Difference*. Princeton, NJ: Princeton University Press.

Sex and Not the City? The Aspirations of the Thirty-something Working Woman

Joanna Brewis

Introduction

Many of the available empirical data clearly suggest that British women's public and private lives have changed markedly during the past few decades. While these developments are not necessarily any more significant than social transitions in earlier eras, and space does not allow for a comprehensive review here, it is nonetheless hard to ignore claims like the assertion that women's participation in employment since 1979 has risen roughly in proportion with a similar decline in men's, and that it is now more likely for women to work for money than not to do so (McDowell, 2001, p. 351; Bristow, 2002a, p. 16). Moreover, although workplace gender equality is still a long way off and the evidence for any progress in this regard is mixed (see—for example, McDowell, 2001; Brewis and Linstead, 2004), some British women do appear, at entry level at least, to be beating men 'at their own game'. For example, between 1981 and 1991, numbers of female general managers increased by 61 per cent, as compared with an increase of 9 per cent in the same occupational category for men (Walby, 1997, p. 37). Similarly, London Chamber of Commerce data show that, of the 450 000 professional jobs created in the UK between 1981 and 1996, 69 per cent were taken by women, and those in the capital did especially well in this regard (McDowell, 2001, p. 351; Charles, 2002, p. 26).

In terms of the private sphere, since the late 1960s there has been a marked rise in divorce rates and a decline in first marriages; more couples are cohabiting; the birth rate is falling (especially amongst professionally qualified or graduate women); children are

Joanna Brewis is in the Management Centre, University of Leicester, University Road, Leicester, LE1 7RH, UK. Fax: 0116 252 5520. *Thank you to the 11 amazing women who made this paper possible and who reminded the author how enjoyable data collection and analysis can be. The author is also grateful to Alan Collins, Ronan Paddison, Edward Wray-Bliss and the three anonymous Urban Studies reviewers for their constructive and patient comments. Finally, sincere thanks to Nikki Rogers for her fantastic transcribing, and to the University of Essex for a grant supporting the purchase of recording equipment and the transcription.*

more likely to be born to older and/or non-married parents, and so on (Bristow, 2002a, 2002b, paras 5–6; Charles, 2002, p. 46, p. 48; Woodward, 2002, p. 188; Elliott, 2003b). Although most British households are still based on the traditional pattern of heterosexual marriage, the arena of the 'personal' does seem to be gradually permitting more varied lifestyle choices. Indeed, given that conventional domestic arrangements have long been criticised by feminists, we can suggest that trends in this area perhaps allow women more latitude than was historically the case.

Moreover, these two sets of issues are interconnected in fairly obvious ways, in that

> demographic changes indicate how education, employment and family no longer constitute a simple linearity in women's lives. Increasing numbers of women form part of the paid labour force and combine family and employment (Hughes, 2002, p. 2).

Charles (2002, pp. 67–69) is one author who provides empirical substantiation of these claims, including the suggestion that between 1984 and 1994 mothers entered paid employment at twice the rate of other women.

In terms of the theme of this Special Issue, then, the relationship between 'sex' (read here as a trope for intimate relationships, marriage, having children and so on) and 'the city' (taken to stand for paid employment, careers, etc.) has altered fairly substantially for British women over the past 30 years or so. But there is of course another more literal interpretation of the city—as large-scale urban/metropolitan space, as concentrated mass of inhabitants, as regional or national centre of political, economic, cultural and intellectual activities (Pimlott and Rao, 2002, p. 5). Moreover, the globe continues to urbanise such that, according to Castells (2002, p. 394), between two-thirds and three-quarters of the world's population will live in cities by the middle of this century. Given the latter claim, the city could be argued to be looming ever larger in the collective consciousness.

In terms of wider social indicators, it therefore seems to be an auspicious time to explore women's experiences of work, personal life and the urban expanse under the aegis of the 'sex and the city' theme. Also noteworthy is an apparent shift in cultural representations of such experiences. As I have argued elsewhere, working women during the 1980s and early 1990s were typically fictionalised as work-obsessed 'career bitches'—in films such as *Working Girl* and *Disclosure*—for example, (Brewis, 1998). A later argument (Brewis, 1999a) focused on two more recent cultural texts—the US television series *Ally McBeal* and the novel *Bridget Jones's Diary* (Fielding, 1996). In both, the eponymous heroines (quite unlike the protagonists in the earlier texts) typically put work a poor second to romantic entanglements, and long for stable relationships with men. And Ally and Bridget may well be just the tip of a substantial cultural iceberg: certainly Elliott (2003b, p. 14) identifies "a chorus of seductive new female voices ... warning [women] of what they might be missing while they are at the office". He refers in particular to Allison Pearson's international bestseller *I Don't Know How She Does It* (2003) in which the central character Kate Reddy eventually leaves her stressful City job and relocates away from London with her family. Indeed, even *Sex and the City*, its apparent celebration of assertive female sexuality and career success notwithstanding, sends decidedly mixed messages in this regard. Bristow (2002a, p. 21), for instance, highlights the ambivalence in Carrie's "screw[ing] around" whilst simultaneously yearning to settle down with her on-again-off-again lover 'Big'.

It is also worth making the point that all of these texts are set in cities—Chicago, London, New York, etc.—and that this is often central to their storylines. Bridget and her three closest friends triumphantly assert—for example, that they are an 'urban family' and Kate habitually uses taxis in order to arrive at her chosen destination on time, given the vagaries of London public transport. She even pumps breast-milk in the back of one cab, such are the demands of her densely packed urban existence.

The concerns and priorities of the women in these cultural artefacts, moreover, seem to me to resonate with a series of themes to do with work, the personal arena and city life which surfaced during my previous empirical work with female professional urbanites—and many of these were in any case familiar from informal conversations with friends.[1] Such discussions implied that these women's jobs, to which they have devoted considerable effort, do not fulfil them sufficiently. They appear therefore to be contemplating paths that 10 years previously they would have rejected (moving away from cities and towns, having a family, giving up work, going part-time or changing career to do something less demanding). These changing aspirations also explain the title of the present paper, given that these women had seemingly begun to place 'sex' (relationships/settling down/having a family) above 'the city' (work/career/the urban setting).

To summarise, several issues have been identified which seem to warrant further investigation as a bundle of social phenomena, especially in terms of their relevance to the experiential reality of everyday life. There is the increasing entry of women into paid employment, set against a relatively new collection of cultural portrayals of this group and the anxieties apparently besetting them, as well as empirical data suggesting that these texts may reflect a more or less accurate portrayal of the 21st century city-dweller back to herself. The changing nature of intimacy is also noteworthy, as well as the increasing global importance of cities. On this basis, what follows is a detailed analysis of what a small group of female professionals feel about their personal relationships, their careers and their lives in the city—or London, to be more specific.[2] Overall, it seeks to contribute to debates concerning urban (dis)content, the legacy of 'second wave'/liberal feminism and life-stages by suggesting the ways in which the respondents' experiences map onto issues discussed in the relevant literatures.

Methodology

To begin with, this research project suggested the use of semi-structured interviewing in order to allow some grasp of how my respondents experience the world around them via the collection of qualitative data; to allow these women to control the content of what was discussed as far as possible, to get at what was important to *them*; and to facilitate an open, informal data-gathering process given that the subject matter required a measure of frankness on their part. In the event, I chose to run focus groups, given that they offer a number of additional advantages (Corbetta, 2003; Miller and Brewer, 2003). First, they are less resource intensive than one-on-one interviewing; an important criterion here in terms of ensuring that an adequate and meaningful data-set was collected in the time available. Secondly, focus groups are suitable when gathering data from a relatively homogeneous group of people who share certain experiences (in this case, female city-dwellers of a certain age-group, pursuing certain occupations) in which the researcher is interested. Hence the descriptor *focus* group. Thirdly, the focus group is not just a group interview—it also allows for the capturing of interactions between respondents and thus permits the researcher to compare and contrast their comments in 'real time'. This was something I was especially interested in doing, given the ostensible similarities between my participants, coupled with a tendency in the more influential (liberal) feminist texts to represent women as an undifferentiated social category. As Hughes (2002, p. 33) suggests, for example, a key liberal feminist tenet is that all women *can* 'make it' and, equally, that all women *want to*.[3]

Moreover, instead of striving for a representative sample from which generalisation would putatively be possible, I adopted the approach described by Sanger as follows:

> Rather than observing people and objects as samples of larger groups in some presupposed classificatory system … examine them in their complex singularity (Sanger, 1996, p. 20).

My sampling was therefore guided both by convenience and purpose. The respondents were all existing contacts but, more importantly, shared key characteristics—being aged between 27 and 35, having experience of living and/or working in London and doing jobs which could be described as professional or managerial—which facilitated exploration of the relevant topics. The appendix contains brief biographical details for each woman, who are all referred to using pseudonyms throughout.

But we should not overstate what focus groups can provide as regards understanding others' thoughts and experiences. As Alvesson and Deetz (2000, p. 71) argue, interview data are always overdetermined by their context: the researcher's (perceived) preoccupations and lines of questioning will unavoidably produce certain data and not others. Given that all my respondents were friends, or friends of friends, and thus had some idea of 'what to expect' from me, such 'channelling' of data was perhaps especially likely in this instance. And there is the fact that focus group participation entails being interviewed alongside others, such that their comments and reactions influence one's own (and vice versa). Further, interview data are "affected by the available cultural scripts about how one should normally express oneself on particular topics" (Alvesson and Deetz, 2000, pp. 71–72), so that the unvarnished 'truth' of how someone thinks and feels is never available through the interview mechanism (or indeed any other form of communication) because everything we express is constructed according to discursive rules. Indeed, Silverman (1993, p. 96) asserts that we often mistake 'cultural script following' for authenticity in talk, because what is most familiar and contextually appropriate typically has the most resonance.

Relatedly, Alvesson and Deetz note the issue of the researcher 'hailing' specific respondent identities during interviewing. As I variously and explicitly labelled my respondents as 'thirty-somethings', 'metropoles', 'professionals' and 'females', their answers and interactions were in all likelihood filtered through these lenses as opposed to alternatives such as 'White'.[4] Overall, then, I am not claiming here that I have gained uninflected access to my participants' subjective worlds. Instead, I contend that what was said during the focus groups in many ways reflected the setting, as opposed to directly corresponding to an experiential reality 'out there'.

However, this is not to suggest that it would have been possible to gather *un*biased data had I only used the most apposite method in the most rigorous way. The dynamics of social research ineluctably mirror interaction in the wider context, in that they are mediated by and based on our own ways of being-in-the-world, themselves irrevocably socially located. Some intersubjectivity is of course possible and qualitative methodologies are, I suggest, better suited to achieving this in research. Nonetheless, any stronger claim to have 'broken through' the self–other boundary is misguided.

I ran three focus groups during the latter half of 2002, consisting of three, five and three participants respectively. All were held at my home, with the intention of putting respondents at their ease as far as possible (Denzin, cited in Silverman, 1993, p. 97; Easterby-Smith *et al.*, 2002, p. 106; Miller and Brewer, 2003, p. 122). The literature recommends that focus groups should consist of between 6 and 10 participants, to facilitate a range of views and to ensure that all members can participate in the discussions. I was also keen to involve respondents working in varying occupations, who had differing personal lives (for example, married, single, cohabiting; mothers and non-mothers). When arranging the sessions, however, it became apparent (as is often the case in empirical research) that the size and composition of each would be dictated by expedience as opposed to anything else. In bringing together working women who live in a geographical area spanning some 60 miles, who work across a larger area and who all have busy lives (also of course a key factor in their being approached at all), I was unable to achieve either the numbers or diversity in each group for which I had hoped. Nonetheless, this in itself seems significant, as it speaks of the juggling of personal and professional lives in which my target group of participants engages on a daily basis.

Each group was tape-recorded and data fully transcribed. Analysis was done manually, on a quasi-grounded basis. I used a combination of codes already embedded in the interview schedule and emergent '*in vivo*' codes to cluster data together, subsequently consulting a range of relevant literature in order to theorise emerging issues. A draft of the paper was also sent to all 11 respondents for their comments. The analysis, which follows, is presented according to the three central themes of the city, lives and loves, and work. These are to some extent artificial divides as the themes overlap with each other: connections are therefore drawn wherever feasible. Within each theme, data are ordered according to the dynamics of discussion within (and across) the focus groups: in line with the intention of allowing my respondents to dictate what was discussed as far as possible, I have also attempted to let the data speak for themselves.

Sex and Not the City?

The City ...

At the outset of her discussion of gender and urban space, Parsons avers that

> There are infinite versions of any one city ... cities have aggregate and multiplicitous identities, made up of their many selves (Parsons, 2000, p. 1).

Perhaps predictably, my 11 respondents likewise describe London in a multitude of ways. But what is also noticeable is the homogeneity of their accounts of metropolitan life.

First, all 11 respondents initially chose London because of the opportunities it was perceived to house. I referred earlier to an understanding of cities as the centre of activities in a region or, in the case of capitals like London, a nation, and this is very evident in these data. Megan—for example, returned to the UK from Germany to be with her partner, but insisted that they live in London because for her it had much more to offer than other British cities. Julia, on the other hand, initially moved to the city because it provided the best postgraduate course in journalism. She goes on to say that a more predictable route following her graduation would have been to work in local radio, but that "I was like 'No, it has got to be London, that is where the national broadcasters are, that is where all the news is'. I was very snobbish about it". Julia's account also connects London to another of our key themes—work—in that she suggests it was the only logical choice in terms of the career she wished to pursue. Similarly, Bella moved south to go to bar school—at the time, one could only do this qualification in London—and stays in the city partly because it is the one place in the UK where she can practise her branch of the law. Wendy likewise, although bemoaning the fierce competition in London for parts, says that this is because it "is the cool place to be" for actors and other performers—which is also why she herself lives and works in the city.

Moreover, although only Victoria was born and raised in London, most of the women self-define as locals and distinguish their urban experiences from those of visitors. Judith recounted an anecdote about being on the London Eye with American, Swedish and Dutch colleagues and "suddenly" seeing the city through their eyes. She describes it as "amazing, [a] completely different perspective". Relatedly, Wendy suggests that

> when I go home everyone thinks "Oh", you know, "Wendy's doing this, it's really exciting' ... they go "Oh, you go out to Soho, you do all this and that" but when you're actually doing it it's just *ordinary, normal,* and it's just like anywhere else ... [my family] love [London] cos it's so different whereas for us it's our everyday lives ...

Melanie also remarked on how much she enjoys having friends to stay, because it allows her on occasion to "do the tourism thing" and "really makes you appreciate what you have got in London". She also implies that people are more likely to visit *because* she is in London—reflecting again the sense of the capital-as-national-hub as well as connecting to this paper's 'lives and loves' theme. Megan agrees, saying that if she lived in the city where she grew up "I don't think I would get as many visits".

These comments, moreover, suggest that living in London may generate an immunity to the city's appeal, which is periodically awakened when one spends time with people who live elsewhere and experience it differently. The participants also complain about not 'using' London to its fullest extent—again illustrating the motif of its centre in particular being 'where it all happens'. Julia—for example, notes that she never goes to nightclubs but used to do so regularly when living elsewhere. Simmel's analysis of the "modern personality as constantly buffeted by the speed and ever-changing variety of the urban experience" (Parsons, 2000, p. 30), such that one becomes dulled by overstimulation and withdraws psychologically from the city, is perhaps apposite here. Some respondents—like Victoria—did say they thrived on London's 'buzz', but Megan, Catherine and Madeleine feel instead that they never have enough time and say this is extremely stressful. Relatedly, Castells (2002, p. 122) suggests that the urban experience of "tension and rush" is brought about by the fact that collective consumption patterns in large cities—of key life provisions such as transport—solidify and become restrictive, such that a mass of people are simultaneously subject to the same inexorable rhythms. However, we should not overlook Simmel's counterpoint that the urbanite continues subconsciously to crave the chaos of city life, which feeling is arguably apparent in the remarks regarding the fresh perspective provided for some of my respondents by visitors to London.

What also emerges from these data, despite the nagging impression that everything revolves (/should revolve) around its axis, is that London is to some extent actually experienced as 'centreless'. As Julia puts it, "Within London we all create our own villages". Melanie describes herself as becoming something of an East London 'snob', being reluctant to socialise beyond her immediate locale. She also comments that

> the thing that makes London localised I think more than anything is that, when you're a local in London, you know the late-night off-licences and you know the late-night drinking dens. That makes you local to your area.

Catherine, similarly, says that she rarely sees friends in north London "cos it's a million miles away" (she lives in the south-east of the city) and wonders whether she might in fact spend more time with these people were she to move away.

This perception of the metropolis as a disparate collection of communities resonates with Patrick Keiller's assertion that "The true identity of London is in its absence" (cited in Robins, 2001, p. 473). Likewise, Wirth (cited in Castells, 2002, p. 37) defines heterogeneity as one of the defining characteristics of the urban space. Moreover, these respondents' experience of London seems to echo Raban's (cited in Stevenson, 2003, p. 69). He argues that it is 'officially' some 25 miles long and 20 miles wide, whereas 'his' city is just 7 miles in breadth and length, bounded variously by Highgate Village and Hampstead Heath, Brompton Cemetery and Liverpool Street station. Raban goes on to say that "I hardly ever trespass beyond these limits, and when I do I feel I'm in foreign territory".

London in these empirical and theoretical sketches is depicted as the city of the multitude;

> a multiplicity, a plane of singularities, an open set of relations, which is not homogeneous or identical with itself (Hardt and Negri, cited in Robins, 2001, p. 489).

Moreover, it is precisely this diversity which

the focus group members value about the city in which they work and/or live. Georgie and Elizabeth—for example, both referred specifically to London's multiculturalism; Elizabeth comparing the area where she works to London in terms of the former's blandness, and Georgie agreeing that the same area—also her home county—is "so White". Diversity is seen as equally significant in terms of the choices it affords. Wendy, Julia, Elizabeth and Georgie all claim there is nothing that one cannot find in London, whether it is shops, food, entertainment, green spaces or waterways. In accounts like these

> London is regarded as a huge cultural reservoir and resource—valued for its numerousness, its complexity, and its incalculability (Robins, 2001, p. 491).

Moreover, Simmel argues that the features of city life which create overstimulation also signal the abundance of opportunity in this context and are exactly those which rural areas do not provide (Stevenson, 2003, p. 25). My respondents certainly seem to prise this very highly, again reflecting their sense of the city-as-centre-of-everything.

But London's heterogeneity was not only described in positive terms. Madeleine mentions its lack of a sense of community and Megan concurs. She says that one is "very much aware that you are always an individual when you are out and about ... it's always a very negative vibe ...". Julia, likewise, contrasted her life in London to Brighton, where she lived for five years and "knew just about everybody". These women, then, do not value the anonymity of the urban crowd—rather than finding it a safe haven into which they can escape, they actually dislike being part of an inchoate group and say it is isolating. Here, they echo Wirth's (cited in Stevenson, 2003, pp. 20–21) negativity regarding the fractured and impersonal metropolitan space, and express nostalgia for the more solid bonds apparently to be found elsewhere.

Both Catherine and Wendy also refer specifically to other cities in their comments on the dynamics of friendship in London. Wendy suggests that at home (in the Midlands), during her studies in the north-west and in her first years in (south) London where she relocated to attend drama school, she lived within a tight network of friends who visited each other regularly. Now, however, she does not have others coming round for coffee in this way and suggests this is "a real London thing ... I think now I've kind of discovered that ... it's not just the fact that I live in [the East End]. I think it's a bit of a London thing anyway". Wendy has also found this transition hard to make. Catherine, also regretfully, says she has found it impossible to "recreate" her student experience, in the same city where Wendy studied, of "just pop[ping] round and you'd see your mates" while living in London. Of course the 'popping round' for which Wendy and Catherine express nostalgia may be related to their student status at the time and the relative freedom of this lifestyle. Nonetheless, Wendy is quite clear in her assertion that her initial experience of London, even as a drama student, was unusual: she identifies her existence now, in another part of the city, as far more commonplace.

Moreover, such comments imply that the respondents in focus group 2 especially sense something specific about London as compared even with other cities they know, such as Brighton or Manchester. They talk of its fragmentation and its scale as having depressive effects on the rhythms of personal relationships—also of course the second of our key themes here. Indeed, Pile (2002, p. 8) argues that the city undermines conventional human associations because it brings so many different social categories together that new and typically more impersonal forms of interaction develop. My participants seem to suggest that this is especially true of London, except maybe within each of Julia's 'villages', also evoked by Georgie, who talks about how the inhabitants of her street all "look out for each other", despite being a heterogeneous mix of ethnicities and occupations.

... Lives and Loves ...

This collection of issues provoked slightly less consistent responses from the participants than was evident with regard to the city, or indeed work, but some interesting underlying patterns nonetheless suggest themselves. For example, several respondents apparently attach a good deal of importance to having an intimate who is also a confidant. Julia remarks that

> since I've been single, you know those moments where today I saw a mad caterpillar which was huge and it had a turquoise blue spike on its bum. And we were ... trying to move it to safety and it was getting really violent towards the piece of paper [we were using]. It was really funny and, you know, you tell a boyfriend about that and I *really* miss that.

Moreover, although she acknowledges that empathetic friendship with a boyfriend does not substitute for sexual attraction, Madeleine says that for her it is still very important. Likewise, Melanie can now talk "completely unselfconsciously" to her partner Frank about "when we have children" (etc.) as opposed to "if this happens", because there is no longer any question that they have a future with each other.

The kinds of 'togetherness' described here seem to chime with discussions of contemporary priorities in sexual relationships. Giddens describes the possibility of the 'pure relationship', which he sees emerging from relatively recent developments like increasingly voluntary choice of partners on the basis of romantic love, relationships enduring only insofar as those involved find them satisfying and contraception freeing women from the "chronic round of pregnancy and childbirth" (Giddens, 1992, p. 26). Moreover, although she does not idealise these relationships quite as Giddens does, Jamieson agrees that there is now a privileging of

> an intimacy of the self ... [which] typically requires a relationship in which people participate as equals ... across genders, generations, classes and races (Jamieson; cited in Woodward, 2002, p. 190).

Wilkinson (2002, pp. 41–42) also remarks on the various research studies which have arrived at similar conclusions regarding the changing nature of intimacy. Closeness and egalitarian companionship appear to be the watchwords here, in relationships characterised by more negotiation, individualism and openness than was historically the case. Nonetheless, Bristow (2002b, para. 11) identifies the accompanying anxieties, such that we may now expect much more from sexual liaisons, even coming to view finding a meaningful personal relationship as an "apparently insurmountable challenge". This motif is certainly apparent in both *Bridget Jones's Diary* and *Ally McBeal*, and is also something which Julia's contributions in particular suggest she is concerned about.

Interestingly, moreover, Madeleine introduced the idea that relationships in the city are often subject to particular kinds of pressures—also a theme in the above cultural texts. She remarks that "everyone" says she should move in with her boyfriend, as they spend so much time together but actually live on opposite sides of the Thames. As Madeleine points out, living together for any London couple would be cheaper, given house prices in the capital and transport costs. On the other hand, as she also implies, it may be that the couple concerned are not ready to take this step in their relationship.

There was less agreement amongst the respondents on the issue of children. However, given that all women can now—at least in theory—choose whether and when to become mothers, this is not unexpected. Wendy—for example, does not wish to have a family and says "it really annoys me as soon as someone says 'You're gonna get married and you're gonna have children'. It just *really* gets my back up and I get *really* defensive". She also recalls experiencing one episode of broodiness and finding it very frightening. This was echoed by Judith, who again does not want children but worries about "waking up one morning, seeing one and thinking '*God!* I want one!', when it may be impossible biologically.[5] Most of the other respondents feel differently—such as Georgie, who is trying for a baby with her husband Callum, and Bella, who says it was the reason she got married. Indeed, she recalls telling the man she is now married to, who was living in the US at the time, that if he did not return to the UK " 'the job as the father of my children will be filled and this is your opportunity to get in while the going is good' ". Similarly, Victoria is feeling "maternal" for the first time in her life.[6] Moreover, and given arguments that middle-class women in particular have "to *become* the sort of persons who could properly have children" (McMahon, cited in Hughes, 2002, p. 66), it is noteworthy that these women acknowledge their broodiness as a recent development. The literature identifies several factors here, also visible in the data: economics (Megan's partner would earn the same anywhere in the world, so when they decide to start a family she can leave her job and they will move away from London); psychological readiness (Bella was nearly 30 when she started to want children after years of feeling the opposite); believing one would be a good parent (Victoria is confident she would have a successful relationship with her baby); and being in the right relationship (neither Megan or Melanie can imagine having children with anyone other than their partners).

But the respondents in focus group 3 especially also appear to subscribe to what Hughes (2002) calls the 'doing it all' discourse as regards combining family and another of our key themes, work. Bella suggests it is impossible simultaneously to be "brilliant" at work, motherhood, managing a home and being a wife. She quotes a friend of hers who says "you end up feeling like an inadequate woman and an inadequate barrister as well. You don't give quite enough for your family, you're not quite enough for your husband and not quite enough [for] … your career". Georgie and Elizabeth on the other hand argue that women need to redefine their expectations around wanting to excel and instead aim to be "good enough", as Elizabeth puts it, at the various aspects of their lives.

Judith, in focus group 1, offers a similar argument regarding her own situation, saying that since she is not going to have children or marry she and her girlfriend will always be in the ideal "double income no kids" situation. She suggests that they therefore have a "lifestyle with the minimum compromise" but also points out that it depends on neither partner starting to want a family. The theme here of the need to juggle work and family life, and how challenging it can be, is also a key motif in contemporary cultural representations of working women. The guilt that Kate Reddy experiences in attempting to manage the rival claims of home and work—for example, is writ large across the pages of *I Don't Know How She Does It*— as its title suggests.

Moreover, the acceptance by these respondents that 'having it all' necessitates a reappraisal of individual hopes and dreams as regards marriage, parenting and career echoes empirical evidence regarding mothers who engage in paid employment. As Hughes puts it,

> Paid work and education become additions to the seemingly intransigent nature of women's responsibilities for domestic care (Hughes, 2002, p. 113).

Similarly, all the available data point to a 'lagged adaptation' in that men do not take on a proportion of childcare and/or household labour which reflects the hours their partners spend at work. Then there is the on-going debate about the effects of having a 'working mother' on children's development and, indeed, whether mothers make reliable employees (Charles, 2002, p. 72; Hughes, 2002, p. 63, following Kaplan)—again something which Kate Reddy and her ilk worry about routinely.

Georgie's suspicion that, when she does have children, it will affect the way others perceive her is also noteworthy here in terms of the links the data suggest between personal life and work. She says

> The whole prospect of having children just *won't affect* Callum's life externally the way in which it will mine … I will first be a mother and then a lecturer and, God, I have got to prove that lecturing comes first and that I am the academic … and not the mother. And Callum will *never* have to do that.

Jamieson (1998/2002, p. 218) agrees that career is still seen as a crucial component of being a man, but not of being a woman, and Nicolson (1996, p. 10) puts the related argument that women are usually seen "in some relation to [their] motherhood rather than professional success". Likewise, Charles (2002, p. 49, pp. 70–71) points to empirical data which demonstrate that, although women's paid employment is now widely accepted, the conviction persists that mothers should put their families first and that they should not go out to work if they have preschool children. Two more recent research projects suggest, similarly, that women would prefer either to leave work or to go part-time after having children and that they deliberately choose more flexible, albeit lower-paid, jobs in order to combine paid work and motherhood (Elliott, 2003a; Elliott and Chittenden, 2003).

But according to my respondents it is not just a matter of balancing work against the demands of offspring. They concur—for example, on the need to "invest" in other relationships, as Elizabeth puts it—reflecting a claim made by Woodward (2002, p. 202) in her summary of Giddens' analysis of intimacy, that "it does begin to look like hard work". Judith recounted an anecdote where a colleague told her she did not take her relationship seriously because she spent so many hours away from home during the week. Indeed, her previous employment in west London involved her leaving the house at 5.30am and not returning until 8pm or so. This resonated with her to the extent that she changed jobs to allow her to work outside the city and nearer her rural home. Bella says that, although she routinely lets her friends down due to work commitments, she feels unable to do the same to her husband. Contrastingly, Georgie suggested that until recently she has tended to take her husband for granted, but has worked much harder on maintaining her friendships. These women, then, insist that both partnerships and friendships can be negatively affected by one's job. Interestingly though, this is much less obvious in the current crop of relevant cultural texts, where personal life is either prioritised *over* work anyway (*Ally McBeal* and *Bridget Jones's Diary*), personal life and work are difficult to disentangle (Carrie the relationship columnist in *Sex and the City*) or the emphasis is on juggling *children* and career (Pearson's novel and its like).

What my respondents seem to suggest, then, is that work can limit the ability to take part fully in *every* other area of life. As Charles points out with regard to the family, this in some ways

> is counter-intuitive because of the generally-held assumption that women's participation in paid employment is limited by their domestic responsibilities; it may in fact be the other way round (Charles, 2002, p. 61).

This observation leads us into the next section of the analysis, as it also hints at the sense amongst many of the focus group participants that work for them is becoming less of a priority.

… and Work

All of my respondents have, at some stage, more or less defined their self-worth in terms of their careers. As Elizabeth puts it,

> When I first left college and I was unemployed for a short time, that was really terrifying and I think that was where I really learned that … it is really important for my identity to have a role, a public role.

Similarly, Judith realised when she started her current job that "there is something in me that gets what it needs, there's an itch that gets scratched by working hard". Hughes refers to constructions such as these as 'Women have made it'. She argues that, through this sort of lens, an individual woman "may be authorised to say I am successful if I have a well-paid job. As a mother without paid work, it is far more difficult to claim success as part of my identity" (Hughes, 2002, p. 4). From its liberal feminist stand-point, the 'Women have made it' discourse therefore defines 'successful' femininity in the same terms as 'successful' masculinity—the only 'good' woman is a working woman, as it were. It makes available a different kind of female subject position from the more traditional one identified by Jamieson and Nicolson above, wherein the womanly ideal is motherhood and career takes a definite backseat to children. Indeed, in order to live out the demands of 'making it', women have to work long hours, avidly seek promotion, derive a sense of enjoyment from their jobs—and evidence also suggests that those who achieve success within these somewhat narrow parameters are very often childless (Hughes, 2002, p. 40).

However, an understanding of the implications of opting for such a life trajectory is obvious in the focus group data. We have—for example, already seen references to the detrimental effects of work on relationships, but participants in focus groups 2 and 3 suggested that it may well also undermine psychological and physical well-being. Julia—for example, told of how nine months on night shifts from 10pm until 10am "*literally* drove me mad. And I, I'm still on medication [laughs] kind of three years later". Similarly, Georgie referred to recently "collapsing from exhaustion and being taken to a hospital".

Further, many of the respondents agreed that their jobs are gradually declining in personal significance. Victoria says "work's not the be all and end all and actually ... I think I'd probably get *bored* if I gave up work but I don't think it *defines* me now". Bella is even more emphatic, saying that her job "has cost me over the years family life, personal life, my social life, my figure, my sense of humour at various times". In fact, if she "had a choice, if I could become a lady who lunches tomorrow, oh, in a *heartbeat*! Oh God, yeah!". Georgie also acknowledges that, in the wake of a difficult year personally, she has begun to reassess her career, realising that she wants more in her life. She thinks this is probably the reason why she and her husband now hope to start a family.

Similarly, the participants in focus group 2 talk resentfully of what they see as a London tendency to define others using their jobs. Catherine mentions a friend who

> doesn't do the 'Where do you live? And what do you do?'. He somehow, and he's just a lovely man, he somehow just manages to engage with people [in other ways] which I just can't imagine how he does it [laughs].

This comment provoked a chorus of agreement from the other women. Megan—for example, expostulated

> I so envy [that]! ... I hear myself doing the 'Where do you come from? Yerh yerh yerh' [mocking tone]. And I *hate it, and I absolutely hate it*, because I don't wanna answer those questions and nobody wants to hear them and I am *desperately* searching for something else to say—and then you ultimately come out with 'So ... ?'

As suggested, the speculation amongst this group of women is that the routine categorisation of others based on occupation is a phenomenon peculiar to London. So while "everybody in London asks ... 'Oh, where do you live? What do you do?': that is how London works" (Julia), elsewhere the tendency is to ask whether someone is in a relationship, especially if they are female. Perhaps, the participants conclude, women in particular are understood to have moved to London for work,[7] and are therefore defined in that way—whereas in smaller places it is presumed that, as Megan suggests, "the first priority" is relationships because "the work can't be that good". This again evokes the idea of the capital city in particular being the commercial and economic hub of a nation, as well as bringing all three of our central themes together.

Bauman's concept of the increasing 'aestheticisation of work' may also be useful here. He suggests that

> An entertaining job is a highly coveted privilege ... 'Workaholics' with no fixed hours of work, preoccupied with the challenges of their jobs 24 hours a day and seven days a week, may be found today not among the slaves, but among the elite of the lucky and successful (Bauman, 1998, p. 34).

The most desirable forms of work, then, are all-consuming, in terms of the enjoyment they provide and with regard to the levels of commitment required. Although this is an ideology which many of my respondents now seem to be resisting, maybe the focus group 2 comments also imply that 'What do you do?' continues to serve as a convenient metropolitan indicator of whether someone else is "among the elite of the lucky and successful". As we already know, there is a tendency on the part of the same participants to identify an absence of community in the city, and to argue that this is again unique to London. Overall, an emerging inference from this group in particular is perhaps that interpersonal dynamics in London are instrumental as opposed to being based on a knowledge of others as having personal lives, friends and so on.

The above data, moreover, seem to testify to Crompton and Harris' (cited in Charles, 2002, p. 38) argument that women's orientations to work shift during their lives. However, although work may be receding in importance for most of my respondents, they agree that it is still very necessary in a practical sense. Bella, for instance, cannot give up her job because she is the household wage-earner; Julia recently turned down her "perfect job" because the salary was £4500 lower than what she currently earns; and Melanie says quite emphatically "Real life, in London, not working—can't imagine it!". She therefore also introduces the idea that being unemployed in the capital would be especially financially challenging, which Megan echoes in remarking that it is difficult not earning wherever one lives in the UK, but this is "multiplied" in London.

In other accounts, the monetary significance of employment was more bound up with its psychological import. Wendy was initially reluctant to move into her partner Paul's house because of her desire to be financially independent. She went on to comment that, although their life together is "great", she does feel the need to work and contribute as much as she can to the household in terms of food shopping and domestic labour to avoid the sense that somehow she is reliant on him. This she says distracts her from what she wants to do career-wise. Relatedly, Victoria mentions that her sister Virginia had to depend on her partner while studying and "felt it was a bit like him dishing out money", although this was not in fact the case. Victoria comments that this was very difficult for Virginia, indeed that "it can feel quite trapping, um, and, and not very pleasant and it probably can drive you to some sort of madness ultimately".

Autonomy is a key issue in Wendy and Virgina's accounts, but an earlier motif also seems to recur across these data in the sense that working for a living (or not) is seen to impinge on one's personal life. Some of the respondents therefore continue in their jobs because of the effects that leaving would have on their lifestyles, while others note the tensions that money can create in relationships. Pahl's classic study of household money management suggests that such connections are far from uncommon—like the indication that, even in families where incomings are pooled,

> especially if they are temporarily out of the labour market women can feel as if they are supplicants. Many regard the money coming into the household as not theirs and feel that they have no right to money they have not earned (Pahl, cited in Charles, 2002, p. 56).

But pragmatism and an emphasis on independence aside, the desire on the part of many respondents to 'downshift' points towards another cluster of issues around what is required to succeed at work. Gender differences are implied here and it seems to me that these women's comments may have been informed by what Hills Collins (cited in Marshall, A., 1994, p. 108) calls their "insider–outsider" status. Judi Marshall's (1984) claim that women managers are 'travellers in a male world' is of course well known—and it is on this basis that I suggest my respondents perhaps have a different understanding of prevailing workplace cultures because they are women. They are, then, simultaneously 'part of the organisational gang' in their professional status and 'outsiders' because of the masculinist qualities of organisations. It is also possible to speculate that many of these women, after a decade or so of paid employment post-graduation and attaining well-paid and responsible positions, have not only begun to question the idea that work is what Victoria calls the "be all and end all" but are also becoming cynical about 'making it' *per se* and the behaviours involved. A similar phenomenon has been documented by other researchers: Judi Marshall's follow-up study *Women Managers Moving On* (1995)—for example, suggests that several of her apparently successful respondents left their jobs because they had grown so disillusioned with their organisation's climate and I have already referred to Crompton and Harris' (cited in Charles, 2002, p. 38) conclusion that women's work orientations alter as time passes.

In any case, there was a marked consensus amongst my focus group participants that being good at what one does for a living is not the same as, or does not always lead to, being successful. Many suggested that ruthlessness, selfishness, politicking, arrogance and sycophancy are often traits which generate success, if defined in the conventional way of moving up the career ladder. But these do not, they assert, make one a good manager, an expert in one's profession or a supportive colleague. Neither, perhaps obviously, are they behaviours these women aspire to. Catherine, for instance, says that those who do well in her area are often "ostentatious characters" who have a good deal of "brav[u]ra" and "charisma", but are not necessarily good osteopaths. Georgie is even more blunt in this regard, arguing that "I think actually who, who gets to the top of academia are usually incredibly ruthless, self-motivated, selfish, uncollegiate people".

These comments are also underpinned by the sense that workplaces encourage certain behaviours which some individuals (mainly men) find it easy enough to adopt, but which others (mainly women) find alienating. Moreover, there has of course been a great deal of recent academic work on precisely this issue, the concerns of which Charles sums up as follows:

> The informal networks in which power is located ... the cultural processes within the workplace which maintain a situation in which men who are white, middle class, and heterosexual predominate in positions of power ... the actions of gendered social actors which maintain gender divisions of paid work (Charles, 2002, p. 43).

It seems here, then, that my respondents are aware of such 'cultural processes' but are not especially interested in buying into them. In the main, they appear to regard being good at their jobs as more important than being successful and there is no particular investment in liberal feminist definitions of 'getting on'. Indeed, Victoria says quite categorically that

> If I do a good job hopefully other people will recognise I'm doing a good job but if they don't I actually don't want to spend my life trying to work the, the ladder. I'm just *not* interested in that at all.

Elizabeth and Georgie, similarly, speak without rancour of their sense that they probably will not become professors. Moreover, although Julia did not take her ideal job because it involved a substantial pay-cut, which may imply that she continues to invest in the 'making it to the top' discourse, she nevertheless found the decision-making process "horrible". Relatedly, Megan, who has similar concerns about changing her employment if she remains in London, commented that

> Either you are underpaid for something you enjoy doing or you're stuck in a job you don't want to do because you have reached a level where they pay you enough to keep you.

Here, both women imply that 'making it'—especially in the city—may well involve certain compromises.

The muted gender theme here—i.e. that men are generally better equipped to succeed at work—was also present when the focus group 3 discussions moved to these women's descriptions of their career-related behaviours. Georgie and Elizabeth agree that they are effective teachers and researchers but say that they 'plod', they are not 'intellectual' and will not "set the world on fire". Similarly, Bella remarks that she is good at her job, but achieves this by being "paranoid. Cos I check everything and I work really hard because I don't have the inbuilt confidence that [my colleagues] have, which is the big difference". She says that she lacks this confidence because she is not from the "narrow social background, generally quite well off, mainly White male, almost exclusively White male, and almost all of them ... practising members of one orthodox religion or another" shared by the majority of her peers.

There is a possible connection here to Bartky's discussion of the gendered nature of shame (cited in Hughes, 2002, p. 42) and Walkerdine's (cited in Hughes, 2002, p. 44) remarks on the gendering of potential. Bartky suggests women are taught from an early age that they are inferior to men—that they internalise a 'judging other' which constantly makes them feel inadequate—and this is then reflected in their behaviours. Differential class-based and ethnic learning patterns may well exist too, as Bella's comments imply. Walkerdine's research in schools concludes that girls are routinely defined as having done well not because of talent, but because they have worked hard or followed 'the rules', whereas boys, even if they perform badly, are usually considered to have potential. Again, such constructions, it would seem, become internalised by the students concerned. Perhaps, then, Bella, Georgie and Elizabeth have learnt to measure themselves against others (their male colleagues in particular), to find themselves lacking in certain characteristics and to understand their achievements as based on hard labour as opposed to natural flair. Contrastingly however, in the relevant cultural texts, Bridget Jones and Ally McBeal do well at work *because* they are irremediably 'feminine'; erratic, emotional, sensitive and so on. Likewise, in *Sex and the City,* Carrie's job requires her to reflect on and analyse relationships—again, a feminine-coded activity.

These data suggest overall that many of my respondents apparently now view work as less of a central life activity, dislike being defined by their jobs, have no especial desire to 'make it to the top' and share some reservations about—or feel that they lack—the characteristics identified as necessary for career success. They also suggest that working for a living often has a negative impact on health, which again echoes earlier discussion about the ways in which paid employment can affect the rest of life.

Conclusion

This paper has attempted to draw down issues around women's paid employment, personal relationships and the city to the micro level; to analyse how they are lived out by a small group of respondents; and to explore the links between their experiences and arguments in academic research. The analysis suggests several interesting conclusions. First, in terms of their lives in London, the respondents focus on its social and physical topographies. There is a sense in which they see the city as both central and rotating around its centre, but an opposing inference that 'real' London existence is lived out elsewhere. London is also depicted as culturally and materially diverse, which the respondents value, but other comments identify this same diversity as translating into a paucity of contact with fellow city-dwellers, such that the urban crowd is both individualised and amorphous. Such constructions are also present in writings on the urban experience. There is, moreover, a belief in focus group 2 in particular that London is to some extent unique—that other spaces, even other cities, foster different interactions. Its scale is certainly seen to create difficulties, especially given poor public transport provision (mentioned many times). Finally, there is agreement for the most part that London is only a temporary resting-place—with the exception of Bella, who loves the city, these women do not regard it as their long-term home or workplace. Judith had indeed already left her London job because of concerns regarding the effects of a long working day plus commuting on her relationship and, since the focus groups were held, Catherine has also moved to another part of the UK.

Overall, London is "at once damned, tolerated, manipulated and celebrated" in these accounts (Stevenson, 2003, p. 140). But other issues (crime/safety, dirt, litter, resource allocation and associated conflicts) did not surface during the discussions. This is perhaps surprising since these are all widely available discourses of the city—and of London in particular. However, it again evokes Parsons' idea of the city as a text which its inhabitants write in idiosyncratic ways, as well as suggesting that those occupying similar subject positions (like the 11 women here) may apprehend the urban expanse in consistent ways. In fact, and to return to an earlier point, it is possible that a more ethnically varied group of respondents, say, would have produced different sorts of data—perhaps with more emphasis on the connected issues of crime and safety, and resource allocation—for example.

Moving to the second theme—lives and loves—and in line with many commentators' claims, these women believe in the importance of confiding and supportive intimate relationships as opposed to more traditional requirements of coupledom such as economic security. Many also construct themselves as having reached the life-stage where their future plans include having children. Even Wendy and Judith, who resist the idea of a family, acknowledged this may change. Again, this reflects claims made in the relevant literature. Moreover, all these respondents insist on the importance of having a full life, including work, a partner, family and friends, whilst some also offer a clear-eyed account of the challenges this entails. The focus group 3 participants in particular suggest that 'having it all' demands a redefinition of precisely what this means, and learning to cope with the concomitant role conflicts—as is also clear in other empirical data.

Finally, with regard to work, the majority of respondents' recent reassessment of priorities perhaps stems from gradually becoming disillusioned with what it has to offer; a sense in which they no longer want to be defined by work; an assertion that the behaviours required to progress at work are harmful to others and may be unachievable in any case; and a belief that working for a living impacts on the ability to enjoy other parts of one's life. What seems to emerge overall is a rejection of the liberal feminist emphasis on 'making it' (Hughes, 2002). While work is still acknowledged to be financially as well as psychologically significant, an emerging emphasis on working to live instead of living to work is noticeable in these data. Many of these women apparently wish to disembark from what Elizabeth calls the "treadmill".

I am not claiming that the issues raised in these focus groups are necessarily specific to an urban setting, or that these stories of lives, loves, work and the city in any way represent even the social group to which my respondents putatively belong (thirty-something working women in London). Nonetheless, there are a number of interesting overlaps between the three key foci. These include the idea of London's problematic impact on personal relationships; or the suggestion that Londoners are especially likely to be defined through their work, which relates to its positioning throughout each focus group as a national (in some cases global) centre, here of business and professional activity. What also seems remarkable is the homogeneity in the respondents' accounts; in particular, a widespread sense of being poised at the beginning of a new life-stage as regards the three central issues.

However, despite some similarities, these women also do not appear to be represented adequately by their fictional counterparts—Ally, Bridget, Carrie, Kate *et al.* In particular, it is not that they crave the 'simple comforts' of the private sphere over the hurly-burly of the public city space. Instead, what is apparent is that the realities of being an urban professional female thirty-something—for *these* women at least—present a series of choices and compromises and generate an emphasis on the importance of 'investing' across all areas of one's life. To close, I would also like to note the optimism that I identify here; the implicit motif (rarely present in empirical accounts of working women's lives) that things can (and indeed will) only get better.

Notes

1. Analysis based on parts of this earlier dataset, which focused in the main on women's experiences of their bodies, appears in Brewis (1999b, 2000), Brewis and Sinclair (2000) and Warren and Brewis (2004).
2. Until recently my own city of residence, London was selected in part for respondent accessibility. However, it is an interesting choice for other reasons. As Pimlott and Rao (2002, p. v) suggest, London is widely acknowledged still to be a "pivotal city" in world affairs and is also in a period of transition following the election of Mayor Ken Livingstone and the Greater London Assembly in May 2000. The latter is identified by Pimlott and Rao as possibly the most important transition in the city's governance since the establishment of the London City Council in 1889. Then of course there is the aforementioned London Chamber of Commerce claim about the relative achievements of metropolitan women professionals during the 1980s and 1990s.
3. Interestingly, however, responses tended to be remarkably consistent both within sessions and between them where all three groups covered the same topics (not always the case due to their individual rhythms and the open-ended nature of the questions). Where respondents differed or issues surfaced only in certain sessions, I have attempted to make this clear in the analysis.
4. Nonetheless, the fact that all my respondents are White, as am I, may well have had some bearing on the data generated—an observation to which I will return in my conclusion.
5. Interestingly, Elizabeth, who is 6 months' pregnant, read several articles just after her 35th birthday describing the decline in women's fertility at this age and "started to panic". This is partly why she is now expecting, as opposed to any "strong desire to have kids".
6. However, Victoria is less sure whether she would want to raise a child in London, which comment again connects back to the city theme in this discussion.
7. Of course most of my respondents did exactly this.

References

ALVESSON, M. and DEETZ, S. (2000) *Doing Critical Management Research*. London: Sage.

BAUMAN, Z. (1998) *Work, Consumerism and the New Poor*. London: Open University Press.

BREWIS, J. (1998) What is wrong with this picture? Sex and gender relations in *Disclosure*, in: J. HASSARD and R. HOLLIDAY (Eds) *Organization/Representation: Work and Organizations in Popular Culture*, pp. 83–99. London: Sage.

BREWIS, J. (1999a) *When Ally met Bridget: organization, sexuality and femininity*. Paper presented to the *1st International Critical Management Studies Conference*, UMIST/University of Manchester, July.

BREWIS, J. (1999b) How does it feel? Women managers, embodiment and changing public sector cultures, in: S. WHITEHEAD and R. MOODLEY (Eds) *Transforming Managers: Gendering Change in the Public Sector*, pp. 84–106. London: Taylor and Francis.

BREWIS, J. (2000) When a body meet a body … : experiencing the female body in and outside of work, in L. MCKIE and N. WATSON (Eds) *Organising Bodies: Institutions, Policy and Work*, pp. 166–184. Basingstoke: Macmillan.

BREWIS, J. and LINSTEAD, S. (2004) Gender and management, in: S. LINSTEAD, L. FULOP and S. LILLEY (Eds) *Management and Organization: A Critical Text*, pp. 56–92. Basingstoke: Palgrave.

BREWIS, J. and SINCLAIR, J. (2000) Exploring embodiment: women, biology and work, in: J. HASSARD, R. HOLLIDAY and H. WILLMOTT (Eds)

Body and Organization, pp. 192–214. London: Sage.
BRISTOW, J. (2002a) Maybe I do, in: J. BRISTOW *Maybe I Do: Marriage and Commitment in Singleton Society*, ed. by C. FOX, pp. 14–31. London: Academy of Ideas.
BRISTOW, J. (2002b) *Wedding stress* (available at http://www.spiked-online.com/Sections/LoveAndSex; accessed 16 July 2002).
CASTELLS, M. (2002) *The Castells Reader on Cities and Social Theory*. Malden, MA: Blackwell.
CHARLES, N. (2002) *Gender in Modern Britain*. Oxford: Oxford University Press.
CORBETTA, P. (2003) *Social Research: Theories, Methods and Techniques*. London: Sage.
EASTERBY-SMITH, M., THORPE, R. and LOWE, A. (2002) *Management Research: An Introduction*. London: Sage.
ELLIOTT, J. (2003a) Young mothers say they want to stay at home, *The Sunday Times*, 6 March, p. 23.
ELLIOTT, J. (2003b) Female, clever, highly educated and childless. Are you a member of the anti-baby party?, *The Sunday Times*, 27 April, p. 14.
ELLIOTT, J. and CHITTENDEN, M. (2003) Women 'choose to have lower pay than men', *The Sunday Times*, 6 April, p. 25.
FIELDING, H. (1996) *Bridget Jones's Diary: A Novel*. London: Picador.
GIDDENS, A. (1992) *The Transformation of Intimacy: Sexuality, Love and Eroticism in Modern Societies*. Cambridge: Polity Press.
HUGHES, C. (2002) *Women's Contemporary Lives: Within and Beyond the Mirror*. London: Routledge.
JAMIESON, L. (1998/2002) The couple: intimate and equal?, in: T. JORDAN and S. PILE (Eds) *Social Change*, pp. 216–222. Oxford: Blackwell.
MARSHALL, A. (1994) Sensuous sapphires: a study of the social construction of black female sexuality, in: M. MAYNARD and J. PURVIS (Eds) *Researching Women's Lives from a Feminist Perspective*, pp. 106–124. London: Taylor and Francis.
MARSHALL, J. (1984) *Women Managers: Travellers in a Male World*. Chichester: Wiley.
MARSHALL, J. (1995) *Women Managers Moving On: Exploring Career and Life Choices*. London: Routledge.
MCDOWELL, L. (2001) Changing cultures of work: employment, gender, and lifestyle, in: D. MORLEY and K. ROBINS (Eds) *British Cultural Studies: Geography, Nationality, and Identity*, pp. 343–360. Oxford: Oxford University Press.
MILLER, R. L. and BREWER, J. D. (Eds) (2003) *The A–Z of Social Science Research: A Dictionary of Key Social Science Research Concepts*. London: Sage.
NICOLSON, P. (1996) *Gender, Power and Organisation: A Psychological Perspective*. London: Routledge.
PARSONS, D. (2000) *Streetwalking the Metropolis: Women, the City and Modernity*. Oxford: Oxford University Press.
PEARSON, A. (2003) *I Don't Know How She Does It*. London: Vintage.
PILE, S. (2002) Social change and city life, in: T. JORDAN and S. PILE (Eds) *Social Change*, pp. 1–39. Oxford: Blackwell.
PIMLOTT, B and RAO, N. (2002) *Governing London*. Oxford: Oxford University Press.
ROBINS, K. (2001) Endnote: To London: the city beyond the nation, in: D. MORLEY and K. ROBINS (Eds) *British Cultural Studies: Geography, Nationality, and Identity*, pp. 473–493. Oxford: Oxford University Press.
SANGER, J. (1996) *The Compleat Observer? A Field Research Guide to Observation*. London: Falmer Press.
SILVERMAN, D. (1993) *Interpreting Qualitative Data: Methods for Analysing Talk, Text and Interaction*. London: Sage.
STEVENSON, D. (2003) *Cities and Urban Cultures*. Maidenhead: Open University Press.
WALBY, S. (1997) *Gender Transformations*. London: Routledge.
WARREN, S. and BREWIS, J. (2004) Matter over mind? Examining the experience of pregnancy, *Sociology*, 38(2), pp. 219–236.
WILKINSON, H. (2002) Hot marriage in a cool climate, in: J. BRISTOW *Maybe I Do: Marriage and Commitment in Singleton Society*, ed. by C. FOX, pp. 41–44. London: Academy of Ideas.
WOODWARD, K. (2002) Up close and personal: the changing face of intimacy, in: T. JORDAN and S. PILE (Eds) *Social Change*, pp. 185–213. Oxford: Blackwell.

Appendix. Respondents' Biographical Details

Focus Group 1

JUDITH: 34, works (as a toy market analyst) and lives with female partner in Home Counties. Previously employed in west London for several years.

VICTORIA: 33, single, works (on business ethics for financial services firm) and lives in London.

WENDY: 27, actress and artist, works and lives (with male partner) in London.

Focus Group 2

CATHERINE: 33, osteopath, works and lives in London with male partner.

JULIA: 31, single, journalist, works and lives in London.

MADELEINE: 32, has male partner, publisher, works and lives in London.

MEGAN: 35, manages electronics product group, works and lives (with male partner) in London.

MELANIE: 27, educational mentor, works and lives (with male partner) in London.

Focus Group 3

BELLA: 33, married, barrister, works and lives in London.

ELIZABETH: 35, university lecturer, pregnant with first child, lives in London with male partner, works in another city.

GEORGIE: 29, married, university lecturer, lives in London with husband, works in another city.

All have at least an undergraduate-level education.

Queer as Folk: Producing the Real of Urban Space

Beverley Skeggs, Leslie Moran, Paul Tyrer and Jon Binnie

1. Spatial Frame

We begin with Michel de Certeau's distinction between place and space

> A place (*lieu*) is the order (of whatever kind) in accord with which elements are distributed in relationships of coexistence. It thus excludes the possibility of two things being in the same location (place). The law of the 'proper' rules in the place: the elements taken into consideration are beside one another, each situated in its own 'proper' and distinct location, a location it defines. A place is thus an instantaneous configuration of positions. It implies an indication of stability (de Certeau, 1988, p. 117).

The counterpart of the ordered, stable 'place' is 'space' which, in de Certeau's sense of the word, is defined by vectors of direction, velocities and time variables; thus space is composed of intersections of mobile elements, actuated by the ensemble of movements deployed within it. Space is the effect produced by the operations that orient it, situate it, temporalise it and make it function in a polyvalent unity of conflicting programmes or contractual proximities. One of the central operations that produce the effect of space is representation, enabling space to appear and be experienced as real (verisimilitude). Our intention here is to explore the conventions of living, speaking and producing the representations that produce place as space. Specifically, we focus on how the conversion of place into space intervenes to produce an 'authenticity' that enables political claims to be made and contested. Space is, therefore, not just a passive backdrop to human behaviour but is constantly produced and remade as groups struggle for power (or against powerlessness) to convert their 'reality' into public knowledge, in which a visible presence can be recognised (see debates on recognition politics—for example, Fraser,

Beverley Skeggs is in the Department of Sociology, University of Manchester, Oxford Road, Manchester, M13 9PL, UK. Fax: 0161 275 2491. E-mail: bev.skeggs@man.ac.uk. Leslie Moran is in the Department of Law, Birkbeck College, Malet Street, London, WC1E 7HX, UK. Fax: 020 7631 6506. E-mail: L.moran@bbk.ac.uk. Paul Tyrer is a freelance researcher and can be contacted via Beverley Skeggs. Jon Binnie is in the Department of Environmental and Geographical Sciences, Manchester Metropolitan University, John Dalton Building, Chester Street, Manchester, M1 5GD, UK. Fax: 0161 247 6318. E-mail: j.binnie@mmu.ac.uk.
The authors are grateful to the ESRC for funding the research project 'Violence, Sexuality and Space' (grant no: L133 251031).

1995; Taylor, 1994). What is known and experienced as local place is only a fragmented set of possibilities that can be articulated into a momentary politics of time and space, but can be turned into a political claim if it can be made to appear as real. For a time, the TV series *Queer as Folk* was one of the operations that oriented, situated and temporalised a place called the gay Village in Manchester. As an operation that works on the viewer long after the programme has ended, we were interested in how it located the respondents of the research via sexuality in the making of space.

The paper is organised into three sections. The first explores the significance of representation to the conversion of space into place. The second provides the background to place-use in the TV programme *Queer as Folk* and examines how a television series intervenes in an aesthetic of authenticity. The third section examines the cultural significance of characterisation as another element that enables investments in place and politics to be made.

Most of the literature that engages with people's differential experience to place and space conceptualises space and place in terms of gendered differences (for instance, Bondi, 1998a, 1998b; Pratt and Hanson, 1994; Pain, 1997; Valentine, 1990, 1993; Rose, 1999; Hubbard, 2001), focusing on the masculinity and heterosexuality of public space and their exclusionary impact on and dangers for females (McDowell, 1996). Bell *et al.* (1994) have drawn attention to the centrality of sexuality in the production of space for some time, focusing on the role of space in the production of sexualised identities, for generating visibility, producing communities and sites of resistance (see Bell, 1995; Bell and Valentine, 1995; Bell *et al.*, 2001; Brent Ingram *et al.*, 1997) and for producing citizenship (Bell and Binnie, 2000; Binnie, 1997, 1998). Binnie and Skeggs (2004) have drawn particular attention to the role of consumption in the making of queer space.[1] And Phillips *et al.* (2000) focus on the production of non-metropolitan space in the making of sexualities. This body of literature, a 'geography of sexuality' (Binnie and Valentine, 1999), allows us to think through the differential use and claims on space. It can be deployed alongside analysis of the affective and representational interrogation of space.

Most of the above research envisages places nestling inside spaces, conceptualising place as that endowed with significant meanings by and for individuals and groups, often for the making of identity. Linda McDowell (1993) makes an important distinction between place as relational and place as a location of a structure of feeling centred on a specific territory. This is also explored by Stephen Pile (1996) who reveals how place is an essential component of the process by which individuals and groups map geographical space on a psychic level, involving the construction of emotional defences and boundaries (see also Hoggett, 1992; Hubbard, 2001; and Sibley, 2001). In this paper, we draw on these traditions of analysis, exploring the sexual politics of space as a politics of reality production, affective response and political claims-making.

For this, we draw on the contrast between *Queer as Folk* and our empirical research findings to explore how *Queer as Folk* offered respondents interpretative frames, forms of identification and maps of meaning, or what de Certeau calls spatial operations—the intersection of mobile elements. *Queer as Folk* contributes to an urban imaginary in which elements come together to produce a collective subject and/or bounded space with which identifications can be made. James Donald (1997), for instance, charts how the myriad imaginative cityscapes narrated by literary and cinematic productions offer a variety of subject positions. In our research, it was a specific TV production, and its secondary press, that produced a narrative space through the different visual clues and discourses available for people to be able to relate to and locate themselves within a place. Richard Sennett (1991) argues that representational images of the city are often more legible than those offered to the urbanite using their own unaided sight on the street. These images, he argues, *actively constitute* the city and its narrations, rather than being simply a representation of an external 'real'; they enable the chaos of space to be turned into a knowable place. Constitutive representations engage and entwine with other elements and operations, especially experience: they have to be plausible and to make sense to those who inhabit the place. In this sense, they are dialogical. Representations may also form part of an 'optical unconscious', as Shaviro (1993) puts it "Cinema is at once a form of perception and a material perceived" (p. 41), which can achieve new effects through its modifications of scale and proximity. People and places script each other, as Marback *et al.* propose

> When you hear or read about a particular city, almost automatically you draw upon what you have previously heard or read about that city to judge what you are hearing or reading now (Marback *et al.*, 1998, p. 3).

Queer as Folk had an impact far beyond its representation particularly because it drew on recognisable places and character. We therefore use *Queer as Folk* in the way generated by Benedict Anderson (1983) to understand the production of an 'imagined community'.[2] Once the representation of the Village by *Queer as Folk* circulates, it informs (possibly unconsciously) how people read and experience the place in a dialogical way: our transcripts could have been used to write the scripts and it felt like the script was inscribed in our transcripts—not surprisingly, as anyone spending any time in the Village would hear the same stories. The significant difference is that the TV series visualises these narratives, fixing them in time and place and making them appear as if real. These visual fixes then feed back into narrative circulation.

We do need to remember, however, that representations always take place in an economy of interests, for as Sharon Zukin and DiMaggio (1996) demonstrate, the production of space—just like the production of TV representations—depends on decisions made about what should be visible and what should not. And as Stephen Pile and Nigel Thrift (1995) illustrate, within scopic regimes, visual practices fix the subject into an authorised map of power and meaning, with only some forms of visual presence having legitimate value; others rendered illegitimate. The institutional landscape of the city, therefore, Pile and Thrift (1995) argue, creates a cultural reality that in part defines the frames through which mapped subjects are rendered legitimately visible. And visibility is crucial to the formation of a sexual politics based on visibility, which is about an empirical recognition of being in or out of place that invariably invokes regimes of placement. Claims for political recognition rely on an investment in a future belief of knowing where one's place should be and making claims for that space; the claims for a 'right' always invoke a projection into the future where justice can be found if the 'right' is given. When a significant TV series 'fixes' visible regimes of placement through characters and place, the place is made into space and consolidated in the urban imaginary, working as a primary reference point for those with no other knowledge of the sexual politics of the place. *Queer as Folk* is a powerful visual statement that celebrates visibility in place. But this is not always the same as the visibility that has been campaigned for by interest-groups from the area.

The different investments, interests and struggles in the space mark it differently: community organisations, capitalism, the council and the many different types of users all mark different visualities of interest. The space is therefore marked by both use-value (given by its users) and exchange-value (the different forms of capital investment). Quilley (1997), for instance, details how the spatial coherence of the Village was produced not just from local lesbian and gay community action and political protest, but also from the incorporation of this protest into local government politics through what would now be described as 'new' Labour politics, in which a cultural-led regeneration strategy was put in place which depended on the creation and promotion of dynamic spaces of difference (see also Mellor, 1997). Quilley notes that the overt support for the Village from 1984 culminated in its recognition as a 'planning entity' in 1991. It is now used to promote the city's tourism and regeneration and is also used as a model for other city regeneration in the UK. This is another attempt to institutionalise interest into a particular visuality whose results manifest in the visual city frame for *Queer as Folk*. Cognisance of all these interests became central to how we understood the intervention of *Queer as Folk* in the research process.

2. Research and *Queer as Folk* Background

Before the screening of the first series of *Queer as Folk*[3] on British TV (a programme which uses Manchester's gay Village as its central public location), a team of researchers, funded by the ESRC for a project on '*Violence, Sexuality and Space*', were generating their own representations of Village life.[4] The research project was based on two locations—Manchester's gay Village and Lancaster's 'virtual' gay space—and was designed to explore the significance of symbolically marked gay space (Manchester's Village) in comparison with unmarked space for the making of safety in the lives of three groups (lesbians, gay men and straight women).[5] The project was interested in how the desire for safety and the experience of violence shape the use of space. The project was multidisciplinary (law, sociology, literature and criminology), its researchers multi-identity (queer, lesbian, gay, straight and transgender) and multimethod (key informant interviews, census survey, focus groups, representational analysis, citizens' inquiry; with each method feeding into another).[6]

At the time of the first screening of *Queer as Folk*, the project was already 10 months into its 30-month duration and had, by the time of the screening, conducted 50 interviews with 'key informants' and 'stakeholders'. By the time of the second series (November 1999), we had conducted a further 3 focus groups with each constituency (making 9 meetings with each group in each location). And on the night of Friday 5 February 1999, a team of 12 researchers conducted a census survey of 13 bars and clubs between the hours of 2pm and 4 pm and 8 pm to midnight in the gay Village of Manchester.[7] In this time, 703 questionnaires were returned (from a total of 730 distributed). The project team also conducted background research into the general levels of violence collected through criminal statistics and evidence of reported attacks to the gay and lesbian switchboard, to the council and attended the Lesbian and Gay police liaison committees. Secondary textual materials such as newspaper, magazine and council/tourist publicity materials were also collected for representational analysis. All these different sources provide a contrasting range of research representations of the same space to that of *Queer as Folk*. Yet, whereas *Queer as Folk* is accepted as a representation, a fictional construct, research data are generally understood to constitute evidence, as a representation of the 'real'.[8] Moreover, as the focus groups progressed, *Queer as Folk* was referred to by nearly all our participants, clouding the fine line between empirical research and televisual representation even further. For a time, it was a central topic of conversation not just about sexuality and the Village but about Manchester more generally. By the second series, the filming of the programme became a public event where a network of telephone calls alerted Mancunians to the time and place of filming; the extras did not need paying.

3. Locating *Queer as Folk*

Manchester's gay Village is a small compact area, comprising mainly of bars, restaurants and leisure venues. Its main thoroughfare Canal Street is an almost completely gentrified area, aesthetically pleasing, built on the side of a canal bank framed by large Victorian red-brick warehouses, mainly converted into expensive loft apartments. Canal Street has been partially pedestrianised and during the summer months the pavements are covered with umbrellas, tables and coffee-drinking clientele, resembling more a Mediterranean city (often compared with Barcelona), as opposed to the northern industrial manufacturing landscape that it inhabits. The area is brightly lit. There are some offices mainly housing 'new' media and PR companies and some social housing with controlled rents (possibly the most sought-after in the whole city). There was also a gay shopping centre 'The Phoenix' (which has since closed). The streets behind Canal Street have also been gentrified as a result of the Commonwealth Games (July 2002). The bars are divided into 'old' and 'new' with the 'new' bars offering a queer architectural statement, such as *Manto's* with their 30-foot plate glass windows which promote visibility rather than being hidden underground.[9]

There was only one designated but temporary lesbian space on Canal Street, on the first floor of a male leather bar, but in 1997 this was converted into a bistro. In 1998, the lesbian manager of *Manto* opened the first women-only bar *Vanilla*. The lesbian club *Follies* recently closed and was seen by all our key informants to be resolutely working-class. There is little that marks the Village space as lesbian. It is predominantly represented (as will be seen) as gay = gay men, although our census survey found that, of our gay and lesbian survey respondents who use the Village on a Friday, 13 per cent were lesbians (26 per cent were gay men and 13 per cent were straight women: the rest were a mixture of identifications, such as trans, queer, bi).[10] Lesbians were for the most part found in lesbian-only venues. The campaigns run by 'Healthy Gay Manchester', which decorate the lamp-posts with flags and posters, mark out the space as symbolically gay male. The billboards that advertise the Mardi Gras festival, renamed Gayfest and Europride (held every August Bank Holiday) contribute to the male-gay signification. The creative vandalism that continually erases the C and the S from Canal Street makes it difficult for the area *not* to be recognised as gay male.

Queer as Folk was first screened on British television on Channel 4 on 23 February 1999 and a further five episodes were broadcast on a weekly basis. Following its success, a second series was commissioned and screened on 15 February 2000. The first episode attracted 2.2 million viewers. As the series was broadcast, it generated a large and vocal response. Following the first episode, the Independent Television Commission, after receiving 160 telephone complaints, announced that the programme would be subject to a formal investigation. It has since received more complaints than any other programme broadcast on UK TV. The first episode in particular in which underage Nathan (aged 15) has sex with Stuart (aged 29) produced a huge response. Its screening coincided with Parliamentary consideration of the reform of the 'Age of Consent' bill. Both left and right of the political spectrum complained. On Channel 4's TV programme *Right to Reply*, Angela Mason, director of the lesbian and gay law reform and parliamentary lobby group Stonewall, complained about the 'damage' that the programme had done by threatening support for the lobby (*The Guardian*, 22 June 1999). *The Daily Mail* (representing right-wing tabloid middle-England respectability) warned that "any nation which allows this without any voices raised in dissent is lacking in both wisdom and self-respect".[11] The equally conservative *Sunday Telegraph* suggested that the first episode was "a celebration of statutory male rape and criminal pederasty".[12] Notwithstanding this condemnation, *Queer as Folk* sold to 13 different territories, making it the best-selling Channel 4 video ever and the first TV series to make it onto DVD; it has been screened at film and TV festivals throughout the world, winning a gold disk for the soundtrack and producing a request from Channel 4 for another 10 hours of programmes. An American interpretation was also screened on HBO and is, like the British version, available globally on video and DVD.

Queer as Folk is a story of three gay men. It is a narrative of friendship and desire between Stuart, Nathan and Vince. Each figure represents a detailed characterisation of a particular variant of gay male sexuality. Stuart is full-on sexual predator, flash, hard, insensitive and completely committed to hedonism. The space of the Village offers him the potential for sexual conquest. He exists on the screen solely to have sex and represent it. Even his after-sex activity is to watch the recording of it. His sexual activity is the cipher through which the other identity positions (Irish, father, son, brother) he occupies are filtered. He is always placed in environments that exist as props for sex, especially the Village which is referenced throughout, with Canal Street as the main venue for his movement. Paradoxically, his sexual activity (the movement through people, opportunities, places) offers him the possibility of not being fixed in time and space, even though gay male identity as sexual cipher is a relatively fixed historically repeated representation.

This is contrasted with Vince, who is rarely the sign of sex, but rather of geeky domesticity. The only time Vince is characterised as a sexual being is when he is domesticated by a respectable monogamous Australian accountant, Cameron. He does have sexual opportunities, but the character flaws of a 'muscle mary' who takes off his girdle to reveal a flabby beer belly and the BBC producer with Brazilian bugs in his a*** all work to comic effect to deter Vince's sexual activity. Vince is closeted at his work and outed by being recognised and therefore marked as gay by the place of the Village. He represents the ordinary in a not very queer way. In fact, Stuart berates him for being a "straight man who has sex with men".

In between these two opposites is 15-year old Nathan, the David Beckham look-alike, the schoolboy (the underlegal-age gay boy) who is enabled in his 'coming-out' by his association with Stuart, an association which becomes an obsession and so drives the narrative into more than one moral dilemma. For Nathan, the space of the Village represents freedom, escape, liberation, utopia. It is a space free from the homophobic bullying at school (a threat which is depicted with all the homo-erotica that such threats usually contain). His best friend Donna supports Nathan throughout. The Village is used as a transitional space from immaturity to maturity, from schoolboy sex and porn magazines to full-on genital sex. Nathan learns to take control; his youth and beauty enable him eventually to occupy the Village space with ease.[13] Nathan's difficulty and desire to occupy the place of the Village contrast with the ease with which Stuart and Vince take the place completely for granted and contrast it with straight space, as noted by one of our respondents

> Yeah, remember that scene in *Queer as Folk* where Vince is going into a straight bar to meet some people that he works with and so he is talking to Stuart on his mobile phone and he is saying "I am going in, I am going in" and he is just wandering round and one of the points he makes is "you know the people are having a conversation with no punch line. It must be a straight bar!" (Adam, mgmfg1, 417).[14]

In *Queer as Folk*, the place is represented as utopic, safe, interesting, fun and 'cool', a place where gay men can roam freely, kissing each other in the street, using the phone box for passionate sex, and where every passing person is an opportunity for an exchange of sex and/or wit and repartee. The lighting of the street is enhanced in *Queer as Folk* with the use of thousands of candles to create a warm and soft ambiance and it has been cleaned.

Queer as Folk feeds into, feeds off, but also constitutes, the range of dominant and circulated representations. It offers different incitement for use and investment. And it also adds to the initial impetus for visibility, for political recognition. Just like the bar *Manto*'s huge windows, it is a statement of 'we're here, we're queer, get used to it'.[15] Our key informants tell us that bar turnover and tourist rates increased rapidly during and after the screening of the first series. Our focus groups regaled us with tales of *Queer as Folk* tourism

> There was an increase in straight people after *Queer as Folk* came out. There was also most like a curiosity value. People came and gawped at us (Andrew).
>
> Because they'd seen *Queer as Folk* they wondered where all the candles had gone down Canal Street [laughter]. They didn't realise it was part of the TV series. They thought it was really like that (Carl).
>
> It was definitely busier for the next couple of months after *Queer as Folk*, that was definitely apparent and then it calmed down. And everyone thought it's s*** again (Andrew).
>
> But it became more expensive than it ever was. I mean I know the prices have increased so you know for me *Queer as Folk* was the almost antithesis of my life (Carl, all mgmfg1, 1147).
>
> Jill blames it on *Queer as Folk*, the Village is now a freak show (Marina, mswfg2, 433).

The impact of *Queer as Folk* fed into debates that had been circulating in the area (via gay and straight media) about the dilution of the 'gayness' of the space. All our respondents offered narratives of invasion (hen parties in particular, which we explore fully in Moran *et al.*, 2003; Skeggs, 1998). *Queer as Folk* was seen to be partly responsible for this dilution. Paradoxically, even the straight women focus group note the loss of 'essential gayness'

> The gay Village I used to like but I think it has been infiltrated by all the corporates, too many straight people and it's losing essentially what it was in the beginning, that sense of community, it doesn't seem to have it anymore. It's like *Queer as Folk* made it so trendy, it doesn't look like that really but it pulled in the crowds (Marina).
>
> *Queer as Folk* has got a lot to answer for making Canal Street or the Village look like not the Village, but this fantastic place where you want to go and then when you go it stunk (smells) or whatever (Sandra, both mswfg1, 57).

These comments feed into the wider talk of nostalgic communities, uncontaminated by heterosexuality, of spaces under siege, of threat and violence by young straight working-class men (signified as 'rough', 'scally'), or by geographical areas (such as working-class Salford and Wythenshawe), which repeatedly represent the Village as a place under threat. The reiteration of the same narratives of mythical stories of violence, led us to be particularly careful about collecting 'evidence' to counteract the flow of urban mythologies of a community under siege. But these stories are necessary as narratives of interest because, for some—such as council workers—their jobs depend upon the existence of a marginalised space, a marginalised community, a place that needs sustenance and support. Just as the crime survey has been used by campaigning groups to demand resources, the stories of danger justify the maintenance of roles and spaces; marginalised groups have learnt how to deploy marginality in the making of political claims (see Brown, 1995). It is different combinations of elements, interests and investments in place that both constitute and disrupt identifications with, entitlements to and territorialisation of, space.

Marina and Sandra's comments engage with what Celia Lury (1991) describes as the 'aesthetics of authenticity', in which the use and representation of a space that is known is read through the discourses and codes available and compared with the experience of it. Clearly, the experience of the Village is very different from its hygienic-romantic representation. The use of local points of knowledge (the street without candles, the smells, the potential danger) refuses the authenticating of the place, yet the use of a space that is easily recognisable as Manchester's gay Village helps simultaneously to authenticate it. In episode five, Vince names all the bars he uses: *Via Fossa*, *the New Union*, *Metz*, *Napoleons*, *New York*, *Cruz*, *Babylon*, *Paradise* and *Poptastic*. All the bars but *Napoleons* were surveyed in our research. There are enough direct signifiers and references for the TV scene to appear as a realist and authentic representation to anyone in Manchester with even a modicum of knowledge about the Village.

Yet what appears as a 'real' place has been given a different meaning. *Queer as Folk* represents the Village as the site of gay male safety and sexual fantasy, not the site of violence and political struggle. As one of our respondents suggest *Queer as Folk* is changing the way people see the Village

> I think because of the filming and that is changing the way people look at the Village … and em you know, its, *Queer as Folk* is getting the city a reputation. People think Manchester is gay. They get a shock when they come here (Julie, mswfg1, 65).

The comfortable occupation of place in *Queer as Folk*, can be contrasted to our research where negotiation of violence is an ever-present concern. For *Queer as Folk*, it is ontological security—coming out, finding an identity, belonging, fitting-in: issues which, in our research, are always underpinned by violence. This becomes apparent when we compare the different representations of the place. For Julie, not only are the aesthetics of authenticity questioned, but she argues that *Queer as Folk* actually has the potential to change how people see and experience the Village—a change, we argue, that could be potentially dangerous.

Our research demonstrates how movement in and out of the area is prone to danger. Gay men are more likely to be victimised in cruising areas or gay-identified neighbourhoods, whereas lesbians report more violent encounters in heterosexual streets. The crime and disorder statistics, which break down crime into policing areas, identify the Village as a 'hot spot'. However, this needs to be broken down by street, the central thoroughfare—Canal Street—is particularly safe whilst the border zones, the access routes into the centre of the Village are potentially dangerous.[16] As we know from prior research on gay male homophobic violence, struggles for visible identities will often incite danger, for visibility is a threat to the normalised landscapes (see Myslik, 1996). The visualising of the Village makes it *simultaneously* a potential space for the expression of visible identity and also a way of identifying potential targets for homophobic violence. This is not represented in *Queer as Folk*, where we enter the place through the safety of the camera and being visibly queer in visibly queer space is celebrated.

Moreover, in *Queer as Folk*, the Village is not only safe but is also represented as a space of escape: Vince from work; Nathan from school homophobia, father homophobia and prying mother; Stuart from homophobic family; and Donna from school sexism and abusive home. There are three scenes of Nathan and Donna running away, leaving suburbia to enter the free and queer space of the Village.[17] This escape becomes a claim on space, or territorialisation through sexuality—one way by which gay men have been able historically to instantiate their sexual identity. Territorialisation is made apparent through the character of Nathan as he becomes more and more confident about his occupation and inhabiting the space called the Village. In the second series, Stuart and Vince move on and out from the constricting place of the provincial (but cosmopolitan) Village to Arizona—the strongest signal that could possibly be given of unconquered, open, vast expanse of space. It is as if gay male sexuality becomes mature through spatial claiming and territorialisation.

The relation between the TV representation and the actual physical place is not just blurred through the use of 'authentic' known bars and clubs in *Queer as Folk*, but also by secondary media. A national gay magazine, in a feature article on the gay scene in Manchester, written to coincide with the 1999 Lesbian and Gay Mardi Gras festival in Manchester, describes the space in the following terms, "a bustling gay scene most recently featured in TV's *Queer as Folk*". (*Fluid*, 5 September 1999). Other instances of ways in which empirical description enters fictional discourse are to be found in discussions about characterisation. One dominant theme in responses to *Queer as Folk* is that the characters in the drama, "... confirmed every stereotype in the book" (*Time Out*, 10–17 March 1999) and "didn't challenge any stereotypes" (*Pink Paper*, 5 March 1999). Other criticisms pointed to the marginality (and stereotyping) of lesbian characters; "all the lesbians wanted babies" (*Pink Paper*, 5 March 1999). Still others complained about the absence of other facets of lesbian and gay culture. One respondent to a *Pink Paper* 'Vox Pop' survey commented

> There's not one brown character even in the background. They're all as White and proper as can be. I've never felt so ashamed to be gay in my life (*Pink Paper*, 5 March 1999).

These criticisms of *Queer as Folk* and demands for a more diverse, inclusive or more complete picture of gay and lesbian *Queer as Folk* life in all its diversity have a common theme—realism. And they are all participating in political claims-making from a particular space.

Yet this realism is never fixed, especially in relation to the use of place: the blurring of knowledge about the sexualised place called the Village is made by the disruption of the different elements that make up the space, in which visible queer is made safe, danger is recoded as safety, dirt as hygiene, bullying into community, immaturity into maturity and boredom as sexual opportunity: these all present (new) ways of reading the place. Participants at a 'Sexuality' conference held in Manchester in summer 1999 were delighted to find all the named bars from the series, but were shocked to find the Village inundated with vagrant drunks, people harassing for money, young women trying to steal cameras and anything else they could lay their hands on. The conference participants, like many other tourists, had imagined a safe gay utopia.

These aesthetics of authenticity become even more intimately insidious and confused when we explore how realism is continually evoked by our research participants in an understanding of characterisation, a characterisation that is always underpinned by the authenticity of the location.

4. Cultural Significance: Mimesis *and* Verisimilitude

> *Paul* (researcher): Did you feel watching [*Queer as Folk*] that it was about you somehow?
>
> *Simon*: No. Certainly not.
>
> *Phillip*: I did, even though I also thought that at the same time it wasn't about me but a little bit of everyone was there somewhere.
>
> *Simon*: I think there are all sorts of elements that are there even though I've never seen myself in any of them (lgmfg 4, 13).

These observations draw attention to the importance of the reader in the generation of meaning and the appropriation of the text. They also demand that we avoid totalising assumptions about the nature of the audiences. As Sara one of the participants in our straight women's focus groups noted

> I talked to a gay man and a lesbian about it [*Queer as Folk*] on the same night it was on. The gay man thought it was a bit over the top, a bit too much and the lesbian thought it was brilliant. ... She just thought it was really entertaining and he just thought it was oh it was a bit too graphic (lswfg3, 37).

One of our gay men's focus group members described the problem in the following terms

> I don't see how you can aim things at gay people really because gay people are from bankers down to tramps in the gutter. How can you aim a TV programme at that kind of range of persons? I don't think really sexuality is anything to have in common with anybody really (lgmfg4, 14).

If places are moments of encounter, as suggested by Amin and Thrift (2002), we can see here an awareness of the different types of encounter beyond the limited representations of the TV programme, but familiar to the research participant, thereby questioning the 'authenticity' of the representation. As the programme attempts to 'fix' sexuality in the place of the Village, resistance is made to this possibility. Another participant noted that

> All my straight friends were going "oh! are you watching *Queer as Folk*?" I was going, "No! I find it really boring". Straight people really like it and apparently young, 15-, 16-year-old girls just think it's fantastic (lgmfg4, 12).

Here is a direct resistance, a refusal to be linked in anyway to the programme and a statement of derisory distinction from it. These observations point to a need to recognise a range of assumptions at play in addressing the question of the cultural and deployment of images in *Queer as Folk* and demonstrate the polyvocal and multiple readings made by media audiences (see, for example, Ang, 1985, 1990; Gillespie, 1995; Hall, 1997; Morley, 1980, 1986; Radway, 1987; Seiter, 1990, 1999; Seiter *et al.*, 1989). The media both link and fragment, and as Amin and Thrift (2002) reveal, enable a broad range of engagements underlined by the fact that place-based media must always interact with the unpredictability of user histories and experience with the characters in place. Whilst the place is literally fixed, named and known, the characterisation is much more ambiguous. We can refer, as many participants did, to the place and challenge its representation, but the generalised characterisation enables more ambivalence: the characters are not known and not fixed, although some of our research participants made strong identifications

> I thought it had a lot of different character types in there so there was someone for everybody if you like (Phillip, lgmfg4, 12).

> One thing I did like about *Queer as Folk* was the fact that you could relate to the character in the fact that 15-year-old Nathan ... I could relate to myself in that situation at 15. And then I could relate to Stuart at another part of my life and I could relate to Vince at another part of my life, so I could see relevant points which I thought were interesting (Mark, lgmfg4, 15).

Identification is transitory, it moves between characters rather than being projected onto a singular being. Each connection represents a particular facet and point of contact, suggesting it is mimesis rather than identification that may produce the connection. Michael Taussig (1993) argues that mimesis is the driving motor of modernity because of the *resonance* and the *perspectival spectrality* that we bring into our own world to explain it and draw on the power before our eyes to absorb it and attribute it to ourselves. He maintains that what is crucial in the resurgence of the mimetic faculty, of copying or imitation, is the connection made between the body of the perceiver and the perceived. Traditional filmic analysis argues that visual representation reproduces a virtual space marked by 'proximity without presence' (see, for example, Fleisch, 1987), yet the proximity of the Village and the presence of our participants within it, often along with the investments that have been made in the space—politically, economically, emotionally—make *Queer as Folk* both proximate and present.

The characters in *Queer as Folk* may work both because they are believable and have aspects to them that cannot fail to connect to some aspects of the lives of gay men: taunted at school and home for being gay; revenge fantasies; being rejected by the family; or offering the potential for fantasy (being able to pick up anybody of one's choosing; being known and loved in a safe and sexually charged space). Recognition and resonance here are key to the readings that are made.

These responses are affective because an emotional identification can be made. In an early audience research study, Ien Ang (1985) identifies 'emotional realism' as one of the primary mechanisms invoked by audiences when watching and discussing TV representations. It is an affective connection that enables the link between the representations of experience and the experience of that person, interpreted through discourses that predispose or coincide with the representations, offering a point of connection. When an affective connection is made via experience to a character representation that is fixed in a place that is known, it is likely that a strong resonance occurs; the space and the character become 'real'.

We can also take into account Walter Benjamin's (1959, p. 255) ideas of 'the flash' wherein "the past can be seized only as an image which flashes up at an instant when it can be recognised and is never seen again". Here, it is the repetition of the flash, rather than its escape, which has the potential of creating recognitions, of resonances, of connections.[18] The flashing through of places as mobile elements underpins the character connection. This relies on the ability to place oneself in the position of the character through an identification of similarity (not difference), whilst also inhabiting a body and space that are known to be different. It is a form of embodied knowing in which experiences become interpreted through the same frame.[19] Yet what enables the connections to be made? First, it must be credible and plausible, and it is here that the place becomes a central reference point for how the characters take shape. The representation must also have reference points that can be understood.

Stephen Neale (1980) maintains that generic convention is essential to this process in which genre and culture come together in the production of 'place'. Cultural verisimilitude refers to the norms, mores and common sense of the social world outside the fiction (such as the experience of space and place), whereas generic verisimilitude refers to the rules of a particular genre; genre conventions. The genre has its 'own laws of verisimilitude'. They provide the guide by which the credibility or the 'truth' of the fictional world associated with a particular genre is guaranteed. Generic verisimilitude provides for a considerable play with fantasy inside the bounds of generic credibility. Neale (1980) contends that when generic verisimilitude and cultural verisimilitude come together, they produce an experience of representation as the real. For instance

> With '*Queer as Folk*', Vince was like a cardboard cut-out but other than that for me, I don't get out much and do much at the moment and it was like the highlight of my week. It was fab. I mean the last episode where he got his own back on that guy, pointing him out and saying whatever he said. It was fab. It just reminded me of the person who said the most poisonous thing to me in the past—I just wish that was me saying it to them, making them feel like that. I thought it was great. I'd love a life like that, the money and all that. *I'd like to be able to walk through the Village like that*, head held high, not worried about what everybody thinks, not hassled. Realistically I'm poor, but I'm happy as well. It's just TV isn't it? (Peter, lgmfg4, 14; emphasis added).

Within this statement, the flash of historical recognition occurs and the generic convention of revenge is used to authorise a fantasy of overcoming marginalisation in a place that is known and experienced. Identification is again *not* made with the singular character but, rather, is made affectively through the mimetic process of recognition, which enables a contact to be made with the past and the present. In a 'flash', the scene before the viewer's eyes becomes a copy of the 'truth' of their own experience powerfully located. The power of *Queer as Folk* here is that it provides a narrative solution that attributes power, revenge and overcoming to a prior negative event in a place that is known. It copies the event but provides the alternative of a different outcome; it makes a substitute interpretation. It changes the morality of the experience and it offers a momentary position of power to be occupied by the viewer in the safety of their own viewing. Basso and Goodwin (1942/1969) suggest that it may be the moral relevance of the narrative that enables a connection to be made between reader and representation. *Queer as Folk* offers a different possibility for interpretation that is brought closer to what is known as reality. By locating that interpretation in a place that is known and recognised, it reinscribes the messiness of place with the fantasy of a safer space.[20]

Our research participants know of the illusory nature of the representation in a way which does not implicate them in the illusion. They are aware of it as an "imaginatively convincing piece of artifice" (Corner, 1992, p. 99). The reference to *Queer as Folk* as "the highlight of my week" draws attention to the programme as a cultural resource. In Peter's comment above, for one moment the representational flash seems to be real, but is swiftly followed by a recognition of the representation as artifice, "It's just TV isn't it". The place location of *Queer as Folk* as somewhere familiar pulls in via recognition and pulls and pushes out by character recognition or resistance. Indeed, as Neale argues, 'realistic' genres do not involve total belief in the accuracy or the reality of their modes of characterisation.

An interesting example that draws attention to the vital importance of suspension of belief in the fictional aspects of the representation is to be found in the following extract

> I had to stop watching the series [*Queer as Folk*] because I got bored with the lack of a decent story line and couldn't handle the terrible accents. I couldn't decide whether they were hideously bad actors or whether that was just Mancunians, it was hard to say really (DavidP, lgmfg4, 12).

Here the deception of truth has failed—no mimesis, no flash, no resonance. The juxtaposition between 'bad actors' and 'just Mancunians' displays a knowledge of the fictional nature of the story, a reference to one of the conventions of verisimilitude, in this instance. Accents and the reference to "lack of a decent story line" suggest that the mode of narrative might also be part of a genre's system of credibility. For David, the series failed to realise satisfactory verisimilitude. There is no historical or emotional recognition, nor any resonances with present position; rather, the lack of any recognisable authenticating character or practice renders *Queer as Folk* without credibility. The aesthetics of authenticity clearly did not work.

This may be because, as Christine Gledhill suggests

> the demand for a 'new' realism from oppositional or emerging groups opens up the contest over the definition of the real and forces changes in the codes of verisimilitude (Gledhill, 1997, p. 360).

These demands, of what can be expected from representations of reality, shift the debate from one about the relationship between representation and reality, to one about interests, stakes and investments.

Our research participants are very well versed in gay and lesbian politics (including the straight women) and know that TV is used in the symbolic struggle for legitimacy and visibility. They are aware of the precariousness of recent political gains. Thus our research participants' emotional and resonant readings are not the only way they respond; they also assess the representation for its value as a moral and political resource, working out the value of investments in different interpretations, presenting their responses as an articulation of a position on lesbian and gay politics

> I'm not quite sure what to make of Nathan. Perhaps he's everyone's worst nightmare, there's the young lad that the older guy is corrupting … I wasn't comfortable with the underage element (Phillip, lgmfg4, 256).

> Yeah, a total cardboard cut-out. I think it made gay people look not very appealing really … It just proved the cynicisms that I have about being a gay man in straight society. It just brought them back to the fore and I just really didn't want to sit through it (Dave P, lgmfg4, 257).

Their responses are an example of an audience versed in the knowledge of the value of representations; they know what could be at stake and they use the interpretation of the fictional text in this battle. This is not dissimilar to accounts of other groups, such as Jacqueline Bobo's (1995) research on Black women as readers of media texts. Or Seiter and Riggs' (1999) account of fundamental Christian readings of children's TV. These groups have a clear awareness of how TV is used both to make and to delegitimate political claims. Moreover, research by Corner *et al.* (1990) into nuclear power debates shows how responses are made to fit into a range of rhetorical frames of understanding, arguing that the 'civic' frame—that which is based on fairness—is the single most powerful regulator of interpretative assessments. The representations are read as stakes in a political battle. The research process then highlights how these stakes are articulated, how positions are taken, how representations provoke reflexivity upon one's own experience and knowledge and how these 'should' be represented to a public (which includes researchers). Reality is not the issue. It is how reality is represented as a moral resource that can either condemn and deny legitimacy via visibility or promote and enable legitimacy that is at stake. Verisimilitude becomes a matter of morality, if what it generates is how value is attributed to those who are represented through it. When these interests are located in a place that has itself been a site of struggle for visibility and legitimacy, the stakes are more significant.

5. Conclusion

Our research suggests that the place of the Village is produced as a stability through the representation of *Queer as Folk*, in which variants of male sexuality are fixed in place as a limited configuration of subject positions. In contrast to *Queer as Folk*, our research reveals not a utopic-hygienic safe community, held together through shared interests and sexuality, but a space occupied very differently by different groups in which distinctions, conflict and struggle proliferate, leading to specific uses, circuits and routes for different groups of people,[21] whose presence disrupts any potential for known, safe, bounded, visible community in which kissing, hand-holding and witty repartee, as depicted by *Queer as Folk*, can take place.

These different interests continually inform the reading of the space and the use of representations in the making of place. It was interest, capital and politics that informed how the Village could be spoken about to us as researchers. The representation of utopia (in *Queer as Folk*) may fit nicely with the capitalist desire to commodify gay space but, as most bar owners/workers know, in a space controlled by local protection rackets, utopia is very rarely a possibility. Yet *Queer as Folk* did make a momentary intervention in the imaginary of the space that could enable a potential reinterpretation of experience. It was also great for tourism. *Queer as Folk* made the rather *passé* Village trendy and interesting once again (on an international as well as a national level). It renewed interest in a space that had been made anew as a place by a different visual representation. The programme drew on and constituted the visual imaginary that is Manchester.

Yet *Queer as Folk* remains a significant marker of gay Manchester (used to market Europride 2003) and it still has currency as a short-hand term for the pleasures of the Village. It did enable people to think about their experiences differently and it did visualise fantasies of possibility otherwise unimagined. It also provided a context for political debate (especially the absence of lesbians within the space; see Munt, 2000; Thynne, 2003). And it made us question the 'evidence' of research about the use and the making of space; in people's imaginaries. The TV programme reproduced almost exactly the dialogue of some of our focus groups and interviews, making us question our hold on 'reality'. Just as the research participants brought together the generic and cultural conventions of verisimilitude, so did we. This makes us argue that we can never understand the transformation of space into place without an understanding of the representations and conventions that enable its conversion. As space is so often used to mark the context, the location, the background to social and cultural practice, we show how that space is made through the practice of underpinning and locating interpretation. This underpinning by relying on an aesthetics of authenticity brings into play interests, investments and experience which both challenge and celebrate the fixed representation, but this is always leaky as a visual place. Representation can never encompass all the constituent elements and subject positions that make a place into a space. As Amin and Thrift (2002) argue, the city is a process in which scripts of it are made through scripts of experience of it. *Queer as Folk* intervened in the space-making of the place, bringing different constituent parts to it, inciting people to enter, creating a different dynamic, in which interests and struggles are re-engaged.

All we can ever know about a space are its mobile elements and how they are deployed to produce a momentary fixity, a temporal configuration of possibilities, which can be deployed by different interest-groups, sometimes territorialised but always the site of struggle. *Queer as Folk* powerfully captured and contained the specific elements of a gay male utopia in a place that was known to be the site of gay politics. The ambiguity between different representations, the similarities between the experience of space and sexuality and the fact that the space has always been the site of different interests made a distinction between representation and reality impossible, although it did not stop people trying to find one (including the research team initially).

As a particular moment in time, *Queer as Folk* offered a fantastical queer visuality that fed into the moment of change in British society and challenged the assumptions about the heteronormativity of space; it was a powerful queer statement, not only 'We're here, we're queer, get used to it'. But 'We're here, we're queer in a space that is close to you that you know'. It was simultaneously a statement of the present by being fixed in a known place and a statement for the future by offering an imaginary utopic of what could be. Paradoxically, it was that simultaneity of present and future that enabled our research participants to challenge its visionary future. Their location in the location meant that they used the codes of authenticity and verisimilitude to challenge the representation of place. The candle-lit utopia was read as the dirty and dangerous place that it could be. The participants did, however, get drawn into the characters in the place and they did see the potential for political claims and challenges to legitimacy. It was the ability offered by the visuality of *Queer as Folk* to reimagine that which is known (either through place or character) that both intervened in our research sense of reality and also enabled us to see how visual relocation can be a powerful mechanism for thinking into the future, without which any sexual politics would be bleak.

Notes

1. Important contributions to the role of space and place in the making of sexuality has also been undertaken outside geography—see Stychin (1998); Moran (1997); Berlant and Freeman (1993). In literature, see Probyn (1990). In Sociology and Cultural Studies, see Cant (1997). See Binnie (1998) for a comprehensive review.
2. Anderson shows how the invention of the printing press and mass literary projects enabled the production and projection of an imagined community.
3. *Queer as Folk* was written by Russell T. Davis, directed by Sarah Harding and produced by Nicola Schindler (for Red Production company).
4. The project on 'Violence, Sexuality and Space' was funded by the ESRC (grant no: L133 251031) as part of the 'Violence Programme' between 1998 and 2001.
5. Straight women were included in the project as previous research had identified how they used gay space as a space of safety (see Skeggs, 1997).
6. Full details can be found at the website (http://les1.man.ac.uk/sociology/vssrp) and from the book of the project (Moran *et al.*, 2004).
7. The Lancaster survey was conducted on two separate occasions (to enable access to gay and lesbian events as no permanent space for gays and lesbians exists) in May 1999. The

available numbers were far smaller, with 250 survey responses returned.

8. Notwithstanding the debates in anthropology on ethnographic representation on which this paper draws.
9. This queer visual statement was requested of the architects by the owners.
10. Our key informants had told us that Friday was the most heterosexualised time.
11. See Manchester's *City Life*, 399, 24 February 2000, pp. 10–12.
12. Quoted in *Pink Paper*, 5 March 1999.
13. Sub-characters are the 'lesbians' (to which Hazel, fag hag mum, responds when they visit "Oh no I'm out of herbal tea"), who are literally played straight: respectable, domesticated and dull. The radical position and political struggles of the lesbian are completely lost. It is straight women who are given the radical outsider position, the two variants of fag hags, depicted as not-dull: Hazel, Vince's mum, and Donna, Nathan's friend, who are able to comment wisely on the antics of their male connections. Yet it is women's general absence from the central narrative that enables the production of the safe, communitarian, liberational representation of space, unencumbered by women, although it is women's labour that underscores the men's lives.
14. Coding is Manchester (m) or Lancaster (l), gay (g) lesbian (l) straight women (sw), focus group (fg) number, transcript reference. Coding was made using NUDIST. Where names are not given, it is because participants wished to remain anonymous. Others chose to use their 'real' names and others chose pseudonyms.
15. *Queer as Folk* offers a representation of a particular queer visibility, but it also incites visibility. It is made at a particular historical moment when *some* of the imperatives to closet and invisibility are being broken down.
16. One street, Minshull Street, is particularly dangerous as it is where young sex workers operate, who are subject to numerous attacks; in 1999, two were murdered.
17. Suburbia operates as the primary representation of the lower-middle-class English, as Roger Silverstone argues in *Visions of Suburbia* (1997).
18. Or, as Taussig (1993, p. 39) argues, using Marx's understanding of revolution as "the subject of a structure whose site is not homogeneous, empty time, but time filled by the presence of the 'now' ".
19. It could also be seen to be part of the process of intercorporeality (see Weiss, 1999).
20. *Queer as Folk* here combines what Corner (1992) describes as the two forms of realism: Realism 1—the project of verisimilitude (of being *like* the real); and Realism 2—the project of reference (of being *about* the real). In both cases, the 'real' is partly a normative construction and disputable independently of any media representation. Corner draws attention to the problems with realism debates that assume the illusory effect of representations. Here, we argue that it is the embedding of the physical, of making a known space recognisable, that makes it appear 'more real' than the representation.
21. This was more usually an account of the avoidance of others (be they class, gender, generational or sexuality based) rather than a coming-together across difference. Shared interests seem to be based around taste cultures: choice of music (usually defined in Bourdieu's (1986) senses of distaste).

References

AMIN, A. and THRIFT, N. (2002) *Cities: Reimagining the Urban*. Cambridge: Polity.

ANDERSON, B. (1983) *Imagined Communities: Reflections on the Origins and Spread of Nationalism*. London: Verso.

ANG, I. (1985) *Watching Dallas*. London: Routledge.

ANG, I. (1990) Melodramatic identifications: television fiction and women's fantasy, in: M. E. BROWN (Ed.) *Television and Women's Culture: The Politics of the Popular*, pp. 75–89. London: Sage.

BASSO, K. H. and GOODWIN, G. (1942/1969) *The Social Organisation of the Western Apache*. Tuscon, AZ: University of Arizona.

BELL, D. (1995) Pleasure and danger: the paradoxical spaces of sexual citizenship, *Political Geography*, 14, pp. 139–153.

BELL, D. and BINNIE, J. (2000) *Sexual Citizenship*. Cambridge: Polity.

BELL, D. and VALENTINE, G. (Eds) (1995) *Mapping Desire*. London: Routledge.

BELL, D., BINNIE, J., CREAM, J. ET AL. (1994) All Hyped up and No Place to Go, *Gender, Place and Culture: A Journal of Feminist Geography* 1(1), pp. 31–47.

BELL, D., BINNIE, J., HOLLIDAY, R. ET AL. (2001) *Pleasure Zones: Bodies, Cities, Spaces*. Syracuse, NY: Syracuse University Press.

BENJAMIN, W. (1959) *Theses on the Philosophy of History*, trans. by H. ZOHN. New York: Schocken.

BERLANT, L. and FREEMAN, E. (1993) Queer nationality, in: M. WARNER (Ed.) *Fear of a Queer Planet: Queer Politics and Social The-*

ory, pp. 193–230. Minneapolis, MN: University of Minnesota Press.

BINNIE, J. (1997) Invisible Europeans: sexual citizenship in the new Europe, *Environment and Planning A*, 29, pp. 237–248.

BINNIE, J. (1998) Re-stating the place of sexual citizenship, *Environment and Planning D*, 16, pp. 367–369.

BINNIE, J. and SKEGGS, B. (2004) Cosmopolitan knowledge and the production and consumption of sexualised space: Manchester's gay Village, *Sociological Review*, 52(1), pp. 39–62.

BINNIE, J. and VALENTINE, G. (1999) Geographies of sexuality: a review of progress, *Progress in Human Geography*, 23, pp. 175–187.

BOBO, J. (1995) *Black Women as Cultural Readers*. New York: Columbia University Press.

BONDI, L. (1998a) Gender, class and urban spaces: public and private space in contemporary urban landscapes, *Urban Geography*, 19(2), pp. 160–185.

BONDI, L. (1998b) Sexing the city, in: R. FINCHER and J. M. JACOBS (Eds) *Cities of Difference*, pp. 177–200. London: Guilford Press.

BOURDIEU, P. (1986) *Distinction: A Social Critique of the Judgement of Taste*. London: Routledge.

BRENT INGRAM, G., BOUTHILLETTE, A.-M. and RETTER, Y. (Eds) (1997) *Queers in Space: Communities/Public Places/Sites of Resistance*. Seattle, WA: Bay Press.

BROWN, W. (1995) Wounded attachments: late modern oppositional political formations, in: J. RAJCHMAN (Ed.) *The Identity in Question*, pp. 85–130. New York: Routledge.

CANT, B. (Ed.) (1997) *Invented Identities: Lesbians and Gays talk about Migration*. London: Cassell.

CERTEAU, M. DE (1998) *The Practice of Everyday Life*, 2nd edn. London: University of California Press.

CORNER, J. (1992) Presumption as theory: 'realism' in television studies, *Screen*, 33(1), pp. 97–102.

CORNER, J., RICHARDSON, K. and FENTON, N. (1990) Textualising risk: TV discourse and the issue of nuclear energy, *Media, Culture and Society*, 12(1), pp. 105–125.

CORTEEN, K. (2002) Lesbian safety talk: problematising definitions and experiences of violence, sexuality and space, *Sexualities*, 5(3), pp. 259–280.

DONALD, J. (1997) This, here, now: imagining the modern city, in: S. WESTWOOD and J. WILLIAMS (Eds) *Imagining Cities: Scripts, Signs, Memory*, pp. 181–202. London: Routledge.

FLEISCH, W. (1987) Proximity and power: Shakespearean and cinematic space, *Theatre Journal*, 4, pp. 277–293.

FRASER, N. (1995) 'From redistribution to recognition? Dilemmas of justice in 'post-Socialist' Age', *New Left Review*, 212, pp. 68–94.

GILLESPIE, M. (1995) *Television, Ethnicity and Cultural Change*. London: Routledge.

GLEDHILL, C. (1997) Genre and gender: the case of soap opera, in S. HALL (Ed.) *Representation: Cultural Representations and Signifying Practices*, pp. 337–387. London: Sage.

HALL, S. (Ed.) (1997) *Representation: Cultural Representations and Signifying Practices*. London: Sage.

HOGGETT, P. (1992) A place for experience: a psychoanalytic perspective on boundary, identity and culture, *Environment and Planning D*, 10, pp. 345–356.

HUBBARD, P. (2001) Sex Zones, intimacy, citizenship and public space, *Sexualities*, 4(1), pp. 51–71.

LURY, C. (1991) Reading the self: autobiography, gender and the institution of literacy, in S. FRANKLIN *ET AL* (Ed.) *Off-center: Feminism and Cultural Studies*, pp. 97–109. London: Hutchinson.

MARBACK, R., BRUCH, P. and EICHER, J. (1998) *Cities, Cultures, Conversations: Readings for Writers*. Boston, MA: Allyn and Bacon.

MCDOWELL, L. (1993) Space, place and gender relations, part II: identity, difference, feminist geometries and geographies, *Progress in Human Geography*, 19, pp. 16–27.

MCDOWELL, L. (1996) Spatialising feminism: geographical perspectives, in: N. DUNCAN (Ed.) *BodySpace*, pp. 28–44. London: Routledge.

MELLOR, R. (1997) Cool times for a changing city, in: N. JEWSON and MCGREGOR, S. (Eds) *Transforming Cities*, pp. 1–34. London: Routledge.

MORAN, L. (1997) *The Homosexuality of Law*. London: Routledge.

MORAN, L. and SKEGGS, B. with CORTEEN, K. and TYRER, P. (2004) *Sexuality and the Politics of Violence and Safety*. London: Routledge.

MORAN, L., SKEGGS, B., CORTEEN, K. and TYRER, P. (2003) The formation of fear in gay space: the 'straight' story, *Capital and Class*, 80, pp. 173–199.

MORLEY, D. (1980) *The Nationwide Audience*. London: BFI.

MORLEY, D. (1986) *Family Television: Cultural Power and Domestic Leisure*. London: Comedia.

MUNT, S. (2000) Shame/pride dichotomies in *Queer as Folk*, *Textual Practice*, 14(3), pp. 531–546.

MYSLIK, W. D. (1996) Renegotiating the social/sexual identities of place: gay communities as safe havens or sites of resistance, in: N. DUNCAN (Ed.) *BodySpace: Destabilising Geographies of Gender and Sexuality*, pp. 156–169. London: Routledge.

NEALE, S. (1980) *Genre*. London: British Film Institute.

PAIN, R. H. (1997) Social geographies of women's fear of crime, *Transactions of the Institute of British Geographers*, 22(2), pp. 231–244.

PHILLIPS, R., WEST, D. and SHUTTLETON, D. (Eds) (2000) *De-centring Sexualities: Politics and Representations beyond the Metropolis*. London: Routledge.

PILE, S. (1996) *The Body in the City: Psychoanalysis, Space and Subjectivity*. London: Routledge.

PILE, S. and THRIFT, N. (1995) *Mapping the Subject: Geographies of Cultural Transformation*. London: Routledge.

PRATT, G. and HANSON, S. (1994) Geography and the construction of difference, *Gender, Place and Culture*, 1(1), pp. 5–31.

PROBYN, E. (1990) Travels in the postmodern: making sense of the local, in: L. J. NICHOLSON (Ed.) *Feminism/Postmodernism*, pp. 176–190. London: Routledge.

QUILLEY, S. (1997) Constructing Manchester's 'new urban village': gay space in the entrepreneurial city, in: G. B. BOUTHILLETTE and Y. RETTER (Eds) *Queers in Space: Communities/Public Places/Sites of Resistance*, pp. 275–295. Seattle, WA: Bay Press.

RADWAY, J. (1987) *Reading the Romance*. London: Verso.

ROSE, G. (1993) *Feminism and Geography: The Limits of Geographical Knowledge*. Cambridge: Polity Press.

SEITER, E. (1990) Making distinctions in TV audience research: a case study of a troubling interview, *Cultural Studies*, 4(1), pp. 61–85.

SEITER, E. (1999) *Television and New Media Audiences*. Oxford: Oxford University Press.

SEITER, E. and RIGGS, K. (1999) TV among fundamentalist Christians: from the secular to the satanic, in E. SEITER (Ed.) *Television and New Media Audiences*, pp. 91–111. Oxford: Oxford University Press.

SEITER, E., BORCHERS, H., KREUTZNER, M. and WARTH, E.-M. (Eds) (1989) *Remote Control: Television Audiences and Cultural Power*. London: Routledge.

SENNETT, R. (1991) *The Conscience of the Eye: The Design and Social Life of Cities*. London: Faber and Faber.

SHAVIRO, S. (1993) *The Cinematic Body*. Minneapolis, MN: University of Minnesota Press.

SIBLEY, D. (2001) The binary city, *Urban Studies*, 38(2), pp. 239–250.

SILVERSTONE, R. (1997) *Visions of Suburbia*. London: Routledge.

SKEGGS, B. (1997) *Formations of Class and Gender: Becoming Respectable*. London: Sage.

SKEGGS, B. (1998) Matter out of place: visibility and sexualities in leisure spaces, *Leisure Studies*, 18(3), pp. 213–233.

STYCHIN, C. (1998) *Nation by Rights*. Philadelphia, PA: Temple University Press.

TAUSSIG, M. (1993) *Mimesis and Alterity: A Particular History of the Senses*. London: Routledge.

TAYLOR, C. (1994) The politics of recognition, in: D. T. GOLDBERG (Ed.) *Multiculturalism: A Critical Reader*, pp. 75–106. Oxford: Blackwell.

THYNNE, L. (2003) Being Seen: 'the lesbian' in British television drama, in: D. ALDERSON and L. ANDERSON (Eds) *Territories of Desire in Queer Culture: Refiguring Contemporary Boundaries*, pp. 202–212. Manchester: Manchester University Press.

VALENTINE, G. (1990) Women's fear and the design of public space, *Built Environment*, 16, pp. 279–287.

VALENTINE, G. (1993) (Hetero)sexing space: lesbian perceptions and experiences of everyday space, *Environment and Planning D*, 11, pp. 395–413.

WEISS, G. (1999) *Body Images: Embodiment as Intercorporeality*. London: Routledge.

ZUKIN, S. and DIMAGGIO, P. (1996) *Structures of Capital: The Social Organisation of the Economy*. Cambridge: Cambridge University Press.

Appendix. Research Project Details

This ESRC-funded research project is entitled 'Violence, Sexuality and Space'. Our central research question was: 'How do groups, vulnerable to social conflict, social instability and social exclusion, produce sustainable security in public spaces?'

We made a comparison between Manchester (with a definable 'gay village') and Lancaster (with only virtual gay space). A comparison was made between three different groups previously identified as 'high risk' (lesbians/gay men/heterosexual women) by crime and victim surveys. A significant feature of the research was that we concentrated on safety *not* violence. We found safety to be just as slippery a concept as violence: it was frequently defined not just as avoidance of violence but also about ways of being, such as sexual security.

We interviewed 55 key informants. Informants were selected on the basis of their commercial, institutional (local government, police) and community expertise, connections and involvement (21 in Lancaster and 34 in Manchester). Each interview lasted between 1 and 2 hours, was recorded (unless requested), fully transcribed and inserted into QR*NUDIST software (see later).

We conducted a census survey to provide a snapshot of the populations using lesbian and gay space. The Manchester survey was conducted on a Friday (lunch time and evening) in February 1999 after key informants identified Friday as the 'most dangerous and heterosexual' time in the Village. The Lancaster survey was carried out on two evenings (Wednesday and Friday when gay events were held) in May 1999. The completion rates in both locations are exceptional: 95 per cent Lancaster and 97 per cent Manchester. We distributed 730 questionnaires in Manchester with a team of 12 researchers 'hitting' each bar and club for 15 minutes. Our Research Associate (Paul Tyrer) managed the whole process with military precision. (See website for full story 'Doing the Village in Military Mode').

We collated all secondary information on the space from media, TV, police reports, criminal statistics, research reports and council documentation to provide as full a 'picture' as possible of the area.

Focus groups were established after the key informant interviews. They were used to explore experiences, perceptions, ideas and policies in relation to the main themes of the research. Each group of gay men, lesbians and straight men met on six separate occasions in both sites, apart from the Lancaster heterosexual women's group that was only able to meet on five occasions. We conducted one transsexual focus group when alerted to their significant presence in Manchester's Village. The statistics from the census survey were fed back to the focus groups to generate engagement between the two different methods of research.

The interviews and focus groups were fully transcribed (and prepared for the ESRC Qualidata Centre to enable secondary investigation). They were initially analysed by each of the team for themes. As this was an interdisciplinary team, a range of themes were identified. Evidence was then collated via QR*NUDIST for themes. These were then read back against initial themes and developed further. Each researcher worked on general themes such as safety, fear, use of space, but also developed specialist interests (for example, Karen Corteen worked on lesbian safety talk (Corteen, 2002), Paul Tyrer worked on heterosexual narratives of invasion, Les Moran on home and Bev Skeggs on straight women and lesbian interactions (see website for full details and book of the project Moran *et al.*, 2004).

Citizens' inquiries were held in each locality to facilitate interaction between our focus group participants and individuals nominated as key local individuals. The events took place in March (Lancaster) and April (Manchester) 2000. Each meeting lasted just over two hours and was fully transcribed (full report available on website).

INDEX